The Microsporidia
of Vertebrates

The Microsporidia of Vertebrates

Elizabeth U. Canning
Department of Pure and Applied Biology
Imperial College of Science and Technology
London, United Kingdom

Jiří Lom
Institute of Parasitology
Czechoslovak Academy of Sciences
České Budějovice, Czechoslovakia

With a contribution by
Iva Dyková
Institute of Parasitology
Czechoslovak Academy of Sciences
České Budějovice, Czechoslovakia

1986

ACADEMIC PRESS

Harcourt Brace Jovanovich, Publishers

London   Orlando   San Diego   New York
Austin   Montreal   Sydney   Tokyo   Toronto

ACADEMIC PRESS INC. (LONDON) LTD.
24–28 Oval Road
LONDON NW1 7DX

United States Edition published by
ACADEMIC PRESS, INC.
Orlando, Florida 32887

British Library Cataloguing in Publication Data

Canning, Elizabeth U.
    The microsporidia of vertebrates.
    1. Microsporidia
    I. Title   II. Lom, Jiří.
    596'.0233   QL368.M5

Library of Congress Cataloging in Publication Data

Canning, Elizabeth U.
    The microsporidia of vertebrates.

    Bibliography: p.
    Includes index.
    1. Microsporidia.   2. Vertebrates—Parasites.
I. Lom, Jiří.   II. Dyková, I.   III. Title.
QL368.M5C36   1986   596'.0233   85-20034
ISBN 0−12−158790−8

PRINTED IN THE UNITED STATES OF AMERICA

86 87 88 89   9 8 7 6 5 4 3 2 1

Contents

2. The Microsporidia of Fish

3. The Microsporidia of Amphibia and Reptiles

Preface

Microsporidia, as parasites of all five classes of vertebrates, have an impact on all work concerning these hosts. There are numerous species in fish, some of which are responsible for mortality or pathological abnormalities, sufficient to make the fish unmarketable. Research using the hosts can be seriously interfered with, sometimes because the parasites cause mortality when the hosts are kept under laboratory conditions, for example the toads kept for pregnancy testing, which died of infectious *Pleistophora myotrophica*. Otherwise infections may be latent but interact with experimental infections and cause anomalous results. *Encephalitozoon cuniculi* could be badly misleading in immunological investigations, as it is possible that a pre-existing infection in laboratory animals could cause immunodepression and enhance the development of a newly introduced pathogen—or *vice versa*. Lastly, the increasingly common reports of serious microsporidian infections in man suggest that they may have to be recognised as of medical importance, certainly in immunocompromised and perhaps also in immunologically intact subjects. It is vital therefore that the infections are recognised and, if possible, identified.

This volume is intended to aid the non-specialist in the general recognition of these parasites and to provide sufficient data and illustrative material to secure an accurate diagnosis of species by the specialist. Diagnoses of species include descriptions of the parasites and of the lesions caused by them at light and electron microscope levels. Data included on the tissue changes associated with the removal of the parasites and repair by the host will enable the reader to assess whether or not the host has previously been infected.

The book is intended to complement the monographs edited by Bulla and Cheng (1976, 1977) on the biology and taxonomy of microsporidia, by updating, providing greater detail and, for the first time in a major work on the microsporidia, illustrating the text with photographs of living and fixed specimens. The authors have worked together to produce a book as far as possible uniform in style and content. However, individuals have taken special responsibility as follows: E. U. Canning—Introduction to the Microsporidia; The Microsporidia of Amphibia and Reptiles; The Microsporidia of Birds and Mammals; and overall editing. J. Lom—The Microsporidia of Fish. I. Dyková collaborated with J. Lom on the section on pathogenicity in Chapter 2 and contributed paragraphs on histopathology throughout the chapter. E. U. Canning and J. Lom together wrote Chapter 5—Techniques.

We would like to express our cordial thanks to many colleagues whose generous assistance made this publication possible. They kindly sent us samples of material, preparations and photographs which they permitted us to reproduce here. Our sincere appreciation is extended to Professor N. Ashton, Institute of Ophthalmology, London, U.K.; Dr. T. Awakura, Hokkaido Fish Hatchery, Japan; Dr. P. Berrebi and Dr. C. Loubès, Université du Languedoc, Montpellier, France; Dr. M. Chen, Saskatchewan Fisheries Laboratory, Canada; Dr. C. Delisle, Montreal Aquarium, Canada; Professor J. M. Doby, University of Rennes, France; Dr. E. Elkan (deceased); Dr. S. Egusa, University of Tokyo, Japan; Dr. A. V. Gajevskaya, Kaliningrad, U.S.S.R.; Dr. G. L. Hoffman, Fish Disease Laboratory, Stuttgart, U.S.A.; Drs. A. H. McVicar and K. McKenzie, Marine Laboratory, Aberdeen, Scotland; Dr. C. Morrison, Fisheries and Marine Service, Halifax, Canada; Dr. S. J. Nepszy, Lake Erie Fisheries Research Station, Canada; Dr. R. F. Nigrelli, Osborn Laboratories of Marine Science, New York, New York, U.S.A.; Dr. J. Ralphs, Plymouth Polytechnic, U.K.; Dr. G. Schubert, Technische Hochschüle, Stuttgart, F.R.G.; Dr. J. A. Shadduck, University of Illinois, Urbana, U.S.A.; Dr. V. Sprague, University of Maryland, College Park, U.S.A.; Dr. C. Summerfelt, Iowa State University, Ames, U.S.A.; Dr. J. Vanderberg, University of Guelph, Canada; Dr. V. N. Voronin, Leningrad Veterinary Institute, U.S.S.R.; and Dr. S. R. Wellings, University of Oregon Medical School, Eugene, U.S.A.

We are also deeply indebted to Mrs. V. Walters for typing the entire manuscript and to Mr. J. P. Nichola and Mr. F. Wright for preparation of the plates.

Elizabeth V. Canning
Jiří Lom

The Microsporidia
of Vertebrates

1. Introduction to the Microsporidia

I. ECONOMIC IMPORTANCE

The subject of this book, a group of organisms sufficiently distinctive to be classified in a phylum of their own, the phylum Microspora, are all intracellular parasites of animals. Commonly known as microsporidia, they are recognised as parasites of considerable importance in fisheries and of increasing importance in veterinary and human medicine. In fish, infections can be widespread and diffused through tissues or localised in cysts, often visible to the naked eye. They can inflict heavy mortality and, under conditions favouring transmission, especially in hatcheries, outbreaks can reach epizootic proportions. With the increasing development of aquaculture, their effects will be ever more widespread and ruinous.

Among mammals their effects are varied. In rodents, infections are often, but not always, mild and many remain undetected. The possibility that latent infections of *Encephalitozoon cuniculi* in stocks of laboratory animals will interfere with experimental research on animals, has been deemed of sufficient importance that serological tests have been devised for screening whole stocks. At the other end of the scale, the same parasite has been shown to cause illness and death in pets and fur-bearing animals and there have been a few records of severe pathological effects in man and other primates. Their occurrence will undoubtedly increase alongside immunosuppressive therapy.

II. GENERAL CHARACTERISTICS AND DEVELOPMENT

A. Diagnostic Features

Microsporidia are small unicellular organisms, all of which are obligate intracellular parasites, with an unique mode of infecting host cells. Resistant spores, normally ingested by the host, hatch under suitable stimuli. A minute tube, the polar tube or polar filament, which lies coiled in the intact spore, is everted through the spore wall with force (Fig. 1.2), enabling the tip to penetrate, but not disrupt, a host cell membrane. The infective agent, known as the sporoplasm, passes from the spore, through the tube and enters the cytoplasm directly. In contrast to those parasites which enter host cells in phagosomes, the microsporidia multiply in contact with the host cell cytoplasm and are protected from lysosomal attack. They must absorb small molecules from the host cell across their plasma membrane, there being no evidence of pinocytotic activity nor of intracellular digestion of particulate matter in vacuoles.

Microsporidia of several genera, including some which parasitise vertebrates, enter into a special relationship with their host cells. The infected host cell absorbs nutrients from the surrounding tissue through its specially modified surface and enlarges to accommodate the proliferating parasites. Concurrently, the host cell nucleus becomes greatly enlarged with multiple nucleoli, presumably to code for and amplify the synthesis of proteins for enlargement of the cell. These hypertrophic cells, which may attain a diameter of several millimetres, are termed xenomas (Chap. 2, Sect. II,A,B).

Microsporidia are unusual in lacking mitochondria, presumably relying on their host cells for chemical energy and in having ribosomes of a size more characteristic of prokaryotic cells. In other respects they are typically eukaryotic, with well developed internal membrane systems including an aggregate of vesicles thought to be Golgi in nature, and a nuclear envelope. Nuclear division takes place within the intact nucleus. In the absence of true centrioles, the spindle microtubules radiate from poles, marked by centriolar plaques, consisting of one or more electron dense superimposed discs and polar vesicles, lying in a shallow or deep depression of the nuclear envelope (Fig. 1.3 insert). In some microsporidia the nuclei are isolated throughout development, whereas in others the nuclei are paired, flattened against one another with membranes abutting in the so-called diplokaryon arrangement. Sexual processes are poorly understood. Fusion of gametic cells is unknown but an autogamous fusion of nuclei and reorganisation of chromosomes within the same cell have been reported in some genera parasitising invertebrates. Meiosis, described only in those genera which characteristically have diplokaryon nuclei, was thought to occur, as a prelude to spore production (Loubès, 1979) in the first division of the sporont (a cell irreversibly committed to spore production) but this view has recently been challenged with regard to the actual timing and sequence of events (Hazard and Broodbank, 1984). Meiosis is unknown, or is certainly unconfirmed in the genera parasitising vertebrates, all of which except *Nosema* (Sect. III,I) *Ichthyosporidium* (Sect. III,E), and *Mrazekia* (Sect.

III,H) have isolated nuclei but the existence of meiosis and karyogamy should be looked for rather than dismissed.

Two phases of development are recognised: merogony also known as schizogony, which is the phase of proliferation, and sporogony which culminates in the production of sporoblasts, which undergo morphogenesis into the highly characteristic spores for transmission. Spores may be released from the skin or in faeces and urine or only on the death of the host and are highly resistant to external conditions, even to a certain extent to drought. The principal variations in the life cycle discussed below are summarised in Fig. 1.1.

B. Meronts and Merogony

Sporoplasms released from spores may develop in tissues far removed from the site of hatching in the gut. Little is known about the transport of early phases but wandering cells, especially undifferentiated mesenchyme cells, macrophages and body fluids probably aid in their distribution. In suitable cells the sporoplasms become meronts.

Meronts are rounded, irregular or elongate cells with little membrane organisation of rough and smooth endoplasmic reticulum (ER) relative to the later stages. The plasma membrane is often simple (Figs. 1.3, 3.21) but some genera parasitising fish show specialisations at the interface with host cell cytoplasm. In the genus *Glugea* there is an electron dense coat and always, a closely associated cisterna of host smooth ER (Fig. 2.18), both of which are dispersed at the onset of sporogony. In the genus *Pleistophora* there is a very thick (0.5 μm) electron lucent amorphous coat penetrated by channels which run between the plasma membrane and a layer of vesicles at the interface with host muscle myofibrils (Fig. 2.59). The whole system appears to provide the parasite with an expanded surface for absorption.

Meronts may have isolated nuclei or diplokarya and they may divide repeatedly by binary or multiple fission or by plasmotomy. Although merogony is almost always in direct contact with host cell cytoplasm, individual meronts of *Encephalitozoon* are closely surrounded by a host-derived membrane (Barker, 1975) and, as the meronts multiply, they come to lie at the edge of an expanded parasitophorous vacuole (Figs, 3.20, 4.8, 4.9). Meronts of *Microgemma* are thought to lie within cisternae of host ER (Fig. 1.4).

C. Sporonts and Sporogony

Sporonts are the stages which divide into sporoblasts. They have at their surface, an electron dense coat which, on the completion of sporulation, will become the exospore layer of the spore wall. Neither the

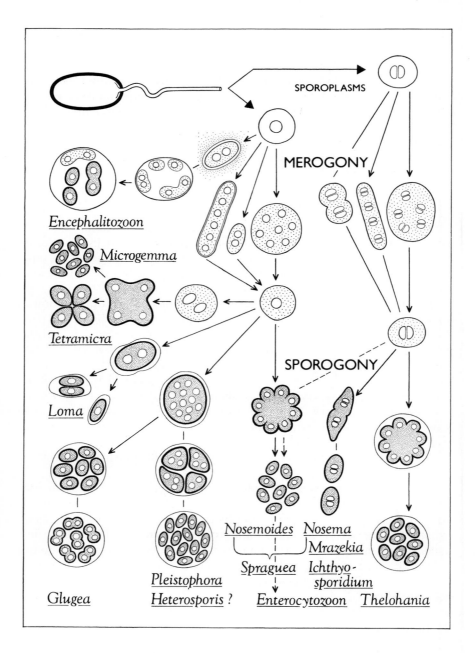

SPOROPLASMS

MEROGONY

Encephalitozoon

Microgemma

Tetramicra

Loma

SPOROGONY

Nosemoides   *Nosema*
                Mrazekia
        Spraguea   Ichthyo-
                    sporidium
Glugea   *Pleistophora*
           Heterosporis ?   *Enterocytozoon*   *Thelohania*

electron dense coat on the merons of *Glugea* nor the thick wall around *Pleistophora* is equivalent to the sporont coat. Once recognised, the sporont coat is a firm indication of the sporogonic process.

Sporonts may have isolated or diplokaryon nuclei. Sometimes sporonts divide directly into sporoblasts by binary fission. Alternatively, they become multinucleate stages (usually called sporogonial plasmodia to indicate the multinucleate state) and pass through a two-phased or sequential series of divisions. In the two-phased division, the cells undergoing the second division are called sporoblast mother cells. The division sequences are varied and highly characteristic for each genus (Fig. 1.1).

Taxonomically the microsporidia are split into two suborders based on sporogonic sequences, in which the spores are either packaged within sporophorous vesicles (SPOVs), also known as pansporoblast membranes (Suborder Pansporoblastina) (Figs. 2.30, 2.57) or are dispersed freely in the host cell cytoplasm (Suborder Apansporoblastina) (Figs. 1.6, 1.7). For the taxonomy of the phylum Microspora, the reader is referred to Sprague (1977, 1982) and Canning (1985).

The majority of microsporidian genera, of which five—*Glugea, Pleistophora, Thelohania, Loma* and *Heterosporis*—are represented in vertebrates, belong to the Pansporoblastina and form SPOVs. The sporont contracts within an envelope laid down on the external surface of the plasma membrane. In *Pleistophora,* the envelope is derived from the amorphous coat of the merons and the SPOV is a thick conspicuous structure (Figs. 1.8, 2.57). In other genera it is formed *de novo* and is fine and membrane like (Fig. 2.20). Within the SPOV, the characteristic coat is secreted at the surface of the sporont (sporogonial plasmodium) before its division into sporoblasts (Fig. 2.20). The resultant number of sporoblasts, and later of spores, is diagnostic for the genera.

In the remainder of microsporidian genera (Suborder Apansporoblastina), no such envelope is formed, the sporont coat being secreted at the surface of the free sporont (Figs. 1.5, 3.24). Representatives of this type of vertebrates are *Encephalitozoon, Enterocytozoon, Nosema, Spraguea, Ichthyosporidium, Microgemma* and *Mrazekia.* The genus *Tetramicra* in

---

Fig. 1.1 Diagrammatic representation of the life cycle of some microsporidia which infect vertebrates. Sporoplasms extruded from the spores, via the polar tube, are represented without stippling. Merogonic stages are represented by light stippling and simple surface membranes. Sporogonic phases are represented by heavy stippling and a dense surface coat. In *Encephalitozoon* all stages develop in a vacuole of host origin. In the rest the merogonic stages are free but the spores are either free or lie within sporophorous vesicles (SPOVs): these first separate as membrane-like structures from the surface of the sporogonial plasmodia.

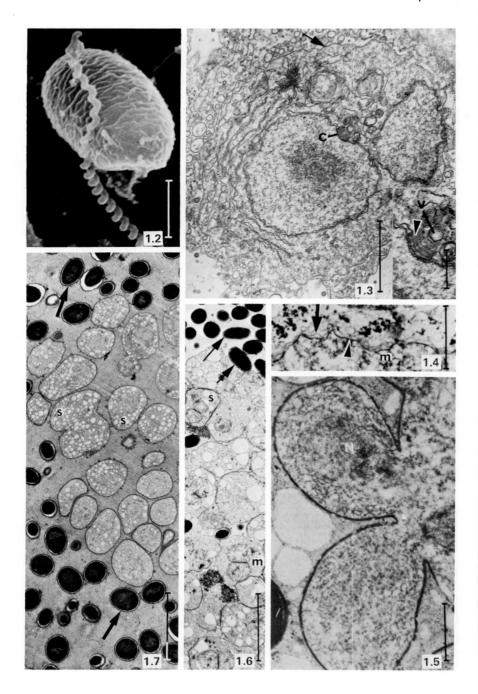

fish is somewhat enigmatic but the vacuole in which the spores lie in the host cell was presumed by the authors (Matthews and Matthews, 1980) to be of host origin and thus not a SPOV.

An interesting feature of the development of some microsporidian genera is that they are able to express more than one sporogonic sequence, in some cases more than one merogonic sequence as well. The course of the two sequences is so different that each was formerly considered to be characteristic at generic level. Sometimes one sequence takes place in a SPOV, the other not. Dimorphism of two kinds is known in microsporidia in vertebrates. The genus *Spraguea* is dimorphic in both merogony and sporogony: one cycle resembles the genus *Nosema* and produces diplo-karyotic spores (Fig. 1.6); the other resembles *Nosemoides,* a genus known otherwise only from invertebrates, and produces uninucleate spores (Fig. 1.7). In the genus *Pleistophora* two kinds of SPOVs are formed: in one

Fig. 1.2 Scanning electron micrograph of a microsporidian spore from which the polar tube has been extruded. Unusually the polar tube has remained coiled in this preparation; it usually straightens to facilitate passage of the sporoplasm. Scale bar = 1 μm. (Photograph by J. Lom.)

Fig. 1.3 *Loma salmonae:* meront from a xenoma in the gills of brook trout; the endo-membrane system is unusually well developed but the simple plasma membrane (arrow) is characteristic of the merogonic stages; the centriolar plaque (c) which is formed at the spindle pole when the nucleus enters mitosis, consists of superimposed electron dense discs (ar-rowhead) and polar vesicles (v) (insert). Scale bar = 1 μm. (Reproduced from Morrison and Sprague, 1981b, with permission of authors and publisher.)

Fig. 1.4 *Microgemma hepaticus:* surface of meront. A fine membrane (arrow), lying close to the surface of the parasite (arrowhead), completely encloses the meront (m); the meront thus appears to lie in an expanded cisterna of smooth endoplasmic reticulum. Scale bar = 1 μm. (Reproduced from Ralphs, 1984, with permission of the author.)

Fig. 1.5 *Microgemma hepaticus:* sporogonial plasmodium. An almost complete electron dense surface coat is laid down on the plasma membrane; sporoblasts are being produced as buds from the plasmodium. N = nucleus. Scale bar = 1 μm. (Reproduced from Ralphs, 1984, with permission of the author.)

Fig. 1.6 Xenoma of *Spraguea lophii:* diplokaryotic meronts (m) and sporonts (s) and cylindrical (*Nosema*-type) spores (arrows); the spores lie free in the host cytoplasm. Scale bar = 5 μm. (Reproduced from Loubès *et al.,* 1979, with permission of the authors and publisher.)

Fig. 1.7 Xenoma of *Spraguea lophii:* sporonts(s) with isolated nuclei which give rise to uninucleate (*Nosemoides*-type) spores (arrows). Scale bar = 5 μm. (Reproduced from Loubès *et al.,* 1979, with permission of authors and publisher.)

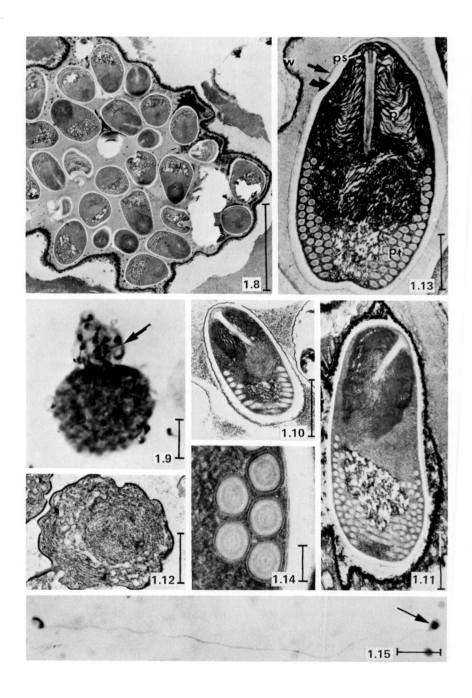

there are numerous microspores (Figs. 1.8, 1.9, 1.10); in the other, much rarer, there are eight macrospores (Figs. 1.9, 1.11). Here the spores are only slightly different in morphology, resembling each other except in size and in minor differences in shape and in the number of coils of the polar tube. The significance of dimorphism and the separate roles of the two spore types is unknown. Some microsporidia of invertebrates, e.g. *Amblyospora,* produce two types of spore in one host. One type is haploid and uninucleate and develops only in an alternate host of a 2-host life cycle, while the other, binucleate type, is involved in transovarial transmission in the primary host (Sweeney, Hazard and Graham, 1985).

Fig. 1.8 *Pleistophora littoralis:* electron micrograph of a microspore SPOV showing sections of microspores in a matrix enclosed by a thick SPOV wall composed of several layers of different electron density. Scale bar = 5 μm. (Reproduced from Canning *et al.,* 1979, with permission of the publisher.)

Fig. 1.9 *Pleistophora typicalis:* adjacent microspore and macrospore SPOVs, the latter with 8 macrospores (arrow). Scale bar = 10 μm. (Reproduced from Canning and Nicholas, 1980, with permission of the publisher.)

Fig. 1.10 *Pleistophora littoralis:* microspore showing 9 coils of the polar tube in one rank. Scale bar = 1 μm. (Reproduced from Canning *et al.,* 1979, with permission of the publisher.)

Fig. 1.11 *Pleistophora littoralis:* macrospore showing similar organisation to the microspore in Fig. 1.10 but with many coils of the polar tube in several ranks and obvious size difference. Scale bar = 1 μm. (Reproduced from Canning *et al.,* 1979, with permission of the publisher.)

Fig. 1.12 *Glugea anomala:* newly formed sporoblast with single nucleus and extensive development of endoplasmic reticulum. Scale bar = 1 μm.

Fig. 1.13 *Pleistophora hyphessobryconis:* mature spore showing wall comprised of electron dense exospore (arrow) and lucent endospore (curved arrow), polar sac (ps), polarplast (p) and polar tube (pt); W = wall of SPOV. Scale bar = 1 μm. (Reproduced from Lom and Corliss, 1967, with permission of the publisher.)

Fig. 1.14 *Pleistophora littoralis:* transverse section of the coils of the polar tube. Scale bar = 0.1 μm.

Fig. 1.15 *Pleistophora typicalis:* microspore after extrusion of the polar tube and emergence of the sporoplasm (arrow). Scale bar = 10 μm. (Reproduced from Canning and Nicholas, 1980, with permission of the publisher.)

D. Sporoblasts (Fig. 1.12)

Sporoblasts are generally ovoid bodies, although they often appear crenated at electron microscope level because of poor fixation. Their development is a process of maturation into spores, i.e. synthesis of spore
organelles. There is a noticeable increase in the amount of smooth and
rough ER. Golgi vesicles are prominent and are responsible, at least in
part, for the secretion of the polar tube and its polar sac (Sect. E,2). A
transparent endospore layer is gradually added to the spore wall under
the dense coat, now called the exospore, which was originally laid down
at the surface of the sporont. When the maturation process is complete
the vesicles of the obsolete Golgi apparatus coalesce to form the posterior
vacuole.

E. Spores

1. External Appearance and Spore Wall

Fresh spores, even at low magnification ($\times 100$) are recognisable to the
practised eye. Usually they have simple outlines and, when viewed with
transmitted light appear greenish and refractile. Hatched spores lose their
refractility and show a distinct wall around a transparent centre. With
phase contrast, intact spores are phase bright and hatched spores are phase
dark. Fresh spores of many species show a posterior vacuole. These are
especially large in species infecting cold blooded vertebrates, sometimes
occupying more than half of the spore volume (Figs. 2.22, 2.23).

The smallest spores found in vertebrate hosts are the ovoid spores of
Encephalitozoon cuniculi from mammals (Fig. 4.14), which measure 2.5
$\times$ 1.5 µm. The largest are those of *Mrazekia piscicola* (Fig. 2.46T) from
Gadus merlangus, which measure 20 $\times$ 6 µm.

Surface ornamentations in the form of filamentous or tubular structures,
as well as a mucous outer layer, which may be present on the spores of
some species from invertebrate hosts, are absent from the species in vertebrates. The exospore is a smooth or corrugated proteinaceous layer
varying in thickness in different species from about 15 nm to 100 nm. The
endospore is chitinous, measures about 150–200 nm thick and is evenly
disposed in the spore wall, except at the anterior end, just off centre,
where it is thinned to 30 nm or less.

The spore is Gram positive. The characteristic reddish-purple reaction
with Gram's stain is of great value in detecting spores in smears and sections, as even single spores are revealed with clarity against an unstained
or lightly counter-stained background. Spores also stain characteristically
with Giemsa's and Goodpasture's stains, retaining a light blue and a red
colour respectively. An autofluorescence has been reported for some

spores (Doby, Rault and Barker, 1975): fresh spores of *Thelohania apodemi*, from the brains of voles, fluoresce strongly, while those of *Encephalitozoon cuniculi*, from mouse peritoneal macrophages, fluoresce feebly.

2. Ultrastructure (Fig. 1.13)

The spore wall encloses an extrusion apparatus, consisting of the polar tube, its anchoring disc (polar sac) and a complex stack of membranes known as the polaroplast, and the infective agent or sporoplasm.

The polar sac is a structure like the cap of a mushroom which fits into the contours of the anterior end of the spore wall. The base of the polar tube fits into the centre of this structure. Spores stained by the periodic acid - Schiff reagent (PAS) show, by light microscopy, a PAS positive dot at this central point (Fig. 2.45). It is referred to as the polar cap and is characteristic in shape and size for a species.

The polar tube runs an oblique course backwards towards the periphery of the spore and falls into a coil in the peripheral cytoplasm. The number of turns of the coil is constant within limits for a species. The diameter of the tube in the coiled region is 100–150 nm. In *Mrazekia*, the basal part of the tube is conspicuously thickened into a rod-like structure, the manubrium, and there is an abrupt constriction at the junction with the coiled part. Fundamentally, the tube consists of a sheath, which acts as a sleeve, through which the tube slides during extrusion, and a complex series of rings surrounding a central core (Fig. 1.14).

The straight part of the polar tube is surrounded by the membranes of the polaroplast. The membranes, arranged singly or in threes, form an anastomosing system known as the lamellar polaroplast. Some species have a second region, the vesicular polaroplast, composed of expanded sacs lying just posterior to the lamellar region.

The remaining space within the spore wall is occupied by the posterior vacuole and relatively undifferentiated cytoplasm containing some rough ER and free ribosomes. This surrounds one or two nuclei which lie centrally within the coil of the polar tube. Although there is no membrane separating this cytoplasm and nucleus from the other spore structures, it is usually regarded as an entity and called the sporoplasm. Spores of *Nosema, Mrazekia, Ichthyosporidium* and the *Nosema*-type spores of *Spraguea* have diplokaryon nuclei; spores of the other genera in vertebrates and the *Nosemoides*-type spores of *Spraguea* are uninucleate.

3. Hatching Mechanism

The eversion of the polar tube and passage of the sporoplasm through it into the host cell have been amply demonstrated since the theory was proposed by Oshima (1937). Kramer (1960) produced light micrographs demonstrating the nuclei within the polar tube and this was confirmed by Ishihara (1968) and Lom (1972) by electron microscopy. Weidner (1976a, 1982) has investigated the nature of the polar tube protein and has suggested a mechanism of tube eversion, effected by molecular rearrangement at the growing tip.

The stimuli causing eversion are not fully understood but suitable pH and the presence of alkali metal ions appear to be essential (Ishihara, 1967). Weidner (1982) demonstrated that external calcium ions were inhibitory but internal calcium was shifted across membranes, probably the polaroplast membranes, during hatching.

When the tube is fully everted, the sporoplasm begins its journey through it, to enter the host cell (Fig. 1.15).

F. Transmission

Transmission of microsporidia normally takes place when spores are ingested by a new host, hatch in the gut and enter epithelial cells, either to develop there or to be transported to their preferred site of development. Several species have been transmitted experimentally to vertebrates. Spores of *Glugea anomala* fed to sticklebacks produced infections (Weissenberg, 1968) as did those of *Pleistophora myotrophica* when fed to toads (Canning, Elkan and Trigg, 1964) and *Encephalitozoon cuniculi* when fed to a variety of laboratory mammals (e.g. Nelson, 1967).

Evidence that *Encephalitozoon cuniculi* may also be transmitted transplacentally has been most convincingly demonstrated in the case of three infected rabbits that had been delivered by Caesarean section and reared in germ-free isolators (Hunt, King and Foster, 1972) and where infection was found in a colony of mice established from four pairs, which had been delivered by Caesarean section and fostered by germ-free rats (Innes, Zeman, Frenkel and Borner, 1962). Reports of infection in puppies, kittens and fox cubs also provide circumstantial evidence for transplacental transmission in nature (Chap. 4, Sect. IIIA,7).

III. DEFINITIONS OF THE GENERA OF MICROSPORIDIA PARASITIC IN VERTEBRATES

A. Genus *Encephalitozoon* Levaditi, Nicolau and Schoen, 1923

Nuclei are isolated throughout development. All stages are in membrane bound vacuoles of host cell origin. No SPOVs. Meronts usually divide by binary fission. Sporogony is disporoblastic. Spores are uninucleate. No xenoma formation. Hosts are usually homoeothermic vertebrates. Type species *Encephalitozoon cuniculi* Levaditi, Nicolau and Schoen, 1923.

B. Genus *Enterocytozoon* Desportes, Le Charpentier, Galian, Bernard, Cochand-Priollet, Lavergne, Ravisse and Modigliani, 1985

Meronts are diplokaryotic cells lying in direct contact with the host cell cytoplasm. Sporogony is polysporoblastic—probably octosporoblastic—by multiple fission of spherical sporogonial plasmodia with isolated nuclei. The development of spore organelles, including the polar tube, is well advanced before fission of the sporogonial plasmodium into sporoblasts. Spores are uninucleate with the endospore layer of the wall poorly developed. Found once, in the enterocytes of a man suffering with AIDS. Type species *Enterocytozoon bieneusi* Desportes *et al.*, 1985.

C. Genus *Glugea* Thélohan, 1891

Nuclei are isolated throughout development. Meronts are cylindrical, with an electron dense coat on the plasma membrane, and are surrounded by a cisterna of host ER. Sporogony is polysporoblastic in SPOVs: multinucleate sporogonial plasmodia divide by multiple fission to give numerous sporoblast mother cells, which in turn divide by binary fission into sporoblasts. SPOV wall is membrane-like. A variable and large number of spores produced within SPOVs. Spores uninucleate. Induces xenomas, encapsulated in sloughed layers of the cell coat, and in which the host cell nucleus is highly branched in the peripheral cytoplasm. Developmental stages of the parasite lie peripherally and spores lie centrally in the xenoma. Hosts are usually fish. Type species *Glugea anomala* Moniez, 1887.

D. Genus *Heterosporis* Schubert, 1969

Nuclei are isolated in all sporogonic stages. Merogony is incompletely known. In sporogony, multinucleate plasmodia within SPOVs, give rise to eight macrospores or more rarely to sixteen or more microspores. SPOV wall is thick. Host cell becomes greatly hypertrophic, finally remaining only as an envelope, which incorporates several persisting host cell nuclei, around the mass of parasites. Spores are uninucleate. Host is a fish. Type species *Heterosporis finki* Schubert, 1969.

E. Genus *Ichthyosporidium* Caullery and Mesnil, 1905

Rounded meronts divide by binary fission; nuclear condition is unknown. No SPOVs. Sporonts are diplokaryotic and disporoblastic. Spores are diplokaryotic. Host cells become syncitial xenomas, apparently arising

by proliferation of connective tissue cells, which coalesce and isolate regions of parasitised cytoplasm within themselves by the formation of fibrous capsules. These cystic stages mature into large xenomas with villous
projections at the surface. Hosts are fish. Type species *Ichthyosporidium
giganteum* (Thélohan, 1892) Swarzewsky, 1914.

F. Genus *Loma* Morrison and Sprague, 1981

Nuclei unpaired throughout life cycle. Uninucleate meronts with a simple
plasma membrane, lie in direct contact with host cell cytoplasm and develop into plurinucleate plasmodia. Sporogony is polysporoblastic within
a SPOV. Sporoblasts numbering up to 8 (possibly up to 16) transform into
spores within the vesicle. Infected cells are hypertrophic, forming xenomas
reaching a diameter up to about 1 mm with a single, centrally-located
hypertrophic nucleus. Xenoma wall is a simple host cell membrane coated
with fibrillar layers, probably of connective tissue orgin. Developmental
stages of the parasite are intermingled with spores throughout the xenomas.
Hosts are fish. Type species: *Loma branchialis* (Nemeczek, 1911) Morrison and Sprague, 1981. Morrison and Sprague (1981b) originally proposed
as the type species *L. morhua* Morrison and Sprague, 1981 which is a
junior synonym of *L. branchialis*. Morrison and Sprague (1981b) found
that spores were isolated or in pairs within vesicles and considered that
sporogony was disporoblastic. Bekhti (1981, 1984) found at least three
spores per vesicle and Loubès, Maurand, Gasc, de Buron and Barral (1985)
found multinucleate cylindrical and round sporogonial plasmodia, producing sporoblasts arranged in chains or in morula-like configurations respectively. The last authors suggested that cross sections of cylindrical
plasmodia provide an explanation for the isolated spores seen by Morrison
and Sprague.

G. Genus *Microgemma* Ralphs and Matthews, 1986

Nuclei are isolated at all stages of development. Multinucleate meronts
which divide by plasmotomy probably lie within expanded cisternae of
host ER and are thus encased in a host membrane. Sporogonial plasmodia,
derived directly from the meronts by loss of the host membrane, lie in
direct contact with host cell cytoplasm. Sporogony is polysporoblastic
starting with the production of sporoblasts by single exogenous budding
and accelerating to produce them by multiple budding and fragmentation
of plasmodium. Spores uninucleate. Infected cells form a xenoma with a
simple plasmalemma extending into microvilli. There is a reticulate host

cell nucleus, situated between a peripheral band of mitochondria and the central region of the cell, occupied by intermingled stages of parasites. Type species *Microgemma hepaticus* Ralphs and Matthews, 1985.

H. Genus *Mrazekia* Léger and Hesse, 1916

Meronts typically divide by binary fission; nuclear condition is uncertain. Sporogony is disporoblastic giving diplokaryotic sporoblasts. Spores are diplokaryotic, long and cylindrical, with a manubrium. No SPOVs. Host cells become xenomas. Hosts are usually arthropods and annelids. Type species *Mrazekia argoisii* Léger and Hesse, 1916.

I. Genus *Nosema* Naegeli, 1857

Nuclei are diplokaryotic throughout most of the life cycle. All stages are in direct contact with the host cell cytoplasm. No SPOVs. Merogony is by binary or multiple fission. Sporogony is disporoblastic. Spores are diplokaryotic. No xenoma formation. Hosts are usually invertebrates, especially insects. Type species *Nosema bombycis* Naegeli, 1857.

J. Genus *Pleistophora* Gurley, 1893

Nuclei are isolated throughout development. Meronts are rounded plasmodia with a thick amorphous wall, external to the plasma membrane. At the onset of sporogony the wall separates from the surface and becomes the thick wall of the SPOV. Sporogony is polysporoblastic. Division of the sporogonial plasmodium is by repeated segmentation. A variable and large number of spores is produced within SPOVs. Spores uninucleate. A rare sequence of sporogony, usually octosporoblastic, gives macrospores. Infections are diffused in tissues without xenoma formation. Hosts are usually fish or arthropods. Type species *Pleistophora typicalis* Gurley, 1893.

K. Genus *Spraguea* Weissenberg, 1976

Dimorphic microsporidia producing two types of spores. All stages are in direct contact with host cell cytoplasm. No SPOVs. In one sequence, all stages of merogony and sporogony have isolated nuclei; sporogony is polysporoblastic giving oval, uninucleate spores. In the other sequence, meronts and sporogonic stages have diplokarya; sporogony is disporoblastic giving slender, curved diplokaryotic spores. Parasites develop in a restricted zone in a hypertrophic ganglion cell, which is bounded by a simple membrane and has a single hypertrophic nucleus. Host is a fish. Type species *Spraguea lophii* (Doflein, 1898) Weissenberg, 1976.

L. Genus *Tetramicra* Matthews and Matthews, 1980

Nuclei are isolated throughout development and all stages develop in membrane-bound vacuoles of presumed host origin. No SPOVs. Meronts are uninucleate or multinucleate cylindrical cells. Merogony ends with the formation of binucleate cells, which enter sporogony. Sporogony is tetrasporoblastic. Spores are uninucleate. Host cells become xenomas, which have a microvillous surface and a reticulate nucleus. Developmental stages and spores are mixed at all levels. Host is a fish. Type species *Tetramicra brevifilum* Matthews and Matthews, 1980.

M. Genus *Thelohania* Henneguy, 1892

Nuclei are diplokaryotic or isolated, according to stage of development. Merogony is inadequately known. Meronts, at least sometimes, are diplokaryotic, dividing by binary or multiple fission. The early sporont is diplokaryotic, giving rise to eight uninucleate sporoblasts within a SPOV. Spores are uninucleate. Hosts are usually arthropods. Type species *Thelohania giardi* Henneguy, 1892.

N. Collective group *Microsporidium* Balbiani, 1884

An assemblage of identifiable species of which the generic positions are uncertain (Stoll, 1961). Sprague (1977) introduced this concept into microsporidian taxonomy.

In the present work the collective group has been used for species, already assigned to genera on insufficient evidence and for species for which a generic placement has never been possible. However, when no new data have been available, the doubtful old generic assignments have been kept and also in some cases when there has been so specific determination (e.g. *Glugea* sp.).

2. The Microsporidia of Fish

I. FISH HOSTS OF MICROSPORIDIA

Abramis ballerus L., blue bream; freshwater
  Glugea sp. of Bogdanova, 1961 (intestine)
Abramis brama (L.) × *Rutilus rutilus* (L.), a hybrid of common bream and roach;
    freshwater
  Pleistophora elegans (ovary)
Acipenser guldenstadti Brandt, Russian sturgeon; freshwater
  Pleistophora sulci (ovary)
Acipenser ruthenus L., sterlet, freshwater
  Pleistophora sulci (ovary)
Aeoliscus strigatus (Gunther), razor fish; marine
  "*Glugea anomala*" Laird, 1956 (subcutaneous; connective tissue)
Alburnus alburnus (L.), common bleak; freshwater
  Pleistophora mirandellae (ovary)
Alosa kessleri volgensis (Berg), Volga black-backed shad; brackish
  Glugea bychowskyi (intestine, testis)
Ammodytes lanceolatus Le Suer, sand-eel; marine
  Glugea caulleryi (liver)
Anarhichas lupus L., Atlantic wolffish; marine
  Pleistophora ehrenbaumi (skeletal muscles)
Anarhichas minor Olafsen, spotted wolffish; marine
  Pleistophora ehrenbaumi (skeletal muscles)
Anguilla japonica Temminck and Schlegel, "Japanese eel"; euryhaline
  Pleistophora anguillarum (skeletal muscles)
Apeltes quadracus (Mitchill), fourspined stickleback; euryhaline
  Glugea weissenbergi (= *Glugea anomala*) (body cavity and
    visceral organs)
Apistogramma reitzigi Ahl, tetra; freshwater
  Pleistophora hyphessobryconis (skeletal muscles)
Asprocottus megalops (Grazianow), humped rough sculpin; freshwater
  Glugea anomala (Zajka, 1965; subcutaneous)
Atherina boyeri Risso, big-scale sand smelt; euryhaline
  Glugea atherinae (intestinal wall and body cavity)

Atherina mochon pontica n. *caspia* Eichwald, Caspian silverside
  Caspian Sea
  Microsporidium sp. of Gasimagomedov and Issi, 1970 (intestine)
Bairdiella chrysura (Lacépède), silver perch; euryhaline
  Microsporidium sp. of Lawler, pers.comm. (liver)
Barbus barbus (L.), common barbel; freshwater
  Pleistophora longifilis (testis)
Barbus capito conocephalus (Kessler), Turkestan barbel; freshwater
  Pleistophora longifilis (testis)
Barbus lineatus Duncker, barb; freshwater
  Pleistophora hyphessobryconis (skeletal muscles)
Blennius pholis L. shanny blenny; marine
  Pleistophora littoralis (skeletal muscles)
Blicca bjoerkna (L.), silver bream; freshwater
  Microsporidium sp. of Gasimagomedov and Issi, 1970 (liver)
Brachydanio nigrofasciatus (Day), spotted danio; freshwater
  Pleistophora hyphessobryconis (skeletal muscles)
Brachydanio rerio (Hamilton-Buchanan), zebra danio; freshwater
  Microsporidium sp. of Kinkelin, 1980 (spinal cord)
  Pleistophora hyphessobryconis (skeletal muscles)
Brachydanio sp., danio; freshwater
  Microsporidium pseudotumefaciens (ovary, organs in the body cavity)
Callionymus lyra L., greater dragonet; marine
  Glugea destruens (skeletal muscles)
Carangoides malabaricus Bloch and Schneider, ajack; marine
  Pleistophora carangoidi (skeletal muscles)
Carassius auratus auratus (L.), fan-tail goldfish; freshwater
  Pleistophora hyphessobryconis (skeletal muscles)
Cepola rubescens L., European bandfish; marine
  Microsporidium ovoideum (liver)
Channa maculatus (Lacepede), snake head; freshwater
  Glugea hertwigi (?) Chen and Hsieh, 1960 (small intestine)
Cheirodon axelrodi Schultz, tetra; freshwater
  Pleistophora hyphessobryconis (skeletal muscles)
Chelon labrosus (Risso), thick-lipped mullet; marine
  Microgemma hepaticus (liver)
Clupeonella cultiventris cultiventris (Nordman) Black Sea sardelle; brackish
  Glugea luciopercae (intestine)
Clupeonella delicatula caspia Swetovidov, Caspian-Pontic sprat;
    brackish to marine
  Glugea luciopercae (intestine)
Colisa lalia (Hamilton-Buchanan), dwarf gourami; freshwater
  Microsporidium pseudotumefaciens (ovary, organs in the body cavity)
Coregonus exiguus bondella Fatio, whitefish of the Neufchatel Lake;
    freshwater
  Thelohania ovicola (ovary)
Coregonus lavaretus pidschian n. *pidschianoides* Pravdin, western Arctic
    whitefish; euryhaline
  Glugea hertwigi (various body organs)
Coris julis (L.) syn. *C. giofredi* (Risso), rainbow wrasse; marine

Glugea depressa (liver)
Nosema marionis (hyperparasite of the myxosporidian *Ceratomyxa coris* parasitic in the gall bladder)
Coryphaenoides nasutus Gunther, rat-tail; marine
  Pleistophora duodecimae (skeletal muscles)
Cottus beldingi Eigenmann and Eigenmann, piute sculpin; freshwater
  Pleistophora tahoensis (skeletal muscles)
Cottus gobio L., sculpin; freshwater
  Pleistophora vermiformis (skeletal muscles)
Cottus sp., sculpin; freshwater
  Microsporidium salmonae (gills)
Crenilabrus melops (L.), corkwing; marine
  Ichthyosporidium giganteum (body cavity)
Crenilabrus ocellatus (Forskal), ocellated wrasse; marine
  Ichthyosporidium giganteum (body cavity)
Crenilabrus tinca (L.) (= the correct name for *C. pavo* of Swarzewski), corkwing, peacock-wrasse; marine
  Ichthyosporidium hertwigi (gills)
Dallia pectoralis Bean, Alaska blackfish; freshwater
  Pleistophora dallii (subcutaneous tissue)
Dentex dentex (L.), dentex; marine
  Glugea machari (liver)
Diplodus sargus (L.), sea bream; marine
  Loma diplodae (gill filaments)
Dorosoma cepedianum (Le Suer), gizzard shad; freshwater
  Glugea cepedianae (visceral cavity)
Dorosoma petenense (Gunther), threadfin shad; freshwater
  Pleistophora sp. of Wellborn, 1966 in Putz and McLaughlin, 1970 (site of infection not stated)
Drepanopsetta hippoglossoides Gill, long rough dab; marine
  Pleistophora hippoglossoideos (skeletal muscles)
Enchelyopus cimbrius (L.), fourbeard rockling; marine
  Loma branchialis (gills)
Entelurus aequorus (L.), snake pipefish; marine
  Glugea acuta (air bladder muscle)
Esox lucius L., common pike; freshwater
  Pleistophora oolytica (ovary)
Fundulus heteroclitus (L.), mummichog; euryhaline
  Glugea sp. of Bond, 1938 (stomach and bile duct)
  Pleistophora sp. of Bond, 1937 (skeletal muscles and spinal cord)
Gadus morhua kildinensis Derjugin, Kildin cod; marine
  Loma branchialis (gills)
Gadus morhua maris-albi Derjugin, White Sea shore cod; marine
  Loma branchialis (gills)
Gadus morhua morhua L., Atlantic cod; marine
  Loma branchialis (gills)
  Pleistophora gadi (skeletal muscles)
Gadus (Pollachius) virens L., pollock; marine
  Glugea punctifera (ocular muscle)
Gambusia affinis (Baird and Girard), mosquitofish; freshwater

Glugea sp. of Crandall and Bowser, 1981 (mesentery, ovary, subcutaneous connective tissue)
Gambusia spp., gambusia; freshwater
  Microsporidium pseudotumefaciens (ovary, organs in the body cavity)
Gasterosteus aculeatus L., three-spined stickleback; euryhaline
  Glugea anomala (subcutaneously and in most of the body cavity organs)
  Thelohania baueri (ovary)
Girardinus caudimaculatus Hensel, caudy; freshwater
  Microsporidium girardini (skin, muscles and intestine)
Gobiodon okinawae Sawada, Arrai and Abe; marine
  Pleistophora sp. of Blasiola, 1977 (unpublished)
Gobiosoma bosci (Lacépède), naked goby; euryhaline
  Microsporidium sp. of Lawler, pers. comm. (gall bladder)
Gobius cobitis Pallas, giant goby; marine
  Glugea sp. of Naidenova, 1974 (intestine)
Gobius niger L., black goby; marine
  Glugea sp. of Naidenova, 1974 (intestine)
  Loma dimorpha (walls of the digestive tract)
Gobius ophiocephalus Pallas, grass goby; marine
  Glugea sp. of Naidenova, 1974 (intestine)
Gobius paganellus L., rock goby; marine
  Glugea sp. of Naidenova, 1974 (intestine)
Gobius platyrostris Pallas, guban goby; marine
  Glugea sp. of Naidenova, 1974 (intestine)
  "Glugea anomala" Naidenova, 1974 (subcutaneously)
Gobius ratan Nordmann, ratan goby; marine
  Glugea sp. of Naidenova, 1974 (intestine)
Gymnocephalus cernuus (L.), common ruff; freshwater
  Glugea acerinae (intestine)
  Pleistophora acerinae (mesentery)
Hasemania nana (Reinhardt), tetra; freshwater
  Pleistophora hyphessobryconis (skeletal muscles)
Hemigrammus erythrozonus Durbin, glow-light tetra; freshwater
  Pleistophora hyphessobryconis (skeletal muscles)
Hemigrammus ocellifer (Steindachner), head-and-tail-light fish; freshwater
  Pleistophora hyphessobryconis (skeletal muscles)
Hemigrammus pulcher Ladiges, tetra, freshwater
  Pleistophora hyphessobryconis (skeletal muscles)
Hippocampus erectus Perry, seahorse; marine
  Glugea heraldi (subcutaneous connective tissue)
Hippoglossoides platessoides (Fabricius), American plaice; marine
  Pleistophora hippoglossoideos (skeletal muscles)
Hyphessobrycon callistus callistus Boulenger, tetra, freshwater
  Pleistophora hyphessobryconis (skeletal muscles)
Hyphessobrycon flammeus Meyers, flame tetra; freshwater
  Pleistophora hyphessobryconis (skeletal muscles)
Hyphessobrycon gracilis (Reinhardt)
  Pleistophora hyphessobryconis (skeletal muscles)
Hyphessobrycon heterorhabdus Ulrey, tetra; freshwater
  Pleistophora hyphessobryconis (skeletal muscles)

Hyphessobrycon rosaceus Durbin, tetra; freshwater
  Pleistophora hyphessobryconis (skeletal muscles)
Hypomesus olidus (Pallas), pond smelt, freshwater
  Glugea hertwigi (intestine and various body organs)
Ictalurus punctatus (Rafinesque), channel catfish; freshwater
  Microsporidium sp. of Herman and Putz, 1970 (heart and intestine)
Leiostomus xanthurus Lacépède, spot; marine
  Ichthyosporidium giganteum (body cavity)
Leuciscus cephalus (L.), chub; freshwater
  Pleistophora oolytica (ovary)
Limanda ferruginea (Storer), yellowtail flounder; marine
  Glugea stephani (chiefly the intestine)
Limanda limanda (L.), dab; marine
  Glugea stephani (chiefly the intestine)
Liopsetta glacialis (Pallas), Arctic flounder; marine
  Glugea stephani (chiefly the intestine)
Lophius americanus Valenciennes, American anglerfish; marine
  Spraguea lophii (ganglia of the central nervous system)
Lophius budegassa Spinola, black bellied anglerfish; marine
  Spraguea lophii (ganglia of the central nervous system)
Lophius gastrophysus Miranda Ribiero; marine
  Spraguea lophii (ganglia of the central nervous system)
Lophius piscatorius L., European anglerfish; marine
  Spraguea lophii (ganglia of the central nervous system)
Lota lota (L.), burbot; freshwater
  Glugea fennica (skin)
  "*Nosema anomala*" Bauer, 1948 (intestine); Izjumova, 1959; Grabda, 1961
  Pleistophora ladogensis (muscles)
Lycodopsis pacifica (Colett), blackbelly ealpout; marine
  Microsporidium sp. (*Glugea?*) of Noble and Collard, 1970 (dorsal muscles)
Macrourus berglax Lacépède, roughhead grenadier; marine
  Glugea berglax (intestine and gall bladder)
Macrozoarces americanus (Bloch and Schneider), ocean pout; marine
  Pleistophora macrozoarcidis (muscles)
Mallotus villosus (Muller), capelin; marine
  Microsporidium sp. of Marchant and Schiffman, 1946 (body cavity)
Melanogrammus aeglefinus (L.), haddock; marine
  Loma branchialis (gills)
Merluccius gayi Guichenot, Chilean hake; marine
  Microsporidium ovoideum (liver)
Merluccius hubbsi Marini, Argentinean hake; marine
  Microsporidium ovoideum (liver)
Mesogobius batrachocephalus (Pallas), froghead goby; marine
  Glugea sp. of Naidenova, 1974 (intestine)
Micromesistius poutassou Risso, blue whiting; marine
  Microsporidium sp. of Gaievskaya and Kovaleva, 1975 (skeletal muscles)
Mollienesia sphenops Cuvier and Valenciennes, molly; freshwater
  Microsporidium pseudotumefaciens (ovary, organs in the body cavity)
Motella tricirrata Nilss., three-bearded rockling; marine
  Microsporidium ovoideum (liver)

Mugil cephalus Risso, golden gray mullet; marine
  Pleistophora destruens (skeletal muscles)
Mullus barbatus L., red mullet; marine
  Microsporidium ovoideum (ovary)
Myctophum punctatum Rafinesque, lantern fish; marine
  Glugea capverdensis (intestine, mesentery, ovary)
Mylopharyngodon piceus (Richardson), black Amur fish; freshwater
  Glugea intestinalis (intestine)
Myoxocephalus quadricornis labradoricus (Girard), Arctic fourhorn sculpin; marine
  Pleistophora typicalis (skeletal muscles)
Myoxocephalus scorpius (L.), European horned sculpin; marine
  Pleistophora typicalis (skeletal muscles)
Nemipterus japonicus (Bloch), thread-fin bream; marine
  Glugea nemipteri (liver, gonads, visceral muscles)
  Microsporidium bengalis (gills)
Neogobius caspius (Eichwald), "Hyrcanian goby"; marine
  Glugea shulmani (intestine)
  Pleistophora tuberifera (subcutaneous musculature)
Neogobius cephalarges (Pallas), mushroom-goby; marine
  Glugea sp. of Naidenova, 1974 (intestine)
Neogobius fluviatilis (Pallas), sandy Black Sea goby; marine
  Glugea sp. of Naidenova, 1974 (intestine)
Neogobius fluviatilis pallasi (Berg), monkey-goby; marine
  Glugea shulmani (intestine)
Neogobius kessleri gorlap (Iljin), gorlap goby; Caspian Sea
  Pleistophora tuberifera (subcutaneous musculature)
Neogobius melanostomus (Pallas), round Black Sea goby; marine
  Glugea sp. of Naidenova, 1974 (intestine)
  "*Glugea anomala*" Naidenova, 1974 (site of infection not stated)
Neogobius melanostomus affinis (Eichwald), Caspian round goby; marine
  Glugea shulmani (intestine)
  Pleistophora tuberifera (subcutaneous musculature)
Neogobius syrman (Nordmann), Black Sea goby; marine
  "*Glugea anomala*" Naidenova, 1974 (subcutaneous)
Noemacheilus barbatulus (L.), common stone loach; freshwater
  Pleistophora macrospora (skeletal muscles)
Noemacheilus malapterus longicauda (Kessler), eastern crested stone loach;
     freshwater
  Pleistophora sp. of Dzhalilov, 1966 (site of infection not reported)
Notemigonus crysoleuces (Mitchill), golden shiner; freshwater
  Pleistophora ovariae (ovary)
Odontogadus merlangus (L.), whiting; marine
  Mrazekia piscicola (pyloric caeca)
Oncorhynchus gorbuscha (Walbaum), pink salmon; freshwater
  Microsporidium takedai (heart and skeletal muscles)
Oncorhynchus keta (Walbaum), chum salmon; euryhaline
  Microsporidium takedai (heart and skeletal muscles)
Oncorhynchus kisutch (Walbaum), coho salmon; freshwater
  Microsporidium rhabdophilia (rodlet cells in the skin, gills and intestine)
Oncorhynchus masu Brevoort, masu salmon; euryhaline

Loma salmonae (gills)
Microsporidium takedai (heart and skeletal muscles)
Oncorhynchus nerka (Walbaum), sockeye salmon; euryhaline
Loma salmonae (gills)
Oncorhynchus nerka var. *adonis* (Walbaum), kokanee; euryhaline
Microsporidium takedai (heart and skeletal muscles)
Oncorhynchus tschawytscha (Walbaum), chinook salmon; freshwater
Microsporidium rhabdophilia (rodlet cells in skin, gills and intestine)
Microsporidium takedai (heart and skeletal muscles)
Opsanus beta (Goode and Bean), gulf toadfish; marine
Nosema notabilis (hyperparasite of the myxosporidian *Ortholinea polymorpha* parasitic in the gall bladder)
Opsanus tau (L.), oyster toadfish; marine
Nosema notabilis (hyperparasite of the myxosporidian *Ortholinea polymorpha* parasitic in the gall bladder)
Osmerus eperlanus eperlanus (L.), European smelt, euryhaline
Glugea hertwigi (intestine and various body organs)
Osmerus eperlanus eperlanus n. *ladogensis* Berg, European smelt; euryhaline
Pleistophora ladogensis (skeletal muscles)
Osmerus eperlanus eperlanus m. *spirinchus* Pallas, Baltic smelt; euryhaline
Glugea hertwigi (intestine and various body organs)
Osmerus eperlanus mordax (Mitchill), rainbow smelt; euryhaline
Glugea hertwigi (intestine and various body organs)
Paracheirodon inesi (Myers), neon tetra; freshwater
Pleistophora hyphessobryconis (skeletal muscles)
Paralepis elongata (Brauer); marine
Glugea sp. of Reimer, 1975 (pers.comm.) (stomach wall)
Parophrys vetulus Girard, English sole; marine
Glugea stephani (chiefly the intestine)
Perca flavescens (Mitchill), yellow perch; freshwater
Glugea sp. (this publication) (skeletal muscles)
Perca fluviatilis L., common perch; freshwater
Glugea sp. of Roman, 1955 (gills)
Perca schrenki Kessler, Balkhash lake perch; freshwater
Pleistophora acerinae (mesentery and intestinal wall)
Percottus glehni Dybowsky, bigmouth sleeper; freshwater
Thelohania peponoides (subcutaneous tissue)
Pholis gunnellus (L.), rock gunnel; marine
Microsporidium sp. of Lom and Laird, 1977 (subcutaneous tissue)
Phoxinus phoxinus (L.), elritze; freshwater
Glugea sp. of Pfeiffer, 1895 (site of infection not stated)
Pimephales promelas Rafinesque, fathead minnow; freshwater
Glugea pimephales (body cavity)
Platessa platessa L., winter flounder; marine
Glugea stephani (chiefly the intestine)
Platichthys stellatus (Pallas), starry flounder; marine
Glugea stephani (intestinal wall)
Platypoecilus maculatus var. *pulchra;* freshwater
Microsporidium pseudotumefaciens (ovary, organs in the body cavity)
Plecoglossus altivelis Temminck and Schlegel, ayu; freshwater

Glugea plecoglossi (various organs of the body cavity, gills, iris, etc.).
Pleuronectes flesus Pallas, flounder; marine
   Glugea stephani (chiefly the intestine)
   Microsporidium sp. of Raabe, 1935 (skin)
Pleuronectes flesus bogdanovi Sandenberg, White Sea river plaice; marine
   Glugea stephani (chiefly the intestine)
Pleuronectes flesus trachurus Duncker, Baltic river plaice; marine
   Glugea stephani (chiefly the intestine)
Poecilia (Lebistes) spp., guppies; freshwater
   Microsporidium pseudotumefaciens (ovary, organs in the body cavity)
Pomatoschistus minutus (Pallas), sand goby (freckled goby); marine
   Glugea anomala (?) (subcutaneous)
Poronotus triacanthus (Peck), butterfish; marine
   Pleistophora sp. of Woodcock, 1904 (liver)
Priacanthus macracanthus Cuvier and Valenciennes, big-eye; marine
   Pleistophora priacanthicola (pericardium, stomach, pyloric caecum,
      intestine, liver, spleen, gonads and body wall muscles)
Priacanthus tayenus Richardson, big-eye; marine
   Pleistophora priacanthicola (pericardium, stomach, pyloric caecum,
      intestine, liver, spleen, gonads and body wall muscles)
Proterorhinus marmoratus (Pallas), tubenose goby; euryhaline
   Glugea sp. of Naidenova, 1974 (intestine)
Pseudopleuronectes americanus (Walbaum), winter flounder; marine
   Glugea sp. (this publication) (intestine)
   Glugea stephani (chiefly the intestine)
Pterophyllum scalare (Cuvier and Valenciennes), angelfish; freshwater
   Heterosporis finki (oesophagus)
Pungitius platygaster (Kessler), southern ninespined stickleback; freshwater
   Glugea anomala (intestine)
Pungitius pungitius (L.), ninespined stickleback; euryhaline
   Glugea anomala (subcutaneous and in most of the body cavity organs)
   Thelohania baueri (ovary)
Rhodeus sericeus amarus (Bloch), common European bitterling; freshwater
   Glugea rodei (serose membrane of internal organs and body wall)
Rhombus maximus (L.), turbot; marine
   Glugea stephani (chiefly the intestine)
Rutilus rutilus (L.) × *Abramis brama* (L.), a hybrid of roach and common bream;
    freshwater
   Pleistophora elegans (ovary)
Rutilus rutilus (L.), roach; freshwater
   Pleistophora sp. of Lucky in Lucky and Dyk, 1964 (= *P. longifilis*) (testis)
   Pleistophora oolytica (ovary)
Salmo gairdneri Richardson, rainbow trout; freshwater
   Loma salmonae (gills)
   Microsporidium rhabdophilia (rodlet cells in skin, gills and intestine)
Salmo gairdneri irideus Gibbons, rainbow trout; freshwater
   Microsporidium rhabdophilia (rodlet cells in skin, gills and intestine)
   Microsporidium takedai (heart and skeletal muscles)
Salmo salar L., Atlantic salmon; euryhaline
   Microsporidium sp. of Plehn, 1924 (eggs, yolk sac of the fry)

Salmo trutta m. *fario* L., brook trout; freshwater
  Glugea truttae (yolk sac of newly hatched fry)
Salmo trutta trutta L., sea trout; euryhaline
  Microsporidium sp. of Plehn, 1924 (eggs, yolk sac of the fry)
Salvelinus fontinalis (Mitchill), brook trout; euryhaline
  Loma fontinalis (gills)
  Pleistophora sp. of Putz, 1965 in Putz and McLaughlin, 1970 (gills)
Salvelinus leucomaenis Pallas, char; freshwater
  Microsporidium takedai (heart and skeletal muscles)
Salvelinus malma (Walbaum), dolly warden char; freshwater
  Microsporidium takedai (heart and skeletal muscles)
Sardina pilchardus sardina (Risso), sardine; marine
  Glugea cordis (heart)
Saurida tumbil (Block), greater lizard fish; marine
  Microsporidium sauridae (visceral muscles)
  Pleistophora sauridae (skeletal muscles)
Scatophagus argus (L.), silver cat; marine
  "*Glugea anomala*" Laird, 1956 (site of infection not stated)
Sciaena australis Gunther, perch of the Brisbane river; freshwater
  Microsporidium sciaenae (ovary)
Scophtalmus meoticus (L.), turbot; marine
  Tetramicra brevifilum (skeletal muscles)
Scorpaena porcus (L.), small-scale scorpion fish; marine
  Microsporidium sp. of Jírovec, 1938 (site of infection not stated)
Seriola quinqueradiata Temminck et Schlegel, yellowtail; marine
  Microsporidium seriolae (skeletal muscles and heart)
Silurus glanis L., European catfish; freshwater
  Glugea tisae (intestine)
  Pleistophora siluri (intestine)
Solea solea (L.), sole; marine
  Pleistophora hyppoglossoideos (skeletal muscles)
Sphaeroides maculatus (Bloch and Schneider), northern puffer; marine
  Glugea sp. (this publication) (intestine)
Stizostedion lucioperca (L.), pike perch; freshwater to marine
  Glugea lucipercae (intestine)
Syngnathus acus (L.), great pipefish; marine
  Glugea acuta (swim bladder muscle)
Synodus lucioceps Ayres, California lizard fish; marine
  Encephalitozoon sp. of Jensen, Moser and Heckmann, 1979 (skeletal muscles)
Taurulus bubalis (Euphrasen), buffalo sculpin; marine
  Microsporidium cotti (testis)
  Pleistophora typicalis (skeletal muscles)
Theragra chalcogramma (Pallas), walleye pollock; marine
  Pleistophora sp. of Arthur, Margolis and McDonald, 1982 (skeletal muscles).
Tilapia melanopleura Dumeril; freshwater
  Loma sp. of Bekhti, 1984 (gill filaments)
Trachurus declivis Jenyns, mackerel; marine
  Microsporidium sp. of Jones, 1979 (pericardial cavity, brain nerve trunks)
Trisopterus luscus (L.), whiting pout; marine
  Glugea shiplei (skeletal muscles)

Valamugil sp., a grey mullet; euryhaline
  Microsporidium valamugili (intestine)
Vimba vimba persa (Pallas), Caspian vimba; Caspian Sea
  Microsporidium sp. of Gasimagomedov and Issi, 1970 (intestine, kidney)
Xiphophorus helleri Heckel, swordtail; freshwater
  Pleistophora hyphessobryconis (skeletal muscles)
Xiphophorus maculatus var. *pulchra* (Gunther), platy; freshwater
Xiphophorus spp., swordtails; freshwater
    both host species: *Microsporidium pseudotumefaciens* (ovary, organs in the
      body cavity)
Zosterisessor ophiocephalus (Pallas), marine
  Loma dimorpha (walls of the digestive tract)

II. PATHOGENICITY

A. The Cellular Level

The growth of microsporidia and the proliferation of merogonic and sporogonic stages bring about progressive degradation of cell cytoplasm and organelles, which culminates in the total destruction of the cell and its replacement by mature spores. The exact mechanism by which the injury is inflicted upon the cell remains a mystery. Even at the ultrastructural level abnormalities at the interface between the parasite and cell cytoplasm are not usually apparent. Where cell ultrastructure is disorganised, unaltered mitochondria persist and concentrate around the parasites, presumably meeting the parasite's energy requirements in the absence of their own mitochondria. A rare example of a destructive influence on the surrounding cytoplasm is provided by *Pleistophora hyphessobryconis*. Here, the parasite stages are surrounded by a halo visible by light microscopy (Fig. 2.2). Ultrastructurally the halo consists of unstructured sarcoplasm, without myofibrils, containing small vesicles, cisternae of smooth ER and groups of ribosomes. This zone abuts directly onto intact myofibrils (Fig. 2.1).

Hypertrophy can be recognised in many more instances than actually recorded in the literature. A cell, packed full of parasites, and hence greatly enlarged in size, is not just mechanically distended but is in fact hypertrophied. The definition and terminology of host cell-parasite complexes date from Chatton (1920) and Weissenberg (1922). Chatton coined the term xenoparasitic complex or xénoparasitome (= "complexe xénoparasitaire" or "xénoparasitome") for an intimately close union of a parasitic dinoflagellate *Neresheimeria paradoxa* with its tunicate host *Frittilaria*. The term xénoparasitome conveyed the idea of an elaborate complex in which the host cell changes its structure and size to become physiologically and morphologically integrated with its parasite. It becomes a separate entity with its own development in the host, at the expense of which it grows. Chatton & Courrier (1923) applied this term to the formations induced by *Nosema cotti* (see *Microsporidium cotti,* Sect. III,L) in *Cottus bubalis.* Weissenberg (1922), used the term "xenon" for this host-parasite cellular union. Since it was a rather unfortunate synonym with the name of a chemical element, he later changed it in his later publications to "xenoma" or "xenom." This term, thanks to its brevity has become generally accepted.

The host cell component of the xenoma has a strongly hypertrophic nucleus sometimes branched or fragmented amitotically into numerous separate nuclei and the cell surface becomes modified for increased absorption. In *Microsporidium cotti, Ichthyosporidium* and *Tetramicra* the cell surface is expanded as microvillus-like structures, whereas, in *Glugea* and *Loma,* the cell surface compensates for the absence of microvilli by mass formation of pinocytotic vesicles.

There are several different kinds of microsporidian xenomas in fish (see Fig. 2.6). In *Heterosporis,* the connective tissue cell enlarges but maintains a simple plasma membrane at its surface (Sect. III,E). In *Pleistophora longifilis,* the epithelial cell of the seminiferous canaliculus is simply hypertrophied (Fig. 2.64; p. 112). In the *Glugea* type of xenoma, the elaborate parasitic formation (Sect. IIIA,1) can reach up to 13 mm in size. The *Ichthyosporidium* type of xenoma can reach dimensions of centimetres. The structure of individual types is known up to a point (see descriptions of species) but the physiology, biochemical interaction and immunological phenomena are completely unknown.

The biological meaning of xenomas seems to be to secure advantage for both host and parasite. For the parasite, it provides a suitable environment for proliferation, while protecting it against host attack by masking it within a host component. The host benefits by confining the parasite and ensuring that free spread of the parasite does not take place. The ultimate fate of the infected cell is of course death, by its complete replacement by the parasite that it harbours.

B. The Tissue Level

A knowledge of the histopatholgical changes caused by microsporidia is fundamental to a real assessment of the pathogenicity of individual species. Even so, there has been no comprehensive study of the tissue re-

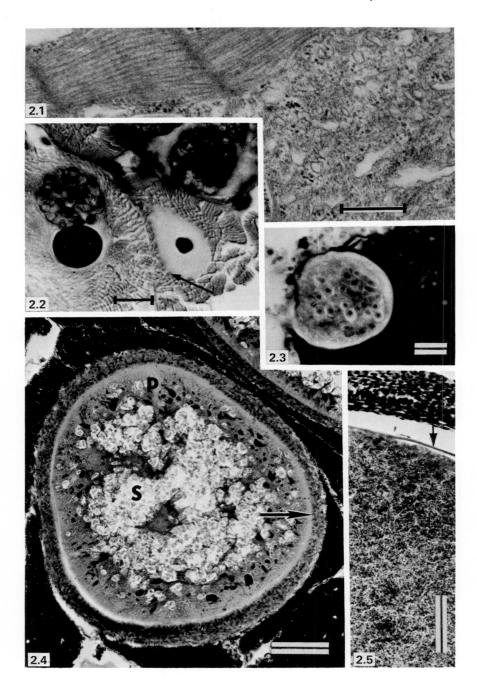

sponses of fish to these parasites, in spite of some early pioneering works. As early as 1899, Mrázek correctly recognized that Doflein (1898) had misinterpreted stages of *Spraguea lophii* as early developmental stages, when they were in fact late stages of infection. Early workers sometimes mistook phagocytosis and host reparative reactions for early stages of parasite development within numerous cells.

After recognition of the type of responses to microsporidian invasion by fish, descriptions of new species have sometimes included observations on tissue reactions (e.g. Drew, 1910; Debaisieux, 1920; Jírovec, 1930; Nigrelli, 1946; Thieme, 1954; Kabata, 1959; Lom and Laird, 1976; Matthews and Matthews, 1980). Mercier (1921) correctly recognized the inflammatory character of the tissue reaction to *Ichthyosporidium giganteum*, noticing at the same time, that the onset of host reaction coincided with the moment of spore formation. Sometimes errors were perpetuated, as when Reichenow (1952) suggested that *Glugea stephani* xenomas originated from several parasitised cells, which nowadays would be interpreted as phagocytes.

Full descriptions of tissue changes have been recorded on occasion: by Awakura (1974) for *Microsporidium* (*"Glugea"* or *"Nosema"*) *takedai* and by Dyková and Lom (1980a) for *Glugea plecoglossi*, both from natural infections. Surveys by Sindermann (1970), Lom (1970) and Canning (1976b)

Fig. 2.1 Muscle bundles of neon tetra infected with *Pleistophora hyphessobryconis:* electron micrograph of the interface between the still intact muscle fibrils (upper left) and the halo of disintegrated sarcoplasm (area shown in Fig. 2.2 by an arrow); scale bar = 0.5 μm.

Fig. 2.2 Histological section of neon tetra muscle showing the striking "halo" of disintegrated sarcoplasm, which occurs especially round the early developmental stages of *Pleistophora hyphessobryconis*. H & E; scale bar = 10 μm.

Fig. 2.3 Early stage of *Glugea anomala*-xenoma, developing within an oocyte of the stickleback, *Gasterosteus aculeatus:* there is no host reaction; xenoma contains globular and cylindrical meronts. H & E; scale bar = 10 μm.

Fig. 2.4 *Glugea plecoglossi:* a well developed xenoma in the testis of *Plecoglossus altivelis:* S, central space with mature spores; P, periphery with merogonic stages and fragmented host cell nuclei; the xenoma wall is encased by a layer of host connective tissue (arrow). H & E; scale bar = 100 μm.

Fig. 2.5 Part of periphery of a mature xenoma of *Glugea anomala:* closely packed mature spores occupy the whole xenoma, now reduced to a "cyst", consisting of the wall (arrow) and cytoplasm packed full of spores. H & E; scale bar = 50 μm.

of pathological changes produced by microsporidian infections in fishes, were comprised of records of macroscopical lesions and estimates of the probable mortality rate caused by the parasites in question but did not deal with reactions at tissue level.

Existing data suggests that all fish tissue reactions to microsporidian invasion involve essentially identical phenomena. However, two patterns may be distinguished, the manner in which these reactions are deployed differs according to whether the species eliciting the response are localised

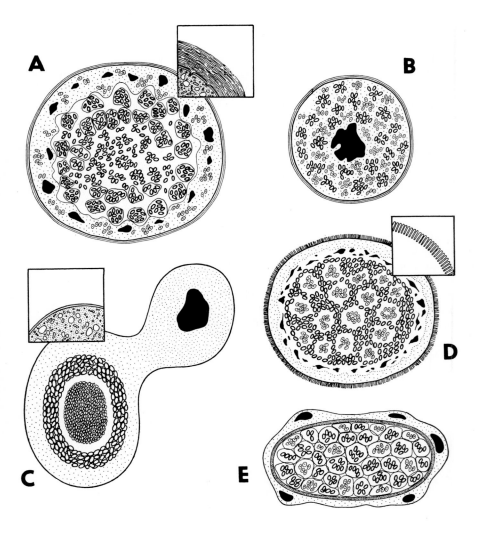

in a xenoma (a hypertrophic cell) or are diffused in the tissues. The former is exemplified by *Glugea* species and the latter by *Pleistophora* species.

1. Tissue Reaction to Xenomas

This pattern is characteristic for species of the genera *Glugea, Ichthyosporidium, Loma, Tetramicra* and *Spraguea,* which develop within the cytoplasm of a single host cell. *G. anomala,* a parasite of sticklebacks, serves as a model for this type of reaction. It has been studied in detail by Weissenberg (1967, 1968) and Debaisieux (1919a, 1920). Sprague and Vernick (1968a) studied the same phenomena in *Glugea weissenbergi,* a closely related species. *Glugea plecoglossi,* a parasite of ayu, is similar. There are three principal stages of development of the xenoma: at an early stage the host cell cytoplasm is uniformly occupied by developmental stages of the parasite (Figs. 2.3, 2.7), when fully grown the xenoma is encased in a refractile wall and the centre of the cell is occupied by a mass of spores (Fig. 2.4); when mature, xenomas are completely filled with mature spores and there is almost nothing left of the host cell components except for the xenoma wall (Fig. 2.5).

The tissue reactions are directed towards isolation of the parasite from the very first stage. Three recognisable stages can be correlated with the three periods of development.

Fig. 2.6 Diagrammatic representation of various types of xenoparasitic complexes, induced by microsporidian parasites infecting fish. All xenomas are in an advanced stage of development. A. *Glugea anomala* type: xenoma wall consists of a laminated layer external to the plasma membrane; developmental stages of the parasite are confined to the peripheral cytoplasmic layer, which also contains numerous profiles of the host cell nucleus; the central region contains SPOVs and free spores. Inset: detail of the laminated wall at the xenoma periphery. B. *Glugea acerinae* type: there is a single centrally located, hypertrophic nucleus; developmental stages of the parasite are mixed with groups of mature spores in SPOVs throughout the host cell cytoplasm; the overall dimensions of the xenoma are much smaller than those of *Glugea anomala* type. C. *Spraguea* type: xenoma wall is a simple plasma membrane (inset); although the whole neurone becomes hypertrophied, including its nucleus which stays single, the parasite develops in one part of the cell only; the parasite mass, in an advanced stage, consists of a centrally located agglomeration of small, slender *Nosema*-type spores and a peripheral, more darkly stained layer of large, oval *Nosemoides*-type spores. D. *Microsporidium cotti* type: cell membrane of the xenoparasitic complex is folded into numerous microvillus-like extensions (inset); at the periphery are located numerous host cell nuclei (or their fragments); islets containing developmental stages are interspersed among mature spores. E. *Heterosporis* type: the host contribution may be a single hypertrophic cell, with several hypertrophic nuclei or possibly a syncitium formed by fusion of several hypertrophic cells; the xenoma contains a mass of SPOVs within a common cyst wall, some containing mature spores and some still with developmental stages.

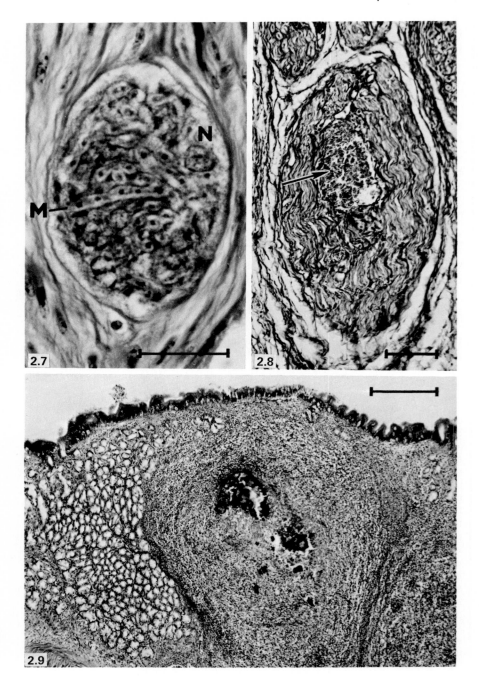

a. *Weakly reactive stage:* The weakly reactive stage usually accompanies young, thin-walled xenomas, which contain only meronts, but in some cases it may be long lasting and overlap with fully grown xenomas; early cell infiltration can be observed. The surrounding tissue is damaged and becomes atrophic due to pressure from the enlarging cell. The damage stimulates proliferation of connective tissue and hyperplasia of collagen fibers, as a result of which a thin concentric layer of connective tissue is laid down around the xenoma. The development of this layer, as well as the duration of the weak tissue reaction, is to a large extent dependent on the amount of interstitial connective tissue in the type of tissue affected.

b. *Productive stage:* When the xenoma is fully grown the host begins its reaction to it by formation of granulation tissue. As a rule, it consists of proliferating fibroblasts, histiocytes and newly formed capillaries (Fig. 2.10; 2.71; p. 128). This reaction may set in as early as 60 days post infection in *G. anomala* (Dyková and Lom, 1978), so that the xenoma is short-lived. Nothing is known of the stimuli initiating the proliferative inflammation although, as Dyková and Lom (1980b) have speculated, it may be elicited by microscopically undetectable changes in the xenoma contents and/or wall. The inflammation may be triggered metabolically or biochemically. The xenoma wall becomes visibly oedematous immediately following the onset of proliferative inflammation and is later invaded by fibroblasts which appear wedged between the stack of lamellae which constitutes the xenoma wall (Lom, Dyková and Canning, 1979). These changes result in complete disappearance of the wall. Simultaneously the appearance of PAS positive substances and calcium deposits within the mass of spores provides evidence of dystrophic changes (Fig. 2.9). At this point, the xenoma turns into a granuloma with a spore mass in its centre, into which grows the newly formed granulation tissue. Secondary xenomas, which have been detected growing amidst the granulation tissue cells (Lom and Dyková, unpublished) probably originate by infection of mi-

---

Fig. 2.7 Early stage of *Glugea plecoglossi*-xenoma, developing in the *lamina muscularis* of the intestine of *Plecoglossus altivelis;* there is not yet any host tissue reaction; the cytoplasm is filled with cylindrical meronts (M); N = host cell nucleus. H & E; scale bar = 20 μm.

Fig. 2.8 *Spraguea lophii:* a mass of mature spores found within a nerve axon which is an unusual site of development. Trichrome; scale bar = 100 μm.

Fig. 2.9 *Glugea anomala* infection; granuloma in the wall of the stomach of stickleback; dystrophic calcification in the centre. H & E; scale bar = 200 μm.

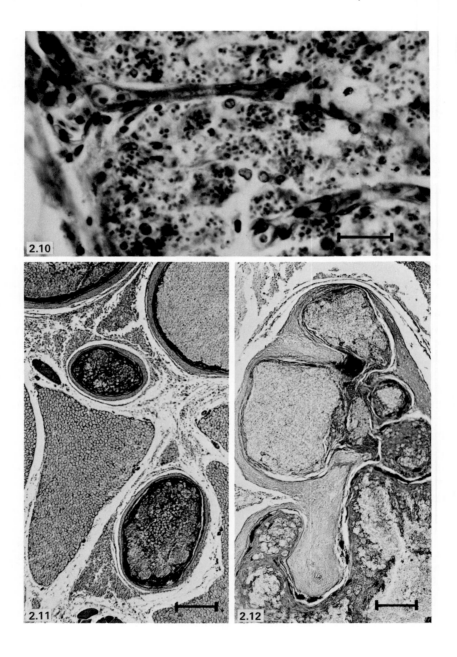

gratory cells which abound in the inflammatory proliferative stage in the granuloma. Secondary xenomas have also been seen in *Glugea capverdensis* (Fig. 2.29; p. 56).

The greatest proliferative changes have been recorded against *G. anomala*-xenomas when they had a subepithelial localisation in the intestine. However, in similarly located *G. hertwigi*-xenomas, changes were minimal. An exudative proliferative inflammation of the type induced by *Glugea* xenomas is also known for *Spraguea lophii* xenomas, when they have reached the spore filled stage (Fig. 2.85; p. 142).

The reaction against the large lobed xenomas of *Ichthyosporidium* when they are filled with spores are somewhat different. In this case the proliferative inflammation is characterised by the presence of epithelioid cells and histiocytes. The wall of the xenoma becomes encased with a single layer of epithelioid cells in a palisade-like arrangement, the long axes of the cells being orientated perpendicularly to the xenoma wall. Other epithelioid cells, further away from the wall, are arranged irregularly.

c. *Granuloma involution:* In *G. anomala,* the mass of spores is gradually eliminated by phagocytosis and subsequent necrotic disintegration within macrophages. The granulation tissue matures, the granuloma diminishes and the resulting fibrous connective tissue may undergo hyalinisation. Gradual repair of the tissue lesion takes place.

2. Tissue Reaction to Diffuse Infections

Species of *Pleistophora* infecting muscle fibres or oocytes replace and destroy the contents of the infected cells without inducing hypertrophy. In *Pleistophora* infections of muscle fibres, there are two paths in the formation of muscle lesions. In *P. hyphessobryconis,* the parasite occupies isolated patches in the cells, always with some sarcoplasm left between

---

Fig. 2.10 *Spraguea lophiii* infection of ganglion cells of the anglerfish *Lophius piscatorius;* newly formed capillaries growing into the granuloma. H & E; scale bar = 20 μm.

Fig. 2.11 *Pleistophora macrozoarcidis* infection of muscle fibres of the ocean pout; the sarcoplasm is completely replaced by the parasite; some fibres are surrounded by a connective tissue envelope. H & E; scale bar = 0.2 mm.

Fig. 2.12 *P. macrozoarcidis* infection; a giant granuloma, formed by several demarcated groups of muscle fibres fused together and containing a necrotic mass. H & E; scale bar = 0.1 mm.

them and the muscle fibres are badly destroyed, long before all the available cytoplasm is invaded. In species like *P. ehrenbaumi* and *P. macrozoarcidis,* the contents of the muscle fibres are completely replaced by the parasites, which are packed closely within the sarcolemma with no space left between them and fusion of several infected muscle fibers may take place (Dyková and Lom, 1980b).

During merogony and sporogony of these muscle parasites the host tissue undergoes dystrophic changes, but the only sign of host reaction is minimal cell infiltration into the myosepta (Fig. 2.51; p. 102). The real host reaction, primarily phagocytosis of spores, sets in when the muscle fibres have disintegrated and the mature spores are released from them.

In *P. macrozoarcidis* infections, groups of affected muscle fibres completely filled with mature spores become encased in connective tissue (Fig. 2.11). Giant granulomas may originate after the fusion of several affected muscle fibres which have undergone necrotic changes (Fig. 2.12). In contrast, in heavy *P. hyphessobryconis* infections, there is a minimal response in the muscles. Granulomas only form if the parasites are atypically located, as for example on the serosa or in the intestinal wall. No signs of regeneration of the muscle fibres can be detected irrespective of the extent of the lesions.

In oocyte-infecting *Pleistophora* species (*P. oolytica, P. ovariae*) there is also a minimum of cell reaction until the parasites are released from the oocytes (Dyková and Lom, 1980b). When mature spores escape through the ruptured oocyte wall, they are removed by phagocytosis (Fig. 2.14). However, since the proliferating connective tissue does not tend to demarcate the affected oocytes, these spores may spread into the lumen of the ovary.

3. Phagocytosis of Spores

Phagocytic cells participate in both types of tissue reaction playing a pivotal role in the host defense mechanism.

In *P. hyphessobryconis* infections, at the stage of muscle disintegration with resulting invasion of migratory phagocytic cells, phagocytosis is extensive. Within the phagocytes, not only isolated spores, but also groups of two to three spores still encased within the matrix of the SPOV may be observed (Fig. 2.63; p. 110). The matrix, and then the spores, are gradually digested. The phagocytes are cells from the granulation tissue, including, without doubt, the fibroblasts.

In *P. ovariae,* spores are also ingested by cells of the granulation tissue

(Fig. 2.14). Various degrees of digestion can be observed; some are still intact spores, some are crumpled with remnants of normal spore contents, and yet others are completely devoid of spore contents. In the last stage of digestion the spore wall is folded or completely compressed, then thinned and dissolved completely, including the chitinous electron lucent layer of the wall. These spore residues are usually enclosed in a distinct vacuole. This shows that the granulation tissue cells have a remarkable potential for digestion of chitin. Furthermore, the way in which the spores are digested testifies to the permeability of the spore wall to host cell enzymes, since the spore contents are dissolved before the wall.

Similar phagocytosis takes place in the granulation tissue replacing the xenomas of *Spraguea lophii* (Dyková and Lom, 1980b). Many of the spore-filled macrophages, in the centre of the granuloma, themselves disintegrate and the resulting spore-containing debris may be taken up by other phagocytes. A large number of the phagocytized spores have well preserved shapes but appear empty or filled with flocculent material (Fig. 2.13). As with the *Pleistophora* spores, it is likely that the spore contents are digested by the enzymes of the phagocytes. The spore walls are ultimately crumpled, compressed and digested in the same way. The process of destruction of spores of *Tetramicra brevifilum* phagocytized by turbot macrophages (Matthews and Matthews, 1980) also follows the same pattern. Whether formed in xenomas or not, spores are ultimately disposed of by phagocytes, unless they are discharged to the environment while the host is still alive or liberated only when the host dies and decomposes.

4. The Outcome of the Host Tissue and Parasite Interaction

The efficiency of the tissue reaction depends upon suitable physiological conditions of the host and favourable environmental factors. Of importance is the ambient temperature which affects not only the humoral response (Corbel, 1975; Roberts, 1975) but also phagocytosis. Low temperatures delay the appearance and development of phenomena such as macrophage responses and fibroplasia (Finn and Nielson, 1971). Similarly the activity of fixed macrophages of the reticuloendothelial system is inhibited at lower temperatures (Ferguson *et al.*, (quoted by Roberts, 1975).

On the other hand, sub-optimal temperatures may impede the development of the parasite. Takahashi and Egusa (1977b) described growth inhibition of developmental stages of *Glugea plecoglossi* under such conditions. Clinical and histopathological changes are thus the result of a

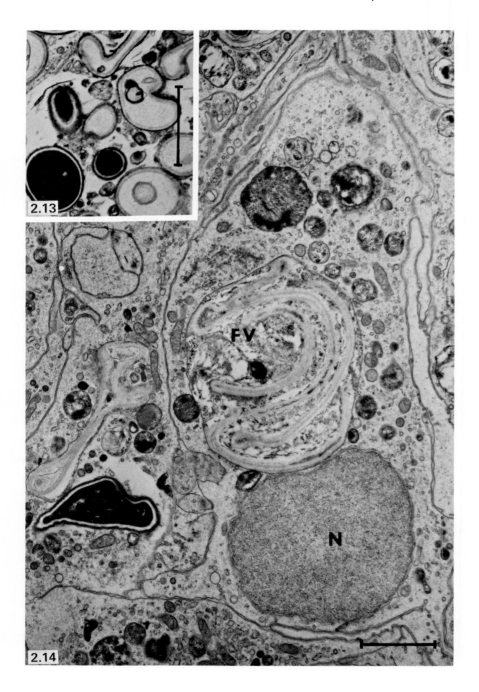

2.13

2.14

complicated interaction of the host's response and parasite's development and both are dependent on an array of environmental factors. Accordingly, differences in the extent of proliferative changes recorded for individual microsporidian infections cannot be interpreted as due exclusively to innate characters of the microsporidian species. The differences may reflect the general dependence of fish tissue reactions on ambient temperature and other factors.

Extensive xenoma development can inflict serious damage on the host in the form of pressure atrophy. Repair of this damage does not necessarily restore the original function of the organ. For example, in heavier cases of *Glugea anomala* and *G. stephani* infections of the stomach, the glandular layer is replaced by connective tissue.

In the xenoma-type of infection the possibility of autoinfection is very limited. In spite of the ability of spores to discharge polar tubes inside intact xenomas (Berrebi, 1978; Dyková and Lom, 1978), secondary xenomas are never found outside its confines.

On the whole, *Pleistophora*-type infections have a worse prognosis than the xenoma-type. Host tissue reaction does not isolate the affected muscle fibres during merogony and these stages are almost certainly responsible for spread of infection. The productive stage of inflammation is only reached when mature spores completely fill the infected muscle fibres. Mawdesley-Thomas and Bucke (1973) assumed that the regenerative potential in muscle injuries is associated with the segmental arrangement of the myotomes and that there is—from the functional point of view—no great demand for regeneration of certain eliminated muscle fibres. In view of this, it is hardly possible that there will be complete tissue repair in *Pleistophora*-type infections.

Although some progress has already been achieved in understanding fish pathology in microsporidian infections, experiments conducted under exactly controlled conditions are urgently needed to trace the exact sequence of tissue changes in various infected organs and to monitor the effect of the intensity of development of infections within the hosts.

Fig. 2.13 Electron micrograph of the final stages of *Spraguea lophii* infections: *Nosemoides*-type spores in various stages of digestion by macrophages; some spores are still intact, others are already reduced to crumpled empty shells; scale bar = 2 μm.

Fig. 2.14 *Pleistophora ovariae* infection in the ovary of golden shiner; phagocytic cell with a food vacuole (FV), containing the remains of partly digested and crumpled spore walls; N = phagocytic cell nucleus; scale bar = 2 μm.

III. DESCRIPTION OF SPECIES INFECTING FISH

A. Genus *Glugea* Thélohan, 1891

The *Glugea*-species complying well with the generic diagnosis given in Chap. 1, Sect. III,C, are *G. anomala, G. atherinae, G. capverdensis, G. fennica, G. heraldi, G. hertwigi, G. plecoglossi, G. rodei, G. shiplei, G. stephani, G. truttae,* and, possibly, also *G. acuta* and *G. cepedianae.* These have large xenomas. A separate group is represented by species forming small xenomas with a different type of organisation (Sect. 2). This group includes *G. acerinae, G. berglax, G. luciopercae, G. machari, G. shulmani, G. tisae* and 3 unnamed species. Future investigations of their life cycles and ultrastructure may require their transfer to a different genus.

Other species formerly assigned to *Glugea* with inadequately known diagnostic features are provisionally left in this genus, pending future revision i.e. *G. bychowskyi, G. caulleryi, G. cordis, G. depressa, G. destruens, G. intestinalis, G. punctifera* and 6 unnamed species. There are two species with characters definitely not pertaining to *Glugea,* but, as they are insufficiently well studied for proper characterisation of new genera, they are transferred, as a temporary measure, into the collective genus *Microsporidium.* These are *Microsporidium cotti* in which the xenoma has a brush border at its periphery (Sect. L) and *Microsporidium* sp. from capelins (p. 167), in which the xenoma wall seems to bear fine fibrils.

In the following systematic treatment of species, the organisation of the xenoma and the developmental stages of the type species, *Glugea anomala,* are described in detail. Subsequently only the special features of other species, which differentiate them from the type species are stressed.

1. *Glugea* Species which Form Large Xenomas

The xenomas exceed 1 mm in diameter and their organisation is typified by *G. anomala.*

Glugea anomala (Moniez, 1887) Gurley, 1893 (Figs. 1.12, 2.3, 2.5, 2.6A, 2.9, 2.15-2.20, 2.22A,B, 2.30, 2.45A, p. 82 2.83A-C, p. 140)

SYNONYMS: *Nosema anomala* Moniez, 1887; *Glugea microspora* Thélohan, 1892; *Nosema anomalum* Labbé, 1889.

Probably also *Glugea weissenbergi* Sprague and Vernick, 1968 (p. 47) and *Glugea gasterostei* Voronin, 1974 (p. 47).

HOSTS: *Gasterosteus aculeatus,* the type host, and *Pungitius pungitius.*

Both hosts are freshwater and euryhaline. Gasimagomedov and Issi (1970) found *Pungitius platygaster* infected with a species forming oval cysts on the intestinal wall, which they identified—most probably correctly—as *G. anomala*.

The *Glugea* species in sticklebacks other than the type host, may be closely related to *G. anomala*. Reports of *G. anomala* in hosts taxonomically remote from sticklebacks, e.g. gobiid fish (p. 47) *Lota lota*, (Bauer, 1948; Izjumova, 1959; Grabda, 1961) are doubtful.

GEOGRAPHICAL DISTRIBUTION: In *G. aculeatus* and *P. pungitius* probably throughout their holarctic distribution, with reports from Europe, Asia and America. Sometimes sporadic, sometimes with more than 50% of the the population infected.

SITE OF INFECTION: Connective tissue of almost all body organs. May be restricted to, e.g. skin or ovary (Stempell, 1904) or infections may be generalised. Weissenberg (1968) claimed that the small intestine, the first tissue to show numerous xenomas in his infection experiments, acted as a focus for the spread of infection.

SIGNS OF INFECTION AND PATHOLOGY: Mature xenomas are opaque white, usually spherical "cysts", several mm in diameter. In subcutaneous sites the cysts are very striking as they bulge from the surface. If located internally large cysts may cause local distortion of the body wall. Xenoma development is asynchronous, even in the same organ, after a single infective dose (Dyková and Lom, 1978).

Serious injury to the host is caused by the growth of xenomas provoking pressure atrophy, and by the ensuing tissue reaction which is described in detail in Sect. IIB,1.

In experimental infections Dyková and Lom (1978) found that proliferative inflammation could develop in as little as 23 days and was always present at 60 days post infection. Granuloma involution could begin at 60 days with necrotic foci and calcium deposits. After 120 days secondary xenomas could be observed in some of the granulomas (Sect. II,B,1).

In some types of tissue, the reparative process never results in complete functional restoration. However, even heavily infected fish apparently tolerate the parasites well and seem to thrive with numerous huge cysts. Information on mortality and growth impairment is lacking.

ORGANISATION OF THE XENOMA: Both the parasite and host cell components of the xenoma of *G. anomala* have been studied at the light microscopic level by Weissenberg (1911a,b, 1913, 1921, 1968) and Debaisieux (1920). The ultrastructure has been described by Lom, Canning and Dyková (1979) and Canning, Lom and Nicholas (1982).

Weissenberg (1968) concluded that the target host cell must be a migratory mesenchyme cell, either a macrophage or histiocyte, to account

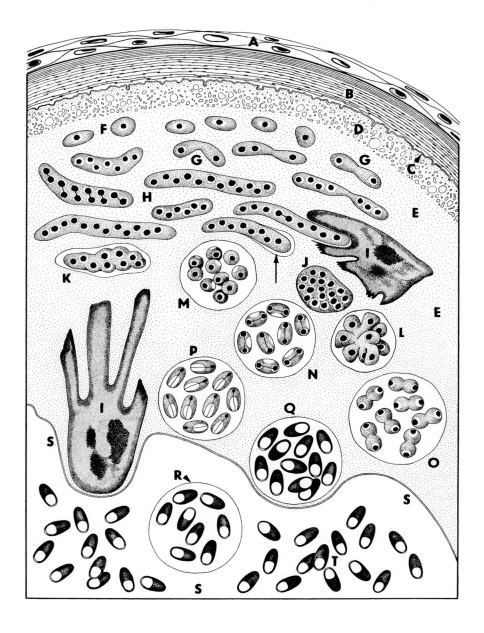

Fig. 2.15 *Glugea anomala:* diagrammatic representation of a part of the xenoma from a stickleback. A, layer of apposed connective tissue, product of the host response; B, refractile xenoma wall, composed of layers of the cell coat; C, cell membrane of the xenoma; D, peripheral layer of the xenoma with increased pinocytotic activity; E, host cell cytoplasm; F, uninucleate, initial meronts; G, their division; H, cylindrical meronts, some with nuclei in the process of division and one with the SPOV being formed around it (arrow); I, host cell nucleus; J, rounded meronts; K, elongate sporogonial plasmodium within an SPOV starting to segment into sporoblast mother cells; L, radial segmentation of sporogonial plasmodium into sporoblast mother cells; M, SPOV with sporoblast mother cells; N,O, sporoblast mother cells in stages of division; P, SPOV with sporoblasts; Q, SPOV with spores; R, SPOV preserved even after breakdown of many of the vesicles with release of mature spores into the central space S; T, agglomeration of free spores in the space occupying the centre of the xenoma.

for the almost ubiquitous distribution of cysts in body organs. He traced the development of parasitised host cells in experimental infections, reporting on the cell hypertrophy, nuclear proliferation and deposition of an external capsule, which halted the migration of the cell.

The capsule is a layer 5 μm thick which reacts positively in the periodic acid/Schiff reaction (PAS) and appears refractile by light microscopy. Ultrastructurally its shows a laminated structure (Fig. 2.17). The laminae, spaced by electron lucent layers, are produced by the sloughing off of successive layers of the cell coat from the xenoma plasma membrane, which is itself pinocytotically active.

In very young xenomas the parasites are randomly distributed (Fig. 2.3) but a definite organisation is seen as the xenomas enlarge. The outermost zone of cytoplasm is parasite-free and contains an abundance of rounded vesicles, flattened cisternae and mitochondria. The merogonic stages occupy the adjacent zone, together with highly branched profiles of the host cell nucleus, while the stages of sporulation tend towards the centre (Fig. 2.15). As more spores are formed, the host cell cytoplasm between spores becomes reduced to narrow tracts. In full-grown, ageing xenomas, the cell cytoplasm and nuclear branches almost completely disappear and the

Fig. 2.16 *Glugea anomala* infection in *Gasterosteus aculeatus:* swellings due to large subcutaneous xenomas (after Möller and Anders, 1983).

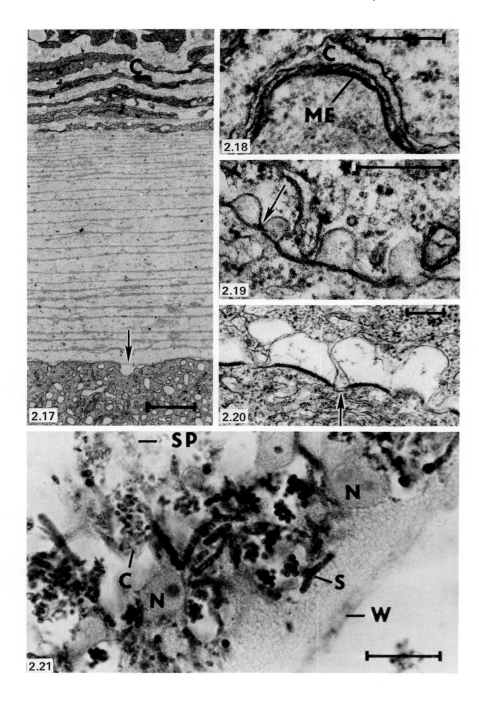

SPOVs break down leaving the cell virtually a bag of spores encompassed by the capsule.

STRUCTURE AND LIFE CYCLE (Figs. 2.17–2.20)

Merogony: The earliest stages of infection, found in mesenchyme cells, 7 days postinfection were only observed by Weissenberg (1913, 1968). He described fusiform stages, 2 × 4 μm large, uninucleate or multinucleate, which he called "primary cylinders". They lay isolated in vacuoles in the host cell (Figs. 2.83A-C, p. 140). If it is true that they lie in vacuoles, this is completely at variance with observations of meronts in more advanced xenomas.

In more advanced xenomas, apparently fixed in position, multinucleate cylindrical meronts ("secondary cylinders" of Weissenberg (1913), measuring up to 2 × 47 μm, lie in direct contact with the host cell cytoplasm. Their plasma membrane is bounded by an electron-dense surface coat, to which there is closely applied a flat cisterna of host cell smooth endoplasmic reticulum, which completely encases the parasite. The cylindrical meronts have up to 32 nuclei in a linear or zig-zag arrangement. The nuclear division products remain connected for some time by a narrow isthmus. The meronts divide by constriction along their length, ultimately into uninucleate daughter cells.

Sporogony: Sporogony is initiated by multinucleate plasmodia, actually the sporonts, derived from the cylinders by rounding up and by a subsequent loss of the electron dense surface coat and dispersal of the endoplasmic reticulum cisterna. The plasmodia develop a second membrane around themselves, external to the plasmalemma. The space between the two membranes of the parasite inflates by the raising up of the outer membrane as a layer of blisters, which swell and ultimately push away as an

---

Figs. 2.17–2.20: Electron micrographs of *Glugea anomala*. Fig. 2.17. Cross section of the xenoma wall showing layers of the cell coat sloughed from the surface; C = flat connective tissue cells of the host; at one place, the cell coat, as it peels off (arrow), overlies an invagination of the surface membrane. Fig. 2.18. Part of the meront surface (ME) bearing an electron dense deposit and surrounded by a cisterna (C) of host cell endoplasmic reticulum. Fig. 2.19. The first stage of the formation of the SPOV around the sporogonial plasmodium; the enveloping cisterna of host endoplasmic reticulum is no longer present; a series of blisters have arisen from the plasmodium surface which maintain contact in places (arrow). Fig. 2.20. Blisters, produced external to the plasmalemma of the sporogonial plasmodium, form the SPOV but still adhere, at narrow points of contact (arrow), to the sporogonial plasmodium surface; between these points of contact, the plasmodium membrane thickens by deposition of a new surface coat. Scale bars = 1 μm (Fig. 2.17), 0.3 μm (Figs. 2.18–2.20).

Fig. 2.21 *Glugea cepedianae:* cortical zone of the xenoma with pycnotic host cell nuclei (N), xenoma wall (W), cylindrical, often branched meronts (S), clusters of sporoblast mother cells (C) and spores (SP). H & E; scale bar = 20 μm.

SPOV envelope. The SPOVs are indentical with the "sporogony vacuoles" of Weissenberg (1913) and other workers.

The sporogonial plasmodium within the vesicle acquires a new electron dense surface coa. and cleaves into uninucleate sporoblast mother cells ("vacuole cells" of Weissenberg), which separate like grapes in a bunch. The sporoblast mother cells, 4 μm in diameter and with a nucleus of 2.5 μm, divide into sporoblasts, first assuming a dumb-bell shape and exhibiting a long-lasting nuclear telophase with the two halves of the nucleus linked by a conspicuous isthmus. Sporoblasts (Fig. 1.12), 3 μm in size, mature into spores. Thus, the plasmodium divides into sporoblasts through a 2-stage process, via sporoblast mother cells, in a polysporoblastic sporogony. There is no solid evidence, ultrastructural or otherwise, of autogamy or meiosis, in spite of earlier assumptions by Debaissieux (1920) and Sprague and Vernick (1968a). The latter authors studied *G. weissenbergi* (see following discussion), a closely related, if not identical species.

The SPOVs measure from about 10 to 30 μm in diameter with 10 to 100 spores per vesicle (Fig. 2.30). The SPOV wall is rather delicate, does not survive for a long time in the xenoma and tends to break when the xenoma contents are teased out.

Spores (Fig. 2.22A,B; 2.45A, p. 82): Oval to elongate-oval, slightly broader posteriorly. The posterior vacuole extends more or less to the middle of the spore, its anterior limit being vaulted, sometimes oblique. Single nucleus and dot-like PAS positive polar cap; 12–14 coils of the polar tube.

Spores from *G. aculeatus* and *P. pungitius* do not differ in size or shape. Dimensions (means and/or ranges) given by various authors are:
Gasterosteus aculeatus (fresh spores): 2 × 6 μm (Stempell, 1904); 2.3 × 3.5 μm (Weissenberg, 1913); 2.2 (1.9–2.6) × 3.5 (3 × 4.1) μm (Lom and Laird, 1976); 2.3 (2.1–2.6) × 4.3 (3.4–5.6) μm and 2.3 (1.9–2.7) × 4 (3.4–4.6) μm (Lom, unpublished data on spores from 2 populations).
Pungitius pungitius (fresh spores): 2.1 (1.9–2.4) × 4.5 (3.5–5) μm (Lom and Weiser, 1969); 2.3 (2.1–2.6) × 4.6 (3.5–5.1) μm (Voronin, 1974).

TRANSMISSION: Transmission has been achieved *per os* (Weissenberg, 1921, 1968), by exposing *G. aculeatus* fry to water fleas (*Daphnia*) previously kept in suspensions of spores, and intraperitoneally (Lom, unpublished). There is no evidence of transovarial transmission.

TAXONOMY—Varieties: Voronin (1974) split *Glugea gasterostei* (p. 47) from *G. anomala* on the basis of spore size and shape and tissues invaded, and further (1976) split *G. anomala* into three "categories" without giving the precise status. These were *G. anomala limnosa*, which invades the surface of *P. pungitius*, *G. anomala baltica*, which invades the surface of *G. aculeatus* and *G. anomala apansporoblastica*, which differs by the absence of SPOVs. This absence, however, was almost certainly due to the age of the xenomas examined and in view of the similarity of spore shape and size, the three varieties probably have no significance.

Species Possibly Identical with *Glugea anomala*

1. *Glugea weissenbergi* Sprague and Vernick, 1968 (Fig. 2.22C, 2.24). SYNONYM: *Glugea* sp. Sprague and Vernick, 1966. Found in *Apeltes quadracus* in the Patuxent river at Solomons Island, Maryland, U.S.A. About 50% of fish were infected. Xenomas were located subperitoneally in the body wall in visceral organs, and, rarely, subcutaneously. Spores are of a more elongate shape and are longer, measuring in the fresh state 3 (2.5–3.6) × 6.5 (6–7) μm (Sprague and Vernick, 1968a) or 2.8 (2.6–3) × 6 (5.6–6.5) μm (Lom, unpublished).

G. weissenbergi was established as an independent species chiefly on account of its different spore size.

2. *Glugea gasterostei* Voronin, 1974 (Fig. 2.41A).

Voronin (1974, 1976 and personal communication) found this species in about 1% of specimens of *G. aculeatus* in the euryhaline areas of Finland Bay near Leningrad, U.S.S.R. It has elongate ellipsoid spores measuring 2.6 (2.1–2.8) × 5.6 (4.9–6) μm when fresh, somewhat larger than those of *G. anomala*. It infected exclusively the internal organs and was rare in *Pungitius*.

Voronin restricted the species *G. anomala* to microsporidia with spore length ranging from 3.5 to 5.1 μm, the species being found both in *G. aculeatus* and *P. pungitius*. Unless future findings corroborate the constancy of size range and shape of spores and the tissue preferences given as characters of *G. gasterostei* and, if no overlapping exists which would indicate the existence of local strains, the species *G. gasterostei* has to be considered synonymous with *G. anomala*.

3. Doubtful reports of *Glugea anomala* from fishes beyond its verified host-range.

Henneguy (1888) and Stempell (1904) each reported finding *G. anomala* from a specimen of *Gobius minutus* (for a discussion of the host identification by Henneguy, see Gurley, 1894). Zaika (1965) found subcutaneous xenomas 1.5–2 mm in *Asprocottus megalops* from Lake Baikal. These contained ovoid spores averaging 2.6 × 5.2 μm and he identified the parasite as *G. anomala* in *Neogobius melanostomus, N. syrman* and in *Gobius platyrostris* from the Black Sea. No further details were given.

Laird (1956) quotes 2 Malaysian fishes *Aeoliscus strigatus* and *Scatophagus argus* as harbouring *G. anomala*, with spores the size of 2 × 5 μm but again, gave no reasons for his identification.

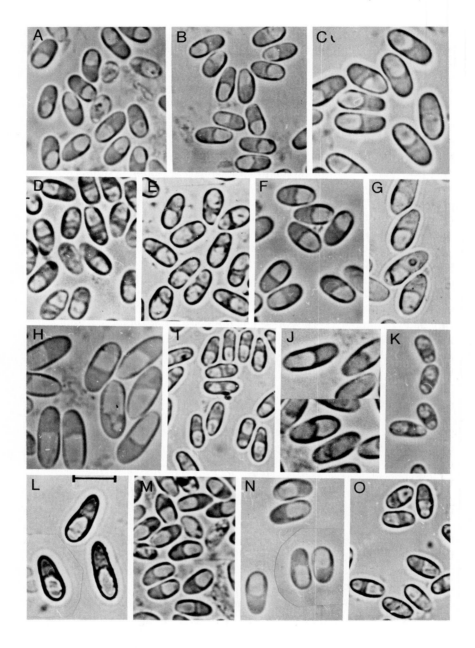

REMARKS: The life of *G. anomala,* although representing a unique model of host parasite interaction at the cellular level, has been virtually unexploited in view of its importance.

Glugea acuta Thélohan, 1895 (Fig. 2.46A)

HOSTS: *Syngnathus acus, Nerophis (Entelurus) aequorus.* Marine.
GEOGRAPHICAL DISTRIBUTION: The Atlantic (Roscoff, Concarneau) and Mediterranean (Marseille) coasts of France.
SITE OF INFECTION: Connective tissue of the dorsal fin muscle.
SIGNS OF INFECTION AND PATHOLOGY: Elongate xenomas of a size measurable in millimeters.
STRUCTURE AND LIFE CYCLE
   Spores (Fig. 2.46A): Ovoidal, $5 \times 3\text{--}3.5$ μm, the anterior end pointed, the posterior end greatly rounded and occupied by a globular vacuole.
REMARKS: Not reported since Thélohan's original finding. Should be reexamined.

Glugea atherinae Berrebi, 1978 (Figs. 2.22G, 2.25, 2.27)

HOST: *Atherina boyeri.* Euryhaline.
GEOGRAPHICAL DISTRIBUTION: Brackish lagoons along the French Mediterranean coast from Marseille to Rousillon, the prevalence in separate lagoons ranging from 1.2 to 21%. The most probable reason why the host is infected in only part of its distribution range in the Mediterranean, is the limited migration of the host populations, which live in "closed" brackish lagoons.
SITE OF INFECTION: Connective tissue of various body organs.
SIGNS OF INFECTION AND PATHOLOGY: In infected yearlings, each fish has a conspicuously swollen body wall, concealing the presence of a large xenoma in the body cavity. Two types of xenoma are formed within the

---

Fig. 2.22 Photomicrographs of spores, fresh if not stated otherwise: A, *Glugea anomala* from *Gasterosteus aculeatus;* B, *G. anomala* from *Pungitius pungitius;* C, *G. weissenbergi* from *Apeltes quadracus;* D, *G. capverdensis* from *Myctophum punctatum* (formol fixed); E, *G. cepedianae* from *Dorosoma cepedianum* (formol fixed); F, *G. stephani* from *Platessa platessa;* G, *G. atherinae* from *Atherina boyeri* (reproduced from Berrebi and Bouix, 1978 with permission of authors and publisher); H, *G. fennica* from *Lota lota:* I, *G. hertwigi* from *Osmerus mordax* (formol fixed); J, *G. plecoglossi* from *Plecoglossus altivelis;* K, *Glugea* sp. from *Sphaeroides maculatus* (formal fixed); L, *G. berglax* from *Macrourus berglax;* M, *G. luciopercae* from *Stizostedion lucioperca;* N, *G. tisae* from *Silurus glanis;* O, *Glugea* sp. from *Pseudopleuronectes americanus.* N.B. the change in internal structure that the spores undergo when fixed. Scale bar for all figures = 5 μm.

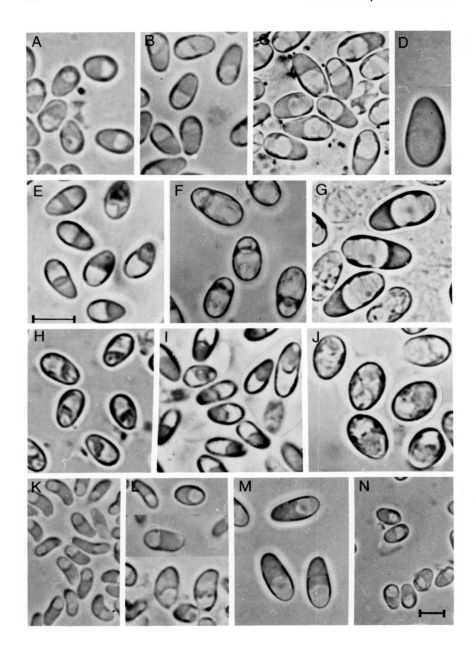

same animal (Figs. 2.25, 2.27). Small ones, numerous in one host 0.1–0.2 mm in diameter, with a wall thickness of up to 5 μm occur throughout the digestive tract except the oesophagus. Large xenomas, not more than 4 in one fish, each up to 13 mm in size, with a wall thickness of 70 μm in various organs and tissues: mesentery and fat tissue (85.1% of all large xenomas observed), swim bladder (6.1%), peritoneum (5.0%) and liver or gonads (3.5%).

The large xenomas are hyaline when young, then milky white and chalky, when packed full of spores.

Both large and small xenomas have the same basic organisation, including the laminated wall. Their structure corresponds to xenomas of *G. anomala*. In large xenomas, the laminated structure of the thick wall can easily be seen by light microscopy.

Advanced xenomas are encased within a layer of host connective tissue which is vascularised. Intestinal xenomas, if numerous, can cause partial obstruction of the digestive tract. Large xenomas cause pressure atrophy of affected organs and impair the viability of the host. Infection may impair the host's resistance to bacterial infection and this alone may be an important mortality factor.

COMMERCIAL IMPORTANCE: *Atherina boyeri* is an important commercial fish and hence the damage by the microsporidia may be a cause for concern.

STRUCTURE AND LIFE CYCLE

Merogony: Rounded meronts with a single nucleus transform into cylindrical meronts 3 × 25 μm with up to 16 nuclei which divide again into uninucleate meronts. All meronts are surrounded, at a distance of 50 nm from their cell membrane, by a continuous cisterna of host cell endoplasmic reticulum.

Sporogony: Multinucleate plasmodia arise from uninucleate meronts.

Fig. 2.23 Photomicrographs of spores, fresh, if not stated otherwise: A, *Pleistophora duodecimae* from *Coryphaenoides nasutus* (formol fixed); B, *P. hippoglossoideos* from *Hippoglossoides platessoides;* C, *P. hyphessobryconis* from *Paracheirodon inessi:* D, *P. gadi* from *Gadus virens;* E, *P. ladogensis* from *Osmerus eperlanus* (photograph by Dr. V. Voronin); F, *P. macrozoarcidis* from *Macrozoarces americanus* (formol fixed); G, *P. ovariae* from *Notemigonus chrysoleucas;* H, *P. oolytica* from *Esox lucius* (formol fixed); I, *Loma branchialis* from *Gadus aeglefinus:* J, *Ichthyosporidium giganteum* from *Leiostomus xanthurus (formol fixed);* K, *Spraguea lophii* from *Lophius piscatorius,* cylindrical, *Nosema*-type spores; L, *S. lophii* from *L. piscatorius,* ovoid, *Nosemoides*-type spores; M, *Microsporidium* sp. of Marchant and Schiffman from *Mallotus villosus* (photograph by Dr. J. Vávra); N, *Pleistophora vermiformis* from *Cottus gobio*. Scale bar in Fig. E applies to Figs. A–M = 5 μm; scale bar on Fig. N = 5 μm.

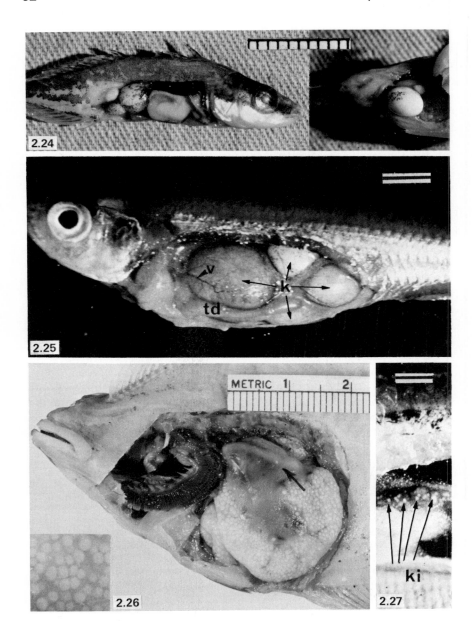

There is vacuolization of the space between the parasite surface and the host cell cytoplasm, then fusion of the vacuoles to produce the SPOVs first 11 μm and later 15 μm in diameter. The production of sporoblasts follows the pattern of *G. anomala*.

The division of sporoblast mother cells was assumed to occur but was not observed. Tubules, 70 μm in diameter, arise from the surface of sporoblasts and extend in the cavity of the SPOVs. Sporoblasts measure 2.3 μm.

Spores (fresh) (Fig. 2.22G): Elongate ovoid, 2.9 (2.6–3.3) × 5.7 (4.5–6.5) μm have a single nucleus (0.7 μm). The posterior vacuole (2.8 μm diameter) occupies one half of the spore length, and has a vaulted anterior border. There are 12–17 coils of the polar tube.

TRANSMISSION: Fresh spores, administered perorally, produced infections in 8.5% of fish. Experimental infections produced xenomas both in the body cavity and subcutaneously. Small xenomas developed "very quickly" and persisted for a long time, while the large ones took at least 6 months to reach the mature stage.

The fish probably acquired infections at the end of summer or beginning of autumn in their first year of life and again in their second year. During winter and spring the xenomas grew and matured. There was a conspicuous drop in the infection rate in summer, which may have been due to migration of fish resulting in "dilution" of the infected population or to death of infected specimens or to elimination of the infection.

Fig. 2.24 Parts of the bodies of 4-spined sticklebacks, *Apeltes quadracus* with large xenomas of *Glugea weissenbergi* in the body cavity: scale divisions = 1 mm (Photograph by Dr. V. Sprague).

Fig. 2.25 An extreme case of parasitism in the body cavity of *Atherina boyeri,* showing 4 large xenomas of diameters from 3 to 7.5 mm, due to *Glugea atherinae* (k arrows); v = perixenomic vascularisation; td = digestive tube. Scale bar = 3 mm. (Reproduced from Berrebi and Bouix, 1978, with permission of authors and publisher.)

Fig. 2.26 *Glugea stephani* infection in the English sole: body wall removed to expose numerous xenomas, causing massive thickening of the intestinal wall and responsible for abdominal enlargement; arrow points to a single xenoma in the liver. Scale divisions = 1 mm. (Reproduced from Wellings *et al.,* 1969, with permission of authors and publishers). Inset: enlargement of part of Fig. 2.26 showing the xenomas. (Photograph by Dr. A. M. McKenzie).

Fig. 2.27 Part of a dissected specimen of *A. boyeri* to show small intestinal xenomas (ki, arrows) of *G. atherinae.* (Reproduced from Berrebi and Bouix, 1978, with permission of authors and publisher.)

SEX DEPENDENCE: In general, females were somewhat more commonly infected than males, i.e. 6.3% females compared with 5.15% males out of a total of 2511 fish, consisting of samples from 8 localities. Detailed data on prevalence of infection in separate size groups of the fish in relation to their sex were given by Berrebi (1978).

HOST SPECIFICITY: Preliminary attempts to infect *Solea solea, Gambusia* sp., *Crenilabrus* sp. and *Gasterosteus aculeatus* failed.

REMARKS: This is a well-suited model species for experimental work. All above data were compiled from Berrebi (1978, 1979) and Berrebi and Bouix (1978). Although Berrebi (1979) did not consider the membrane around the spores to be a SPOV, all data from his publications are in favour of this interpretation. Thus the type of sporogony is identical with that in *Glugea anomala*.

Glugea capverdensis Lom, Gaievskaya and Dyková, 1980 (Figs. 2.22D, 2.28, 2.29)

HOST: *Myctophum punctatum*. Marine.

GEOGRAPHICAL DISTRIBUTION: The Atlantic, region of Cape Verde; one infected specimen, 5.8 cm in size, was found.

SITE OF INFECTION: Intestinal wall, mesentery, ovary.

SIGNS OF INFECTION AND PATHOLOGY: Mature xenomas, rounded or oval, up to 2 mm with a wall up to 3.3 μm thick, were extremely plentiful in the organs affected. In some places, only small islands of the original tissue were left unchanged. They caused considerable pressure atrophy by their sheer numbers. In the ovary, they reduced the number of oocytes. In the intestine, conglomerates of xenomas were so closely packed, that they often bulged one into another. They were localized subepithelially or were situated beneath the visceral sheet of the peritoneum. Many xenomas protruded into the intestinal lumen so that they seriously obstructed passage through it. Quite often, formation of secondary xenomas (Fig. 2.29) similar to those in *G. anomala* (Sect. IIB,1b) could be observed.

STRUCTURE AND LIFE CYCLE

Merogony: Globular (1.5 μm), oval (1.5 × 2 μm) or spindle shaped (exceeding 2 μm in length) meronts grow into multinucleate cylindrical meronts (up to 20 × 2–3 μm) which fragment into uninucleate cells. Nuclei are all about the same size, 0.8 μm.

Sporogony: Elongated sporogonial plasmodia with an average of 14 nuclei fragment within SPOVs into uninucleate sporoblast mother cells, either via multinucleate segments or by radial segmentation. Mother cells divide into 2 elongated sporoblasts distinguished by a striking chromophile axis (Fig. 2.28M,N).

Spores: (formol-fixed) (Fig. 2.22D): Ellipsoid or very slightly ovoid, 2.3 (1–2.6) × 4.2 (3.6–4.8) μm; a single nucleus about 0.8 μm; the large posterior vacuole reaches to about the mid-line of the spore.

Glugea cepedianae (Putz, Hoffman and Dunbar, 1965) comb.n.
(Fig. 2.21, 2.22E)

SYNONYMS: *Microsporidia* sp. Bangham, 1941; *Pleistophora cepedianae* Putz, Hoffman and Dunbar, 1965.
HOST: *Dorosoma cepedianum.* Freshwater.

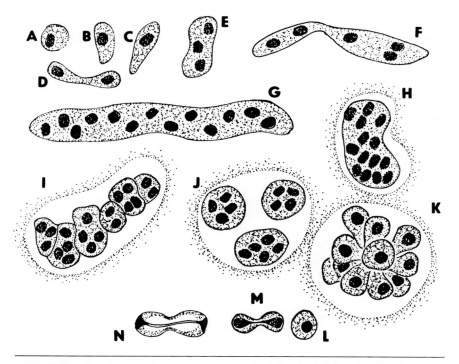

Fig. 2.28 Developmental stages of *Glugea capverdensis:* A,B,C, uninucleate meronts; D, dividing meront; E, trinucleate meront; F, multinucleate meront dividing by plasmotomy; G, multinucleate cylindrical meront; H, sporogonial plasmodium within the SPOV; I,J, fragmentation of the sporogonial plasmodium; K, cleavage of the plasmodium into sporoblast mother cells; L, sporoblast mother cell; M,N, sporoblast mother cells dividing into sporoblasts.

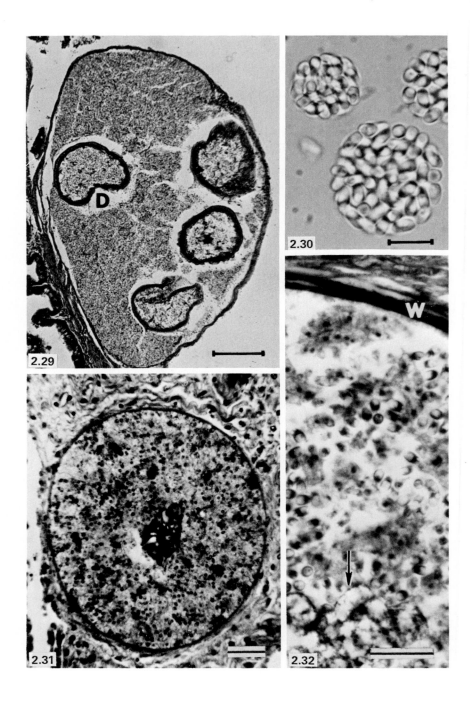

GEOGRAPHICAL DISTRIBUTION: Ohio, U.S.A. Common in some of the fish reservoirs. In their first year, up to 65% of young shad may contract the infection. In their second year, the infection rate slowly decreases and after spawning, in the third year, no infected fish are found, probably as the result of the greater mortality of infected fish (Price, 1982).

SITE OF INFECTION: Visceral cavity.

SIGNS OF INFECTION AND PATHOLOGY: Huge "cyst" (up to 1 cm) protruding conspicuously at the side of the yearling fish; never more than one cyst per fish. The cyst is a large xenoma compressing the visceral organs against the body wall. There is no adjacent cellular reaction in most infections. The connective tissue layer encasing the xenoma may be 4–10 μm or even sometimes up to 300 μm thick. At the periphery of the xenoma, subtending its wall, there is a thick layer of spongy cytoplasm. This layer contains irregular, vesicular host cell nuclei, up to 20 μm in size, and developmental stages. The centre of the xenoma is occupied by a mass of mature spores. The wall of large "cysts" often breaks and large accumulations of mature spores are found within the body cavity, separated from the neighbouring tissue only by a thin layer of melanin pigment. Mortality of young gizzard shads was high (Putz, Hoffman and Dunbar, 1965).

COMMERCIAL IMPORTANCE: Extensive infection may inflict heavy losses on populations of gizzard shad, which have value as bait fish.

STRUCTURE AND LIFE CYCLE (interpreted from Putz, Hoffman and Dunbar, 1965 and from Lom, unpublished)

Merogony: Meronts develop into cylindrical to C-shaped stages measuring 3.3 × 17 μm, with numerous nuclei, and then give rise to spherical bodies about 5.2 μm in diameter with 2 dense nuclei 1 μm in size.

Fig. 2.29 Four secondary xenomas (D) within a *Glugea capverdensis* xenoma pervaded by host's phagocytes: the whole complex is attached to the *lamina muscularis* of the intestine. H & E; scale bar = 100 μm.

Fig. 2.30 Sporophorous vesicles of *G. anomala* from *Gasterosteus aculeatus* released intact from the xenoma; scale bar = 10 μm.

Fig. 2.31 Xenoma of *Glugea acerinae* located within the submucosa of the intestine of *Gymnocephalus cernuus,* and surrounded by connective tissue: mature spores are mixed with clusters of developmental stages; there is a single hypertrophic, centrally-located host cell nucleus. (Photomicrograph by Professor O. Jírovec). Trichrome; scale bar = 25 μm.

Fig. 2.32 Portion of the xenoma of *G. acerinae,* showing the central host nucleus, now almost destroyed (arrow), the xenoma wall (W) and islets of mature spores mixed with developmental stages. H & E; scale bar = 20 μm.

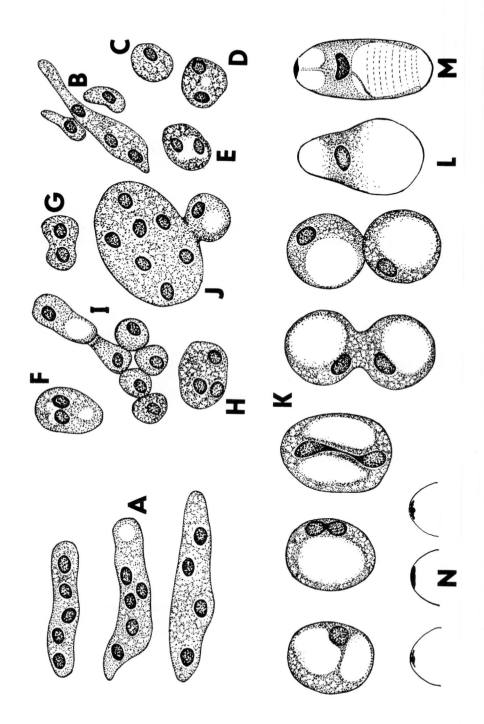

Sporogony: Irregular cells, measuring 9 μm with a 3.2 μm large nucleus, become sporonts. Sporonts grow into oblong multinucleate sporogonial plasmodia, which change to an ovoid shape, 30 μm in diameter. These dissociate within a SPOV into a cluster of 6–20 uninucleate sporoblast mother cells, measuring 3 μm, which later produce spherical sporoblasts less than 3 μm in size.

Spores (fixed) (Fig. 2.22E): Slender ovoid, 4.9 (4.2–5.6) × 2.3 (1.7–2.8) μm, in which the large posterior vacuole has elliptical outlines.

REMARKS: We have transferred this species to the genus *Glugea* because of the presence of a *Glugea anomala*-type xenoma, cylindrical multinucleate meronts and SPOVs containing many spores. However, the large size of the uninucleate cells which produce the sporogonial plasmodium is a rather unusual character when compared with the developmental pattern of *G. anomala*.

Glugea fennica (Lom and Weiser, 1969) Lom and Laird, 1976
(Figs. 2.22H, 2.33)

SYNONYM: *Nosema fennica* Lom and Weiser, 1969.

HOST: *Lota lota.* Freshwater.

GEOGRAPHICAL DISTRIBUTION: Helsinki, Finland. One infected fish examined. According to Voronin (1980), also from Leningrad and Buryat (and probably other) regions in the U.S.S.R.

SITE OF INFECTION: Subcutaneous tissue of the body and fins.

SIGNS OF INFECTION AND PATHOLOGY: The numerous whitish nodules, round or oval, up to 2.5 mm large, on the body surface, represent xenomas with walls 1–3 μm thick. In the one specimen found, the massive occurrence of "cysts" inflicted serious damage to the epidermis.

STRUCTURE AND LIFE CYCLE

Merogony: Oval uninucleate meronts, multiply repeatedly by binary fission and later give rise to multinucleate meronts, mostly irregularly rounded, with up to 11 nuclei and up to 11 μm in diameter; rather rarely meronts are cylindrical, 3 × 13 μm, with up to 5 nuclei.

Sporogony: Multinucleate sporogonial plasmodia fragment into sporoblast mother cells, characterised by the presence of a pale vacuole, which may even appear in the plasmodium before its division into mother cells. These rounded mother cells, measuring 3–4 μm, (Fig. 2.33K) enlarge and

---

Fig. 2.33 Developmental stages of *Glugea fennica*. A, multinucleate cylindrical meronts; B, cylindrical meront, giving rise to uninucleate stages C which grow to larger meronts again; D,E,F,G,H,I, stages in the growth of sporogonial plasmodia; J, rounded sporogonial plasmodium giving rise to sporoblast mother cells; K, division of sporoblast mother cell into two sporoblasts L; M, spore; N, shapes of polar cap after PAS staining.

divide in two spherical sporoblasts, 3 μm in diameter, each with a voluminous vacuole. Sporoblasts grow to 5 μm diameter, elongate and transform in spores.

Spores (fresh, Fig 2.22H): Elongate oval, of a fairly constant shape 2.5–3 × 6.8–8.1 μm (Lom, unpublished); 2.6 (2.3–2.9) × 7.1 (5.9–8.2) μm (Voronin, 1980). Posterior vacuole occupies ½ to ⅔ of the spores, and the oval nucleus is located just above it. The PAS positive polar cap is 0.25 × 1 μm (Fig. 2.33N).

Glugea heraldi Blasiola, 1979 (Fig. 2.41B, p. 72, 2.45C, p. 82)

SYNONYM: *Nosema* sp. Lom, 1972
HOST: *Hippocampus erectus*. Marine.
GEOGRAPHICAL DISTRIBUTION: New York aquarium (Lom, 1972); Florida coast (Blasiola, 1979), both U.S.A.
SITE OF INFECTION: Whitish xenomas up to 0.8 mm in size in the skin, with peripheral developmental stages and central mass of spores.
STRUCTURE AND LIFE CYCLE
Spores (fresh): Elongate ovoid, 4.6 (4.1–5) × 2.5 (2.3–3) μm. Posterior vacuole occupies about ½ of the spore volume, its anterior border is vaulted or slanted. The PAS positive polar cap is dot-like (Lom, unpublished).
REMARKS: Herald and Rakowicz (1951) first recorded the occurrence of infection in this host. Blasiola (1979) used this old material for his description. Lom (1972) described the ultrastructure of the extruded spores and polar tubes.

Glugea hertwigi Weissenberg, 1911 (Figs. 2.22I, 2.41G-I, p. 72, 2.45B, p. 82)

SYNONYM: *Glugea hertwigi* var. *canadensis* Fantham, Porter and Richardson, 1941.
HOSTS: Originally described from *Osmerus eperlanus eperlanus;* common also in *O. eperlanus mordax* according to Schrader (1921) and several Canadian authors. According to Bogdanova (1957) also in *O. eperlanus eperlanus* m. *spirinchus*. Akhmerov (1946) and Vinnichenko, Zaika, Timofeev, Shtein and Shulman (1971), found it in *Hypomesus olidus;* Shulman and Shulman-Albova (1953) in *Coregonus lavaretus pidschian* n. *pidschianoides*. Euryhaline. (See also Remarks.)
GEOGRAPHICAL DISTRIBUTION: In the Baltic at Rügen, an almost 100% prevalence in smelts was recorded (Weissenberg, 1913). In North Russian lakes, the prevalence in *Hypomesus olidus* ranged from 1 to 59% (Anen-

kova & Khlopina, 1920). In North American lakes, in Canada and the U.S.A., the prevalence of infection in *Osmerus eperlanus mordax* was in some places as high as 100% and ranged from 4.6 to 100% and affected young-of-the-year and adult smelts alike. In the lakes of the Canadian province of Quebec, 11 out of 27 water bodies harboured massively infected smelts (Delisle and Veilleux, 1969). There may be a considerable annual fluctuation of infection. Detailed data may be found in Akhmerov (1946), Dechtiar (1965), Delisle (1965), Haley (1952), Nepszy, Budd and Dechtiar, (1978) and Sherbourne and Bean (1979).

SITE OF INFECTION: The principal site in smelts is the intestine but when the infection is massive, it spreads to various other organs and virtually all of them may be affected, including trunk muscle and gills. A thorough analysis of the frequency of infection in separate parts of the digestive tract (of which the anterior part of the intestine was most heavily infected) and other sites, was presented by Delisle (1972), based on examination of 1200 adult smelts. The smallest number of xenomas was found in the kidneys. The prevalence of infection in some organs is sex dependent; Nepszy and Dechtiar (1972) found similar infection rates in the intestine of males and females (97% and 79%, respectively) but widely different rates in the gonads (17% and 77%, respectively).

SIGNS OF INFECTION AND PATHOLOGY: Conspicuous, generally numerous, round or oval, white xenomas. All xenomas are of about equal size, mostly 2–4 mm, exceptionally up to 8 mm, with a laminated wall 2 μm thick and may show through the skin of the host. Typically, a large white knot, composed of many xenomas, bulges out of the abdominal wall in the anal region (Fig. 2.41G, p. 72).

Very large numbers of "cysts", averaging 250 per host in mass mortalities, seriously impairs the viability of the host. Most frequently the effect is on the intestine, in which the tissue may disintegrate completely resulting in general septicaemia and/or intoxication. The lumen may be occluded by the large *Glugea* nodules, resulting in death. Fecundity may be severely reduced by the infection of gonads and/or xenomas may block the gonadal pore and prevent the discharge of sexual products (Sindermann, 1963).

Infected fish survive for a period of time, if organ function is not severely impaired, as was found in one population of smelt in Lake Erie: in 1966, 88.8% of the young-of-the-year were infected and in the following year, the figure was 90% (Nepszy and Dechtiar, 1972). However, if a stress situation intervenes, heavily infected populations may succumb to mass mortality. In the late spring, mortality is due to spawning stress and in other seasons is due to environmental stress.

After some time, xenomas may be destroyed by host tissue reaction.

Delisle (1969) found that some rainbow smelt populations were refractory to infection, even if exposed to rich sources of infection. He interpreted this as a form of physiological resistance.

COMMERCIAL IMPORTANCE: Nepszy, Budd and Dechtiar (1978) confirmed that *Glugea hertwigi* infection plays a leading role in massive mortalities in Lake Erie, as 75% or more of dead fish were heavily infected. In 1972, Nepszy and Dechtiar reported that dead smelts were washed ashore along a 120 km stretch of this lake. The estimated weight of dead fish was 70 tons, amounting to 1% of the total 1969 Canadian catch. Delisle (1972) estimated that the average annual losses of smelts in Lake Erie were as high as 10 million individuals. A catastrophic mortality may result in the complete elimination of a stock of smelts. In addition to direct losses, reduced fecundity limits population growth, so that there are insufficient smelt to support populations of important commercial fish, e.g. lake trout. Epizootics as severe as those in Canadian lakes were reported by Haley (1952, 1953, 1954) from New Hampshire and by Anenkova-Khlopina (1920), Bogdanova (1957) and Petrushewsky and Shulman (1958) from the U.S.S.R.

STRUCTURE AND LIFE CYCLE: Development is similar to that of *Glugea anomala* (Weissenberg, 1913).

Merogony: According to Fantham, Porter and Richardson (1941), young uninucleate meronts are oval, 2.3–3.3 µm and binucleate cylindrical meronts are 2–3 × 3–5.4 µm.

Sporogony: Sporoblast mother cells do not assume the dumb-bell shape but remain in tight clusters after separation from the sporogonial plasmodium. After karyokinesis in the mother cells, the daughter nuclei stay united for a very long time by a thin desmose.

Spores (fresh): *Osmerus eperlanus eperlanus:* elongate oval, 2.3 × 4.6–5.4 µm (Weissenberg, 1913); spores from *O. eperlanus mordax:* 2.2 × 4.4 µm (Haley, 1952); 2.1 × 4.3 µm (Dechtiar, 1965); 2.3 × 3.9 µm (Legault and Delisle, 1967); 2.4 × 4.9 µm (Delisle, 1969). It is unknown whether these spores were measured fresh or fixed. Lom (unpublished, Figs. 2.22I, 2.45B): formol fixed spores 2 (2–2.2) × 4.7 (4.4–5.2) µm, are elongate oval, somewhat wider posteriorly, the posterior vacuole occupying about half the spore length, with a vaulted anterior margin and thin side cytoplasmic walls. PAS positive polar cap is dot-like, nucleus located in the centre of the spore.

Spores (fixed) from *Hypomesus olidus:* 2.2 × 5.9 µm and 2.4–2.7 × 4.6–5 µm (Anenkova-Khlopina (1920) and Akmerov (1946), respectively).

Spores (fixed) from *Coregonus lavaretus pidschian* n. *pidschianoides:* 2.2–2.3 × 5–5.5 µm, length of the vacuole 2.4–2.7 µm (Shulman and Shulman-Albova, 1953).

TRANSMISSION: The infection is contracted perorally by ingestion of spores or by devouring infected individuals (Delisle, 1972).

Weissenberg (1913), studying the growth rate of xenomas in European smelt, concluded that the infection period was limited to late spring. Delisle (1969), however, observed a steady increase in the prevalence of non-lethal infections in young-of-the-year of rainbow smelt from June (6.7%) to September (93.2%). The intensity of infection gradually increased up to 57 cysts per fish. This means that at 4 months of age, almost 100% of smelt fry is infected and this argues in favour of infection over a long, rather than limited, period.

SEASONAL FLUCTUATION OF INFECTION: Regular samplings in Canadian lakes (Skerry, 1952; Delisle, 1972; Chen and Power, 1972) showed an increase in the prevalence of infection as well as an increase in the intensity in the warm seasons of the year, alternating with a decline in prevalence towards winter months. The peak of infection was probably due to the favourable conditions for parasite's growth and its spread in the fish population. The abrupt or slow decline in the prevalence after the spawning period and in autumn was evidently due to mortality of heavily infected specimens, exacerbated by mating-induced or environmental stress (Delisle, 1972).

REMARKS: *G. hertwigi* is a superb model for experimental and epizootiological studies. The only marked difference between *G.anomala* and *G. hertwigi* is in the sizes and shapes of the spores (Weissenberg, 1968). In view of Weissenberg's success in infecting 2 out of 4 *Gasterosteus aculeatus* with *G. hertwigi* spores, the separate identities of both species should be reinvestigated.

G. hertwigi var. *canadensis* described by Fantham, Porter and Richardson, (1941) should be regarded as *G. hertwigi*. Chen (1956) identified with *G. hertwigi* a microsporidium forming "cysts" up to 2–3 cm, in the fat body of about 5% of specimens of *Mylopharyngodon piceus* in China. Fixed and stained spores averaged 1.7×3.1 μm. The identification with *G. hertwigi* is most probably erroneous. The same applies to a species with oval spores, averaging 3.3×1.9 μm when stained, reported under this name by Chen and Hsieh (1960) in the small intestine of a single specimen of *Channa maculatus* in China.

Glugea nemipteri Weiser, Kalavati and Sandeep,1981
(Fig. 2.46B, p. 86)

HOST: *Nemipterus japonicus*. Marine.

GEOGRAPHICAL DISTRIBUTION: Lawsons Bay near Waltair, the Gulf of Bengal, India; 9 out of 82 fish were infected.

SITE OF INFECTION: Liver, gonads and smooth muscles.

SIGNS OF INFECTION AND PATHOLOGY: Xenomas, 8–12 mm in diameter thought to be infected host macrophages, were of the *G. anomala* type. Xenomas were surrounded by a resistant elastic outer wall, 30–50 μm thick.

STRUCTURE AND LIFE CYCLE

Merogony: Uninucleate and binucleate meronts give rise to cylindrical meronts 9 × 25 μm with up to 12 nuclei.

Sporogony: 16 to 64 spores within a SPOV.

Spores (fixed and stained): broadly oval, 4.5–5 × 5.5–6 μm, with a large posterior vacuole.

Glugea pimephales (Fantham, Porter and Richardson, 1941) Morrison, Hoffman and Sprague, 1985

SYNONYMS: *Nosema pimephales* Fantham, Porter and Richardson, 1941; *Microsporidium pimephales* (Fantham, Porter and Richardson,1941) Sprague, 1972.

HOSTS: *Pimephales promelas, P. notatus*. Freshwater.

GEOGRAPHICAL DISTRIBUTION: North America.

SITE OF INFECTION: Mesentery, in the body cavity.

SIGNS OF INFECTION AND PATHOLOGY: Small fry, only 15–30 mm long, show one or two large oval xenomas which measure up to 2–3.3 mm. These distend the abdomen, compress the visceral organs and obliterate the body cavity. It is doubtful that fry with large cysts survive. The xenomas are structurally similar to those of *G. anomala*.

STRUCTURE AND LIFE CYCLE

Merogony: Uninucleate meronts (comparable with "primary cylinders" of Weissenberg, 1913) undergo repeated binary fission, some of them producing elongate multinucleate meronts ("secondary cylinders"). Some of these meronts undergo multiple fission.

Sporogony: Rounded sporogonial plasmodia arise by transformation of elongate multinucleate meronts. They then secrete the SPOV wall. Within the SPOV, smaller plasmodia are produced by plasmotomy, which in turn undergo multiple fission by rosette formation to produce sporoblast mother cells. The latter divide to produce two spores each.

Spores (fixed and stained): Ovoid, 5.2 (4.5–6) × 2.6 (2.5–3) μm; uninucleate; posterior vacuole occupies half of the spore; polar tube in a single rank of 14–16 coils.

REMARKS: All data is taken from the redescription of this species by Morrison, Hoffman and Sprague (1985), who supplied a very good ultrastructural anlaysis of the parasite but no illustration of the spores.

Glugea plecoglossi Takahashi and Egusa, 1977 (Figs. 2.4, 2.7, 2.22J, 2.34–2.40)

SYNONYM: *Glugea* sp. of Sano, 1969 in Putz and McLaughlin, 1970
HOST: *Plecoglossus altivelis.* Freshwater. *Salmo gairdneri* can be infected experimentally (Takahashi and Egusa, 1977a).
GEOGRAPHICAL DISTRIBUTION: Japan; quite common in *P. altivelis* cultures.
SITE OF INFECTION: Body cavity, ovaries, testis, digestive tract, pyloric caeca, spleen, liver, trunk muscle, heart, gills, iris, fat body.
SIGNS OF INFECTION AND PATHOLOGY: Large xenomas (Fig. 2.4) (5 μm) visible through the skin, which may raise the surface into conspicuous bulges. The smallest xenomas, containing a single meront, measure only 8 μm. The structure, development and host tissue reactions are similar to those of *G. anomala* (Sect. IIB,1, IIIA)

In heavy infections, innumerable large xenomas, (Fig. 2.4) in most body organs cause damage by pressure atrophy and there is additional injury due to host reactions.
COMMERCIAL IMPORTANCE: *G. plecoglossi* causes one of the most dangerous diseases of cultured ayu, *P. altivelis,* and hence is of great economic importance.
STRUCTURE AND LIFE CYCLE (Figs. 2.34–2.40)

Merogony: Early meronts are rounded, 3–4 μm with one compact nucleus. They appear at 5 days after experimental infection. They give rise to multinucleate cylindrical meronts, 3.5 × 15 μm, which divide by transverse fission.

Sporogony: Sporonts grow into cylindrical or globular multinucleate stages within a developing SPOV, in which they divide into uninucleate sporoblast mother cells, 3 × 4 μm. Each mother cell produces two slender sporoblasts which mature into spores.

Spores (fresh, Fig. 2.22J): Elongate ellipsoidal 2.1 (2–2.5) × 5.8 (5.1–6.2) μm; posterior vacuole occupies about ½ of the spore (Fig. 2.22J); its anterior border is usually vaulted and the side cytoplasmic walls are rather thin. The nucleus is 1.3 × 1 μm. The PAS positive polar cap measures 0.5 × 0.3 μm.

Development of xenomas was significantly retarded at 16°C but was normal at 18°C and at temperatures above this point (Takahashi and Egusa, 1977b) in experimentally infected fish.
HOST SPECIFICITY: Transmission to rainbow trout was possible but not to stickleback *Gasterosteus aculeatus microcephalus* (Takahashi and Egusa, 1977a).
TREATMENT: Takahashi and Egusa (1976) have tested nine drugs, of which

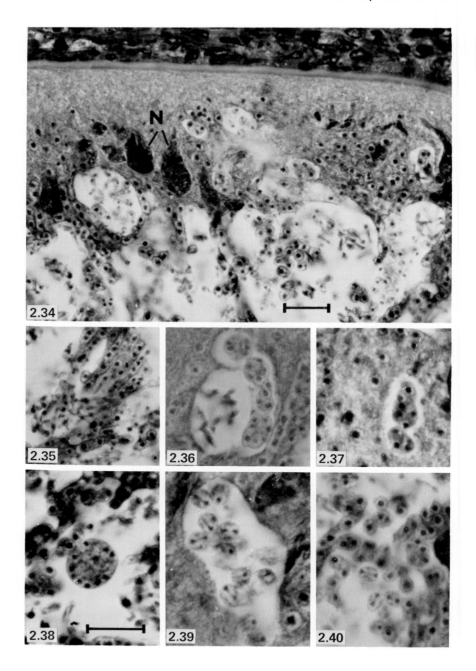

only fumagillin was effective. If applied perorally to ayu fingerlings during the period of 10–25 days after infection, at a dose of 50 mg/kg per day for 3 days or at 12.5 mg/kg/day for 10 days, fumagillin effectively controlled infections. Paradoxically the weight gain of treated fish was not significantly higher than that of untreated fish but the mortality (presumably due to the drug) was sometimes considerably higher. Takahashi (1978) listed as the agents which will inactivate spores of this species high temperature, ultra-violet radiation, cationic surface active agents and chloride drugs.

REMARKS: The above data, except when stated otherwise were taken from Takahashi and Egusa (1977a, 1977b) and supplemented (Lom, unpublished). This species is a good model for studies on fish microsporidia.

Glugea rodei Kazieva and Voronin, 1981 (Fig. 2.46C, p. 86)

HOST: *Rhodeus sericeus amarus*. Freshwater.
GEOGRAPHICAL DISTRIBUTION: Azerbaidzhan S.S.R., Varvarin dam lake, U.S.S.R. In 1974–1978, 40–100% of fish collected were infected.
SITE OF INFECTION: Serose membranes of internal organs, most often intestine, and body wall.
SIGNS OF INFECTION AND PATHOLOGY: In young fish, *G. anomala*-like xenomas measuring 1–3 mm visible through the body wall.
STRUCTURE AND LIFE CYCLE
  Merogony: Meronts, of which no details were given, are located at the xenoma's periphery.
  Sporogony: Round multinucleate plasmodia measure 8–15 μm; SPOVs were rarely observed.
  Spores (fresh): Elongate oval, 2.1 (1.9–2.2) × 5.3 (5–5.5) μm in size. The posterior vacuole occupies ½ to ⅔ of the spore length; its anterior border is straight, obliquely set.

Fig. 2.34–2.40 Developmental stages of *Glugea plecoglossi* within the xenoma cytoplasm. Fig. 2.34. Periphery of the xenoma showing the surrounding connective tissue, the xenoma wall, peripheral cytoplasmic layer of the xenoma, large host nuclei (N), meronts of various size and vacuoles (SPOVs) with sporogonial plasmodia, sporoblast mother cells and sporoblasts. Fig. 2.35. Cytoplasmic islets with cylindrical meronts surrounded by sporoblasts. Figs. 2.36 and 2.37. Sporogonial plasmodia fragmenting into sporoblast mother cells. Fig. 2.38. Globular sporogonial plasmodium. Fig. 2.39. SPOV with sporoblast mother cells dividing into sporoblasts. Fig. 2.40. Dividing sporoblast mother cells, some within a large free space, presumably after breakdown of the SPOVs. H & E; scale bar on Fig. 2.34 = 15 μm; scale bar on Fig. 2.38 = 15 μm; applies to Figs. 2.35–2.40.

Glugea shiplei Drew, 1910

SYNONYM: *Pleistophora shiplei* (Drew, 1910) Sprague, 1977
HOST: *Trisopterus luscus*. Marine.
GEOGRAPHICAL DISTRIBUTION: English coast at Plymouth. Only one infected 10 cm fish was caught.
SITE OF INFECTION: skeletal muscles, muscles of the stomach and intestine.
SIGNS OF INFECTION AND PATHOLOGY: cysts 3 × 5 mm, some visible through the skin. Cysts were encased by a thin layer of connective tissue and contained a colourless gelatinous substance, with numerous circular bodies resembling nuclei and a few large elongate nuclei, suggestive of the host cell nuclei in *G. anomala:* the cyst wall was "very thin and transparent".
STRUCTURE AND LIFE CYCLE: Drew's rather imprecise and hazy account of the circular bodies in the cyst contents describes what could have been developing SPOV. In the centre of the cyst he found mature SPOVs containing large numbers of spores.
  Spores (probably fixed): Pyriform 2.5 × 3.5 μm, with a pointed anterior end, and large posterior vacuole.
REMARKS: Drew's (1910) is the only record of this parasite. Sprague transferred the species to the genus *Pleistophora* because of Drew's observation of SPOV. Since SPOVs are known to develop in the genus *Glugea* as well and Drew's description revealed several other *Glugea*-like features, the species is returned into the latter genus.

Glugea stephani (Hagenmüller, 1899) Woodcock, 1904 (Figs. 2.22F, 2.26).

SYNONYMS: *Nosema stephani* Hagenmüller, 1899; Sporozoan Johnstone, 1901; Protozoan Linton, 1901; *Glugea* sp. Youssef and Hammond, 1972; *Glugea* sp. Olson and Pratt, 1973.
HOSTS: *Pleuronectes flesus* var. *passer* (= the source of material for Hagenmüller's description); common in *Platessa platessa, Pseudopleuronectes americanus,* and *Parophrys vetulus*. Other hosts: *Limanda limanda* (Doflein and Reichenow 1927–1929); *Limanda ferruginea* (Fantham, Porter and Richardson, 1941); *Pleuronectes flesus bogdanovi, Liopsetta glacialis* (Shulman and Shulman-Albova 1953); *Rhombus flesus trachurus* (Shulman 1962); *Platichthys stellatus* (Jensen and Wellings 1972). All marine.
GEOGRAPHICAL DISTRIBUTION: First found in *Pleuronectes flesus* near Marseille, *G. stephani* is presently known as a common parasite of the above named hosts, along the Atlantic and the Pacific coasts of the U.S.A., in the Northern, Baltic and Black Seas and in seas around England. The

prevalence in a given host fish stock may reach 50% or even more. Takvorian and Cali (1981) who examined 1840 fish of 8–33 cm categories, found that apparently healthy flounders from the U.S. Atlantic coast had an overall infection rate of 6.63%. This percentage represents the mean carrying rate of the parasite in this fish population; these are the fish that have survived the initial stages of infection.

No correlation between prevalence and salinity, season or sex of the host has yet been demonstrated. However, the ambient temperature is crucial for establishment and further development of *G. stephani*. McVicar (1975) found that infection was absent in *Platessa platessa* in Scottish inshore waters, where sea bottom temperature rarely exceeds 12°C. Olson (1975) found that the distribution of infection in *Parophrys vetulus* in Oregon was restricted to areas where temperatures of 15°C or higher prevail during summer. Experimental evidence has also been obtained: McVicar (1975) found that infected *P. platessa* kept at 16°C developed lethal infections and succumbed within 80 days, while fish kept at 11°C did not become infected; Olson (1976) failed to infect *Parophrys vetulus* kept at 10° or 11°C but succeeded if they were kept at 15°C. McKenzie, McVicar and Wadell, (1976) found that further development or spread of the parasite did not take place when infected *Platessa platessa* were transferred from heated tanks to open sea cages at cooler temperatures. In this context it is puzzling that Shulman and Shulman-Albova (1953) found *G. stephani* in hosts inhabiting cold waters of the White Sea.

SITE OF INFECTION: Subepithelial connective tissue of the digestive tract. In heavy infections, also mesentery, surface of liver, bile duct, pancreas, mesenteric lymph nodes and ovary.

SIGNS OF INFECTION AND PATHOLOGY: Whitish xenomas about 1 mm in diameter, in the gut wall. In heavy infections (Fig. 2.26), a mass of xenomas may completely replace the structure of the intestinal wall so that it is rigid, thickened and has a chalk-white, pebbled appearance. Sometimes, the enlarged bowel causes the abdominal wall to bulge markedly (Wellings, Ashley and McArn, 1969).

The xenoma arising probably from an infected neutrophil (Bekhti, 1984) has a wall about 8 μm thick, consisting of numerous PAS-positive layers; the xenoma plasma membrane is folded and amplified by fine tubular extensions (Weidner, 1976).

In moderate to heavy infections, the lamina propria, submucosa, and muscularis of the stomach, pyloric caeca and the intestine harbour numerous xenomas. The intestinal wall may reach a thickness of 4 mm even in young fishes, the epithelial layer of the mucosa disappears and the intestine becomes non-functional. Heavily infected plaice lose weight rapidly and intestinal failure results in death. In experimentally infected juvenile

Platessa platessa mortalities occur within 2 months of adminstration of spores, typically with inflammation of the visceral organs (McVicar, 1975). In juvenile *Pleuronectes flesus,* heavy infections also result in a decrease in erythrocyte size and in body weight losses (Bekhti, 1984).

Many of the mature xenomas are destroyed by tissue reactions of the host. Earlier observations of aggregates of host cells containing phago-cytized spore in a "diffuse infiltration" (Hagenmüller, 1899) or in "pseu-docysts" (Woodcock, 1904) led Doflein and Reichenow (1927 to 1929) to postulate that *G. stephani* cysts originated from several invaded cells. In fact, such formations represent the productive stage of host tissue reaction (Sect. IIB,1b) during elimination of the xenoma. Some xenomas burst and the discharged spores are scattered in the tissues of the gut wall, alongside blood vessels (Stunkard and Lux, 1965). McVicar (1975) obtained some evidence that the level of infection in individual fish fell with time and Bekhti (1984) observed that fish above 260 mm in length are resistant to reinfection. The immune system of the host might have contributed to these phenomena.

COMMERCIAL IMPORTANCE: *G. stephani* may prove to be a limiting factor in the growth of natural and cultured populations of *Platessa platessa* and *Pleuronectes flesus*. Bückmann (1952) noted that heavy infections of aquarium-kept *Pleuronectes platessa* by *G. stephani* could be fatal. Stun-kard and Lux (1965), who found heavy infections in young *Pseudopleu-ronectes americanus* but not in fish longer than 95 mm, concluded that the infected individuals did not survive their first year and estimated that there can be 40–50% loss of first year flounders. McVicar (1975), recording an outbreak of microsporidiosis in a *Platessa platessa* farm, observed a mean of 49.4% of infection of which about 10.7% were heavy enough to result in mortalities of the fish.

STRUCTURE AND LIFE CYCLE: The scarce data include mainly ultrastruc-tural observations which agree basically with findings on *G. anomala*: Jensen and Wellings (1972); Youssef and Hammond (1972). Bekhti (1984) observed paramural bodies in sporoblast mother cells.

Merogony: Dividing meronts with 3–8 nuclei lie in direct contact with host cell cytoplasm.

Sporogony: Sporoblast mother cells occur in groups of 8 or more per SPOV.

Spores: Since Hagenmüller (1899) gave neither measurements nor il-lustrations, the description is based on subsequent identification: In *Pla-tessa platessa* (fresh): elongate oval, less frequently ovoid (Fig. 2.22F) 2.7 (2.3–2.8) × 4.7 (4.2–5.2) μm. Posterior vacuole reaches to about the middle of the spore, its anterior limit is straight, sometimes vaulted, its side cytoplasmic borders are quite thin. Oval nucleus, 0.5–0.8 μm, a distinct PAS-positive polar cap (Lom, unpublished). Twelve coils of the polar tube have been recorded by Jensen and Wellings (1972) and 14 coils by Takvorian and Cali (1981).

TRANSMISSION: *G. stephani* is transmitted directly, by the ingestion of spores. Olson (1975, 1976) infected juvenile *Parophrys vetulus* by feeding them brine shrimps and amphipods that had previously ingested spores. McVicar (1975) failed to establish an infection by this method or by adding spores to the water but was successful when he mixed the spore suspension with food. He also established infections by introducing spores directly into the digestive tract and by inoculating them intraperitoneally. The latter way of infection resulted in xenomas formed free in the body cavity. McVicar suggested that free macrophages may have been the site of infection.

About one month after infection, xenomas are visible under the dissecting microscope and spores are present after 2 months. Juvenile fish are particularly susceptible to infection (Stunkard and Lux, 1965; McVicar, 1975).

Infection of *G. stephani* may serve as a guide to the particular nursery grounds used by the fish. In Yaguina Bay in the U.S.A. infection only occurred in the upper estuary (Olson and Pratt, 1973). From the prevalence of infection offshore, the proportion of fish that had utilized the upper estuary as a nursery was deduced.

HOST SPECIFICITY: *Citharichthys stigmaeus* Jordan and Gilbert and *Oncorhynchus keta* Walbaum are refractory to experimental infection (Olson, 1976).

REMARKS: For additional data on the occurrence of *G. stephani* see Radulescu and Vasiliu-Suceveanu, 1956: Stunkard and Lux, 1965; Wellings, Ashley and McArn, 1969; McArn, Ashley and Wellings, (1969) and Moller (1974). Further studies on this excellent experimental material should answer some important questions; for instance whether the microsporidia in populations of *Liopsetta glacialis* and *Pleuronectes flesus bogdanovi* from the White Sea are adapted to the cold water or constitute a separate taxonomic unit; also whether infected juveniles perish or eliminate the parasite. The material is so abundant that it affords the opportunity for studying the course of infection in nature.

Glugea truttae Loubès, Maurand and Walzer, 1981

HOST: *Salmo trutta* m. *fario*. Freshwater.
GEOGRAPHICAL DISTRIBUTION: A trout farm near Geneva, Switzerland; 1 out of 400 fry specimens examined was infected.
SITE OF INFECTION: Yolk sac of a 3-week-old trout fry.

SIGNS OF INFECTION AND PATHOLOGY: No xenoma is formed. Developmental stages occur in the periblast of the syncytial part of the yolk sac, encasing the yolk substance, while spores are concentrated in the yolk.

STRUCTURE AND LIFE CYCLE: *G. truttae* has been described only at the ultrastructural level (Loubès, Maurand and Walzer, 1981).

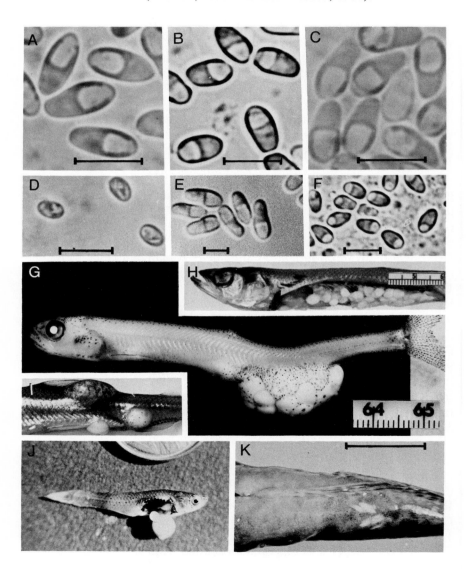

Merogony: Meronts globular, with up to 4 nuclei, closely surrounded by a cisterna of endoplasmic reticulum.

Sporogony: The SPOV forms around a multinucleate sporogonial plasmodium by means of series of blisters detaching the outer covering membrane in a process identical to that found in *G. anomala*. The sporogonial plasmodium divides into sporoblast mother cells which produce 2 spores each.

Spores (measured in ultrathin sections): 5 μm long (= 5.5–6 μm in fresh state according to the authors). There are 22–25 coils of the polar tube.

REMARKS: In spite of the absence of xenoma formation, Loubès, Maurand and Walzer (1981) allotted *G. truttae* to the genus *Glugea* rather than to *Pleistophora,* mainly because of its developmental similarity with previously described *Glugea* sp.. The single case of infection was detected in sectioned material. It may have been an aberrant infection in the yolk sac by a species of *Glugea* which is normally xenoma forming. The cellular reactions which lead to xenoma formation would not be inducible in yolk. Under normal circumstances the xenoma with its laminar wall should still be considered as one of the main characters of the genus *Glugea*.

Glugea sp. of Crandall and Bowser, 1981 (Fig. 2.41E,J)

HOST: *Gambusia affinis*. Freshwater.
GEOGRAPHICAL DISTRIBUTION: Southern California, U.S.A.
SITE OF INFECTION: Mesenteries, liver, ovaries and subdermal connective tissue.

Fig. 2.41 A–F: photomicrographs of spores, fresh, if not stated otherwise: A, *Glugea gasterostei* from *Gasterosteus aculeatus* (photograph by Dr. V. N. Voronin); B, *G. heraldi* from *Hippocampus erectus;* C, *Thelohania baueri* from *Pungitius pungitius* (photograph by Dr. V. N. Voronin); D, *Microsporidium ovoideum* from *Merluccius hubbsi* (formol fixed); E, *Glugea* sp. of Crandall and Bowser (reproduced from Crandall and Bowser, 1981 with permission of authors and publisher); F, *Nosema notabilis* from *Ortholinea polymorpha* infecting *Opsanus tau.* In all figures, scale bars = 5 μm. G–K, lesions due to microsporidian infection in fish; G–I, smelts *(Osmerus eperlanus mordax)* infected with *Glugea hertwigi;* G, a large knot of xenomas in the anal region, scale divisions = 1 mm (reproduced from Delisle, 1969, with permission of author and publisher); H, smelt in spent condition with body wall removed to expose the xenomas in the body cavity, scale divisions = 1 mm (reproduced from Nepszy and Dechtiar, 1972, with permission of author and publisher); I, xenomas showing through the body wall, scale divisions = 1 mm (photograph by Dr. C. Delisle); J, body cavity of *Gambusia affinis* opened to show a large xenoma due to *Glugea* sp. of Crandall and Bowser (photograph by Dr. P. R. Bowser), size of the fish is 2.8 cm; K, spindle shaped foci of infection with *Pleistophora vermiformis* showing under the skin of the ventral side of *Cottus gobio* (photograph by Dr. J. Prouza); scale bar = 1 cm.

SIGNS OF INFECTION AND PATHOLOGY: Xenomas to 2 mm in diameter. Heavy infections resulted in severe abdominal swelling and prevented efficient swimming. Xenomas accounted for as much as 35% of the body weight, and displaced internal body organs which were sometimes atrophied.

STRUCTURE AND LIFE CYCLE: A sporogonial plasmodium, sporoblast mother cells and SPOVs were mentioned.

Spores (fresh): Ovocylindrical, measuring 3.4 (3–4) × 7.4 (6–8) μm. The large posterior vacuole has a straight anterior border and reaches almost the mid-spore length.

2. *Glugea* Species which Form Small Xenomas

Several species of microsporidia are responsible for the induction in fish of xenomas which do not surpass 1 mm in diameter and are usually much smaller. The hypertrophic host cell nucleus remains single and occupies a central position in the cell until it degenerates. Developmental stages, where known, are not limited to peripheral layers but pervade the whole xenoma. The presence of true SPOVs has yet to be proven. At present, these species are assigned to the genus *Glugea*. There is little doubt that a new genus will have to be established for them but, as their biology is poorly understood at present, this step is considered premature. In addition to the named species, 3 unnamed species have been included in this group.

Glugea acerinae Jírovec, 1930 (Figs. 2.6B, 2.31, 2.32, 2.42)

HOST: *Gymnocephalus cernuus.* Freshwater.

GEOGRAPHICAL DISTRIBUTION: The river Labe, Czechoslovakia. A single infected specimen found.

SITE OF INFECTION: Intestinal wall.

SIGNS OF INFECTION AND PATHOLOGY: Small xenomas up to 0.35 mm (Figs. 2.31, 2.32), were located in the *tunica propria,* sometimes in submucosa or *muscularis* of the intestine. The xenoma wall is 0.2–0.5 μm thick. Jírovec supposed that each xenoma was derived from an infected migratory mesenchyme cell. Its nucleus hypertrophied into a single, vacuolated and disorganised structure which disappeared in late-stage xenomas. The mature xenomas frequently elicited a tissue reaction (Sect. IIB,1).

STRUCTURE AND LIFE CYCLE

Merogony: Long, multinucleate, branched cylinders (Fig. 2.42A), which at the end of the merogony cycle dissociate into short chains (Fig. 2.42D).

Sporogony: Uninucleate stages 2.5–3.5 μm produce 2 sporoblasts (Figs. 2.42E-H). In some xenomas, Jírovec (1930) observed large lobed stages with many nuclei at the border (Fig. 2.42C).

Spores (probably fixed): Ellipsoid, 3.5–4.5 × 2.5–3 μm, with a large posterior vacuole and a single nucleus. Spores may be released into the gut contents from broken xenomas and thus discharged from the host.

REMARKS: Of the sporogonic stages observed by Jírovec (1930) the large multinucleate bodies might have been SPOV-producing plasmodia and the uninucleate stages might have been sporoblast mother cells, corresponding to typical stages of *G. anomala.* There are very few characters which can be used to differentiate *G. acerinae* from other *Glugea* species forming small xenomas in the intestine of freshwater fish. The possible identity with *Pleistophora acerinae* cannot be excluded (see p. 93).

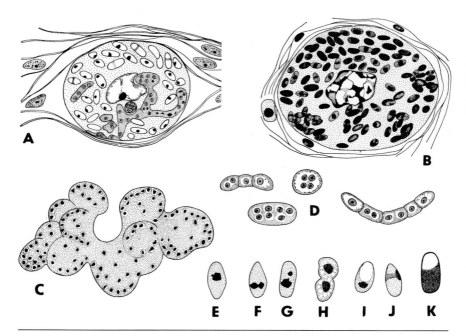

Fig. 2.42 Developmental stages of *Glugea acerinae* (after Jírovec, 1930): A, a small, young xenoma with a centrally located, vacuolated host cell nucleus. The cytoplasm harbours mature spores in addition to rounded, cylindrical or branched meronts; B, xenoma with mature spores and a central host nucleus; C, large multinucleate lobose plasmodium from advanced xenomas; D, meronts and their division; E–H, division of a sporoblast mother cell into two sporoblasts; I, sporoblast; J, stained spore; K, live spore (drawn to a larger scale than Fig. J).

Glugea berglax Lom and Laird, 1976 (Figs. 2.22L, 2.43)

HOST: *Macrourus berglax*. Marine.

GEOGRAPHICAL DISTRIBUTION: northern edge of the Grand Banks of Newfoundland. Out of 7 fishes examined 5 were infected.

SITE OF INFECTION: Mid-intestine, at the boundary between the *submucosa* and *muscularis,* and wall of the gall bladder.

SIGNS OF INFECTION AND PATHOLOGY: Whitish xenoma up to 0.5 mm in diameter with a wall 1–2 μm thick, coated by a thin layer of connective tissue and, on the side facing the *muscularis,* by enlarged muscle cells. There is a single hypertrophic host cell nucleus in the centre, which later disappears. The developmental stages are distributed throughout the whole xenoma.

STRUCTURE AND LIFE CYCLE

Merogony: Rounded uninucleate cells (Fig. 2.43A-D) measuring 3–5 μm with nuclei up to 1.6 μm, often lying in groups, divide or give rise to multinucleate, rather elongate meronts up to 10–15 μm. Multinucleate meronts produce uninucleate stages sometimes arranged in chains. Merogony stages are in close contact with the cytoplasm.

Sporogony: Sporogonial plasmodia (Fig. 2.43E) with about 9 compact nuclei the size of 0.8 μm measure up to 7 μm diameter. They develop within SPOVs and divide into clumps of uninucleate stages (Fig. 2.43F), in fact the sporoblast mother cells which measure 2–2.5 μm. These cells are smaller than those of the merogonic sequence and groups of them are common in mature xenomas where merogony is rare. The mother cells divide into 2 sporoblasts. A distinct vacuole appears in the sporoblast during its formation into a spore. Curiously, the mother cells and sporoblasts are smaller than mature spores.

Spores (fresh, Fig. 2.22L): Elongate ovoid, 2.7 (2.4–3.1) × 6.4 (4.2–8) μm. The wider posterior half is occupied by a vacuole reaching more or less to the mid-body of the spore. The vacuole has a rounded anterior limit and thick cytoplasmic side walls with distinct coils of the polar tube. There is a single oval nucleus, and a dot-like PAS positive polar cap (Fig. 2.43G,H).

TRANSMISSION: Mature spores were found in the bile of infected fish. Exit of spores via the alimentary canal is likely.

Glugea luciopercae Dogiel and Bykhowsky, 1939 (Fig. 2.22M)

SYNONYM: *Glugea dogieli* Gasimagomedov and Issi, 1970.

HOST: *Stizostedion lucioperca*. Doubtful hosts of this species are *Clupeonella delicatula caspia* (in Dogiel and Bykhowsky, 1939) and *C. cul-*

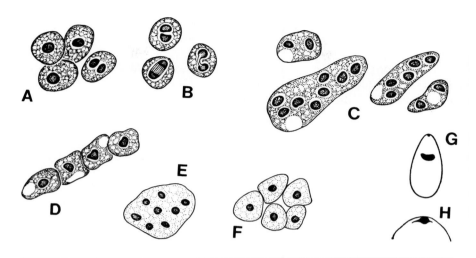

Fig. 2.43 *Glugea berglax,* some of the developmental stages: A, uninucleate meronts; B, their division; C, multinucleate elongate meronts; D, fragmentation of an elongated meront into uninucleate cells; E, sporogonial plasmodium producing (F) sporoblast mother cells, each of which will give rise to two sporoblasts; G, outline of mature spore showing polar cap and nucleus; H, PAS-positive polar cap.

tiventris cultiventris (in Muradian, 1972). All of them freshwater to brackish.

GEOGRAPHICAL DISTRIBUTION: In *S. lucioperca* and *C. delicatula caspia* in the Caspian Sea (Dogiel and Bykhowsky, 1939). In *S. lucioperca* in the Volga river basin (Shulman, 1962); in the Aral Sea, 50% of *S. lucioperca* examined were massively infected (Osmanov, 1971); Kakhowska Dam, Ukraine (Iskov, 1966) and Lake Khoroshee, Siberia (Kashkowsky, Razmashkin and Skripchenko, 1974). Also recorded in *S. lucioperca* from brackish lagoons on the Rumanian Black Sea coast (Radulescu and Vasiliu-Suceveanu, 1956) and Dimitrov Dam, Bulgaria (Grupcheva and Lom, 1980).

SITE OF INFECTION: In *S. lucioperca,* typically in the submucosa of the intestine. Shulman (1962) also gives mesentery, connective tissue of the ovary and gills. In the only specimen of *C. delicatula caspia* which Dogiel and Bykhowsky (1939) found infected, there were 3 round cysts in the gills.

SIGNS OF INFECTION AND PATHOLOGY: Xenomas are tiny (0.2–0.4 mm) whitish nodules which can be very numerous, up to 134 nodules per cm^2 (Iskov, 1966), or may fuse together into a flat layer in the intestine (Dogiel and Bykhowsky, 1939: Osmanov, 1971). The wall, 1–2 μm thick, is surrounded by a 8–9 μm layer of connective tissue cells. In the centre of the xenoma, there is a degenerate host cell nucleus (Gasimagomedov and Issi, 1970).

The lesions in the intestine appear to be extensive enough to impair function considerably. According to Shestakovskaya (1983 and personal communication) the presence of more than 30 xenomas in the intestine may be harmful for small fry.

COMMERCIAL IMPORTANCE: In some regions of the U.S.S.R., it is a serious pathogen for young *S. lucioperca,* an important food fish.

STRUCTURE AND LIFE CYCLE: Developmental stages were not observed.

Spores (fresh): From *S. lucioperca:* elongate oval, 2.1 (1.7–2.4) × 4.5 (4.1–4.8) μm, with a narrower anterior end and a posterior vacuole reaching to the mid-body of the spore. The vacuole has a vaulted anterior border (Fig. 2.22M). There is a single nucleus and a dot-like PAS positive polar cap (Grupcheva and Lom, 1980).

According to Dr. E. V. Shestakovskaya (1983, personal communication) spores retain infectiousness for 6 months.

CONTROL MEASURES: To prevent spreading of the disease, Shestakovskaya (1983, personal communication) recommended banning transfer of fish younger than 5 years, after which they no longer harbour viable spores.

REMARKS: In the brief original description of Dogiel and Bykhowsky (1939), the spores were described as oval or rather pear-shaped, measuring about 3 μm. It was not stated whether these spores were fixed or fresh. Shulman (1962) illustrated oval spores of larger size. He listed *S. lucioperca, Alosa kessleri volgensis* and *C. delicatula caspia* as hosts. He did not specify from which host the material was taken for his description of spores. The differences in *G. luciopercae* spore size between the original description (3 μm) and all subsequent findings is puzzling. A complication was introduced by Gasimagomedov and Issi (1970), who considered *G. luciopercae* to be a mixture of species and ignoring the original name then established *G. dogieli* for the parasite of *S. lucioperca* and *G. bykhowskyi* for the parasite of *Alosa kessleri volgensis* (see p. 85). They did not specify the host of *G. luciopercae.* Of the possibilities that the xenomas in *S. lucioperca* intestine are due to 2 separate species, or that the original measurements were grossly inaccurate we consider the latter more probable. *G. dogieli* is therefore a junior synonym of *G. luciopercae.*

Further studies are needed to show whether the microsporidian recorded from *C. delicatula caspia* by Dogiel and Bykhowsky (1939) is also *G. luciopercae.* They evidently based their opinion on spore size, the aberrant measurements of spores from *S. lucioperca* matching those of spores from *C. delicatula caspia* but differing from measurements of two other species, with which they made comparisons (*G. anomala* and *G. hertwigi*).

Muradian (1972) identified, as *G. lucipercae,* a microsporidium that he found in *Clupeonella cultriventris cultriventris,* in the Danube delta in Rumania. The prevalence was 4.5 to 47%. He gave no description of the parasite, which invaded the intestinal wall, pyloric caeca and mesentery, to a degree that the infection might have been fatal. However, his statement that the cysts markedly protruded into the gut lumen, suggests that they were large. This is at variance with the small size of the xenomas of *G. luciopercae* in the original description.

Glugea machari (Jírovec, 1934) Sprague, 1977 (Fig. 2.46D)

Synonym: *Octosporea machari* Jírovec, 1934
Host: *Dentex dentex.* Marine.
Geographical distribution: Rab Island off the Yugoslavian coast of the Adriatic Sea. Two infected specimens found in which the xenomas were scarce.
Site of infection: Liver.
Signs of infection and pathology: Whitish xenomas, round to ovoid, up to 400 μm, in the liver parenchyma. They have a wall, 0.3–1 μm thick, invested with a layer of fibroblasts 7–15 μm thick. Often found near blood vessels, sometimes apparently lying in blood lacunae which may have been formed around them. Only mature xenomas, without developmental stages and host nuclei were found.
Structure and life cycle
Spores (not stated whether fresh or fixed (Fig. 2.46D)): Variable in shape from rod-like to pyriform or ovoid and often bean-shaped, slightly curved; they measure 3–4.5 × 0.8–1.5 μm and have a large posterior vacuole and a single centrally located nucleus.
Remarks: Jírovec (1934) assigned the species to the genus *Octosporea* Flu, 1911 because of the curved or rod-shaped spores. Sprague (1977) transferred the species to *Glugea* since the spores are uninucleate, in contrast to the diplokaryotic spores of *Octosporea. G. machari* seems to be quite distinct from *Microsporidium ovoideum,* also from the liver of Mediterranean fishes, because of spore size and shape.

Glugea shulmani Gasimagomedov and Issi, 1970 (Fig. 2.44)

Hosts: *Neogobius caspius; N. fluviatilis pallasi; N. melanostomus affinis.* Marine, 10–12% salinity.
Geographical distribution: Southern Caspian Sea, district Samur-Kayakenta at the town of Begdash, U.S.S.R.
Site of infection and pathology: Very small xenomas, 18–80 μm,

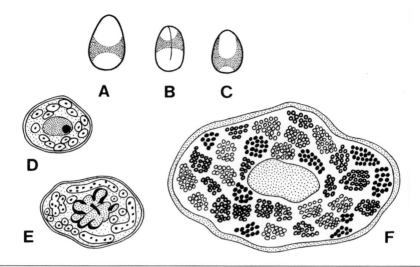

Fig. 2.44 *Glugea shulmani:* spore embedded in glycerin-gelatine (A), stained with Heidenhain's iron haematoxylin (B) and Giemsa (C). D–F, stages of development of *G. shulmani* xenoma: D, an early stage with parasite cells surrounding the host cell nucleus; E, more advanced stage with lobose host cell nucleus; F, fully grown xenoma surrounded by connective tissue cells. (Reproduced from Gasimagomedov and Issi, 1970, with permission of the authors and publishers.)

thought to be derived from a subepithelial cell, have a wall 2.5–5 μm thick (possibly this would have included the apposed fibroblasts) and a centrally located, hypertrophic nucleus which eventually becomes a chromophile mass 25 μm in diameter.

STRUCTURE AND LIFE CYCLE

Merogony: Uninucleate meronts and multinucleate cylinders were illustrated but not described.

Sporogony: Not described and SPOVs were not mentioned but spores were arranged in groups.

Spores (glycerin-gelatin preserved): Ovoid or pyriform, with a large posterior vacuole and measuring 2.2–2.4 × 1.2–1.6 μm; 1.2 × 2.4 μm after additional fixation.

REMARKS: Gasimagomedov and Issi considered that this species closely resembled *G. luciopercae* as described by Dogiel and Bykhowsky (1939) by virtue of its small, pear-shaped spores. They made a taxonomic error in quoting "*G. luciopercae* part." as a synonym of *G. shulmani*. The neogobiids were never reported as hosts of the *G. luciopercae*. The description of *G. shulmani* is incomplete and requires re-examination.

Possibly the *Glugea* sp. invading *N. melanostomus* and *N. fluviatilis,* another gobiid species in the Black and Azov Seas, reported by Naidenova (1974), is identical with *G. shulmani.*

Also the microsporidian, reported by Shumilo (1959) as *G. anomala* from *N. fluviatilis* might well be *G. shulmani.* Shumilo (1959) gave no description of the parasite but stated simply that the gut was covered by a confluent layer of cysts, up to 1000 cysts per cm^3. This heavy infection caused constriction of the gut and emaciation of the host.

Glugea tisae (Lom and Weiser, 1969) Lom and Laird, 1976 (Figs. 2.22N, 2.45D)

SYNONYMS: *Nosema tisae* Lom and Weiser, 1969; possibly *Plistophora siluri* Gasimagomedov and Issi, 1970.
HOST: *Silurus glanis.* Freshwater.
GEOGRAPHICAL DISTRIBUTION: River Tisza near Szolnok, Hungary; light infection in 2 out of 8 catfish.
SITE OF INFECTION: Submuscosa of the intestine.
SIGNS OF INFECTION AND PATHOLOGY: Xenomas, up to 0.6 mm with a wall 3 μm thick or slightly thicker. In the centre of the xenomas lies a degenerate remnant of the hypertrophic host cell nucleus, more than 100 μm in diameter. Sporoblasts and spores which fill the xenoma even occupy the nucleoplasm.
STRUCTURE AND LIFE CYCLE: Merogony was not observed.

Sporogony: Large numbers, possibly more than 100, of sporoblasts and spores were situated within large SPOVs (called sporogonic vacuoles in the original text). The SPOVs appeared as compartments within the cysts and, as all sporoblasts within each compartment were at the same stage of maturity they undoubtedly arose from a large sporogonial plasmodium although this was not seen.

Spores (fresh) (Figs. 2.22N, 2.45D): Ovoid, sometimes bean-shaped or very elongate, mostly with one side vaulted more than the other. They measure 2.2–2.6 × 4–5 μm. A large posterior vacuole reaching into the anterior half of the spore has a round or flat anterior border, and thick cytoplasmic side walls with grooves indicating the polar tube coils. There is a single, crescent-shaped nucleus and a dot-like PAS positive polar cap. Spores often occur in pairs.
REMARKS: The species *G. luciopercae* and *G. tisae* bear strong resemblances to one another but their degree of kinship can only be settled after detailed re-examinations. Sprague (1977) suggested that *Pleistophora siluri* Gasimagomedov and Issi (1970) from *Silurus glanis* from the U.S.S.R. (Caspian Sea) is probably synonymous with *G. tisae. P. siluri,* in the in-

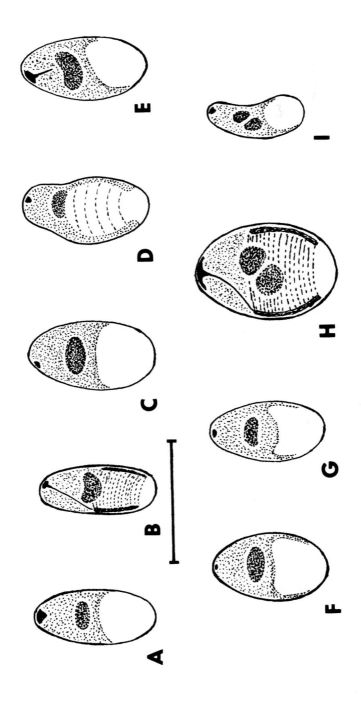

testinal submucosa, forms numerous yellowish cysts up to 410 μm, which contained what appeared to be SPOVs compressed into irregular shapes. Measurements given for SPOVs were 16 × 25 μm and they usually contained over 50 spores, rarely 12–16. The SPOVs although not illustrated very convincingly, could be identical with the compartments of *G. tisae*. No mature spores were found in the centre of the cyst and one explanation given by the authors for this was that the centre might have been the site of the original host cell nucleus.

Data given for *P. siluri* spores were: shape elongate-ovoid and measurements 1.8–2.4 × 4.8–5 μm (fresh) and 2–2.4 × 3.6–4.8 μm (fixed and stained). They were reported as having a small posterior vacuole and 2 nuclei but, as there is a single nucleus in both *Glugea* and *Pleistophora*, this might have been a mistaken observation. In all other respects *P. siluri* closely resembles *G. tisae* and, in agreement with Sprague (1977) they are considered synonymous.

Glugea sp. of Bond, 1938

SYNONYM: *Glugea hertwigi* Weissenberg *pro parte*—in Bond (1938).
HOST: *Fundulus heteroclitus*. Marine.
GEOGRAPHICAL DISTRIBUTION: Chesapeake Bay at Baltimore, U.S.A. Atlantic coast. Only 4 infected specimens found.
SITE OF INFECTION: Mucosa of the stomach and the bile duct.
SIGNS OF INFECTION AND PATHOLOGY: Xenomas, 33–122 μm, which have a large host cell nucleus at the centre and developmental stages at the periphery.
STRUCTURE AND LIFE CYCLE: Spores (fresh) are elongate, cylindrical, narrower anteriorly and have a large posterior vacuole; when fixed and stained spores measure 3–4 × 1–1.5 μm.
REMARKS: Bond (1938) stated that the xenomas were similar to *G. acerinae*, but the spore structure differed from it and from *G. stephani*. Spores were almost identical with those of *G. hertwigi* but the xenomas of *G. hertwigi* are quite different.

Fig. 2.45 Microsporidian spores, showing the outline of the posterior vacuole as seen in fresh state and with the nucleus and PAS polar cap as seen after staining. A, *Glugea anomala;* B, *G. hertwigi;* C, *G. heraldi;* D, *G. tisae;* E, *Pleistophora hippoglossoideos;* F, *P. macrozoarcidis;* G, *Loma branchialis;* H, *Ichthyosporidium giganteum;* I, *Spraguea lophii (Nosema*-type spore). The spores are drawn to scale (bar = 5 μm).

Glugea sp. (Fig. 2.22O)

HOST: *Pseudopleuronectes americanus*. Marine.
GEOGRAPHICAL DISTRIBUTION: The Indian River, Delaware, U.S.A.; one infected specimen found. Material presented to the author by Dr. V. Sprague.
SITE OF INFECTION: Submucosa of the intestine.
SIGNS OF INFECTION AND PATHOLOGY: Xenomas usually 0.2–0.3 mm, occasionally 0.8 mm in diameter, with a wall 1.5–2 μm thick. Each had a central hypertrophic host cell nucleus up to 40 μm large. Only mature xenomas were encountered.
STRUCTURE AND LIFE CYCLE: Merogony was not observed.

Sporogony: Sporoblasts and spores inside SPOVs were found, in addition to free spores.

Spores (formalin-fixed): Oval, uninucleate, measuring 2.2 (2–2.3) × 4.1 (3.7–5) μm, with a large posterior vacuole, reaching to about the midbody of the spore, and a dot-like PAS positive polar cap.
REMARKS: The type of xenoma is consistent with this parasite being a species of *Glugea* but, in the absence of information on developmental stages, no specific name has been assigned to it. The shape and size range of spore are similar to those of *Glugea stephani* and the host is identical. However, the structure of the xenoma is so different that identification with *G. stephani* is precluded. This species provides evidence that it is not just the host reaction and properties of the cell type but also the biology of the parasite that is responsible for the final structure of the xenoma.

Glugea sp. (Fig. 2.22K)

HOST: *Sphaeroides maculatus*. Marine.
GEOGRAPHICAL DISTRIBUTION: Atlantic coast of the U.S.A. Material presented to the author by Dr. J. Couch in 1969.
SITE OF INFECTION: Intestinal mucosa.
SIGNS OF INFECTION AND PATHOLOGY: Xenomas up to 250 μm in diameter, with a wall thickness of 1–2 μm and a centrally located, hypertrophic nucleus. The xenomas found contained only mature spores.
STRUCTURE AND LIFE CYCLE

Spores (formol-fixed): Elongate oval, 1.9 (1.7–2) × 3.8 (3.3–4.3) μm; a large, rounded posterior vacuole and a dot-like PAS positive polar cap at the anterior apex. The small, formol-fixed piece of infected tissue given to the author was unfortunately not satisfactory for an adequate description.

3. *Glugea* Species of Provisional Status

Redescription of these species is required before their generic status can be confirmed. Many will require transfer to another genus. Unless redescribed most of them can be considered *nomina dubia*.

Glugea bychowskyi Gasimagomedov and Issi, 1970 (Fig. 2.46E)

SYNONYM: *G. luciopercae* Dogiel and Bykhowsky, 1939 part. sensu Shulman, 1962.
HOST: *Alosa kessleri volgensis*. Freshwater.
GEOGRAPHICAL DISTRIBUTION: Caspian Sea at the island of Tyulenyevo, U.S.S.R.
SITE OF INFECTION: Intestinal walls and testis.
SIGNS OF INFECTION AND PATHOLOGY: All tissues of the testes were pervaded by microsporidia, with only the envelope, *tunica albuginea* free of spores.
STRUCTURE AND LIFE CYCLE: Development unknown.
    Spores (glycerine-gelatine preserved): Pyriform, 1.8 × 3.6 μm; held together in pairs, which, the authors believed, indicated that it belonged to a *Glugea*.
REMARKS: The only material available to Gasimagomedov and Issi were glycerine-gelatine preserved smears of the intestine and fixed pieces of invaded testis. However, the description and size of the spores and characters such as their refractility stainability and presence of "two vacuoles" is inadequate for a diagnosis.

Glugea caulleryi Van den Berghe, 1940

SYNONYM: *G. microspora* Van den Berghe, 1939 (= *nomen preoccupatum* by *G. microspora* Thélohan, 1892, this being in turn a synonym of *G. anomala* Moniez, 1887).
HOST: *Ammodytes lanceolatus*. Marine.
GEOGRAPHICAL DISTRIBUTION: Wimmereux, France; a "rather frequent" infection (Van den Berghe, 1940).
SITE OF INFECTION: Liver.
SIGNS OF INFECTION AND PATHOLOGY: Parasites are localized in spherical encysted masses, appearing as white spots near the surface or surrounded by liver tissue. What must have been terminal infections appeared as generalised, overwhelming infections of the liver cells.

STRUCTURE AND LIFE CYCLE
  Spores: Ovoid, 1–1.5 μm long.
REMARKS: The data available is indequate even for a generic diagnosis.
Van den Berghe's was the only finding.

Glugea cordis Thélohan, 1895 (Fig. 2.46F)

HOST: *Sardina pilchardus sardina*. Marine.
GEOGRAPHICAL DISTRIBUTION: French coast at Marseille.

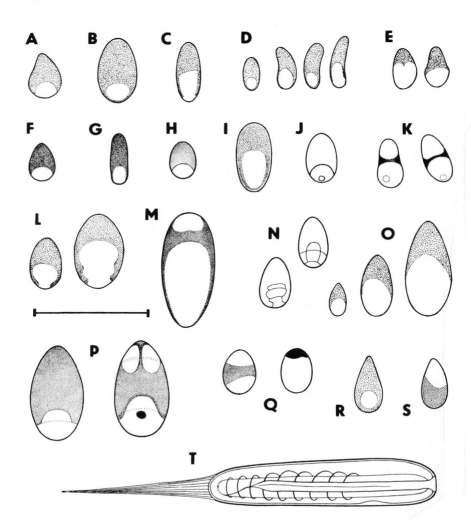

SITE OF INFECTION: Connective tissue and possibly muscle fibres of the heart.

STRUCTURE AND LIFE CYCLE: Irregular spots on the heart, distributed especially on the anterior surface of the ventricle, may have been xenomas.

Spores: Ovoid, 3–3.5 × 2 μm, with pointed anterior and a large posterior vacuole.

REMARKS: Genus uncertain. Thélohan's was the only finding.

Glugea depressa Thélohan, 1895 (Fig. 2.46G)

SYNONYM: *Nosema depressum* (Thélohan, 1895) Labbé, 1899
HOST: *Coris julis*. Marine.
GEOGRAPHICAL DISTRIBUTION: Marseille, France.
SITE OF INFECTION: Liver.
STRUCTURE AND LIFE CYCLE: "Very small white spots" on the surface of the liver.

Spores: Elongate, almost cylindrical, 4.5–5 × 1.5–2 μm, with a relatively small posterior vacuole.

REMARKS: Genus uncertain; not found since Thélohan.

Glugea destruens Thélohan, 1891 (Fig. 2.46H)

SYNONYM: *Nosema destruens* (Thélohan, 1891) Labbé, 1899
HOST: *Callionymus lyra*. Marine.
GEOGRAPHICAL DISTRIBUTION: Atlantic coast of France at Concarneau and Roscoff (Thélohan) and coast of Africa at Cape Town (Gaievskaya and Kovaleva, 1975).
SITE OF INFECTION: Muscles.
SIGNS OF INFECTION AND PATHOLOGY: Invaded muscle fibres undergo vitreous degeneration. Vegetative stages were simply described as a pro-

Fig. 2.46 Microsporidian spores redrawn to scale (bar = 10 μm) from various sources. A, *Glugea acuta* (after Thélohan, 1895); B, *G. nemipteri* (after Weiser *et al.*, 1981); C, *G. rodei* (after Kazieva & Voronin, 1981); D, *G. machari* (after Jírovec, 1934), note the extreme shape variation; E, *G. bychowskyi* (after Gasimagomedov & Issi, 1970); F, *G. cordis*, G, *G. depressa;*, H, *G. destruens;*, (F–H after Thélohan, 1895); I, *G. intestinalis* (after Chen, 1956); J, *G. punctifera* (after Thélohan, 1895); K, *Pleistophora dallii* (after Zhukov, 1964); L, *P. duodecimae* (after Lom *et al.*, 1980); M, *P. elegans* (after Auerbach, 1910); N, *P. macrozoarcidis* (after Nigrelli, 1946); O, *P. priacanthicola,* microspore, normal spore and macrospore (after Hua & Dong, 1983); P, *P. macrospora* (after Léger & Hesse, 1916); Q, *P. sauridae* (after Narasimhamurti & Kalavati, 1972), in one spore, the enormous PAS-positive polar cap is demonstrated; R, *Microsporidium valamugili* (after Kalavati & Lakshminarayana, 1982); S, *M. sauridae* (after Narasimhamurti & Kalavati, 1982; T, *Mrazekia piscicola* (after Cépède, 1924).

toplasmic mass without membrane and cyst envelope (Thélohan, 1891). Gaievskaya and Kovaleva (1975) observed large cysts 1–2 × 5–10 mm which they took for hypertrophic muscular bundles.

COMMERCIAL IMPORTANCE: Unsightly cysts make it impossible to use the fish as food.

STRUCTURE AND LIFE CYCLE

Spores (Fixed?): Oval, 3–3.5 × 2.2–5 μm, with a rounded posterior vacuole (Thélohan), Gaievskaya and Kovaleva give 3.6 × 2 μm.

REMARKS: Thélohan's (1891) description of "vitreous degeneration" and of lack of cyst envelopes might correspond to Pleistophora rather than to Glugea-infections, so that the generic assignment is doubtful. It is also not certain, if Gaievskaya and Kovaleva (1975) dealt with the same parasite.

Glugea intestinalis Chen, 1956 (Fig. 2.46I)

HOST: Mylopharyngodon piceus. Freshwater.

GEOGRAPHICAL DISTRIBUTION: China. In about 3% of the fish in different samples.

SITE OF INFECTION: Mucosa of the small intestine.

STRUCTURE AND LIFE CYCLE

Spores (fixed?) Oval, 6.2(5–6.3) × 3.6(3.1–4) μm with a single nucleus and a large posterior vacuole in which threads of the polar tube are visible.

REMARKS: Genus indeterminate.

Glugea punctifera Thélohan, 1895 (Fig. 2.46J)

SYNONYM: Nosema punctiferum (Thélohan, 1895) Labbé, 1899

HOST: Gadus (Pollachius) virens. Marine.

GEOGRAPHICAL DISTRIBUTION: Atlantic coast of France at Concarneau.

SITE OF INFECTION: Connective tissue of ocular muscle.

SIGNS OF INFECTION AND PATHOLOGY: A xenoma wedged between the muscle fibres, with spores centrally located visible in Thélohan's (1895) illustrations but unmentioned in his text.

STRUCTURE AND LIFE CYCLE

Spores (not stated if fresh): 4–5 × 3 μm, with a globular posterior vacuole; identical in shape, but larger than those of G. anomala (Thélohan, 1895).

REMARKS: Generic status indeterminate. Akhmerov (1951) claimed—without describing it—that he had found G. punctifera as a serious intramuscular parasite of walleye pollock, Theragra chalcogramma from the

Okhotsk and Japanese Seas. He alleged that this microsporidian also infects a "cod" from the Okhotsk Sea.

Glugea sp. of Bogdanova, 1961

Found in the intestinal wall of *Abramis ballerus* (freshwater) in the Volga river, U.S.S.R. No data available.

Glugea sp. of Naidenova, 1974

HOSTS: *Gobius niger; G. paganellus; G. cobitis; G. ophiocephalus; G. ratan; G. platyrostris; Neogobius fluviatilis; N. melanostomus; N. cephalarges; Mesogobius batrachocephalus; Proterorhinus marmoratus.* Marine or euryhaline.

GEOGRAPHICAL DISTRIBUTION: Black and Azov Seas. In some localities, prevalence of infection in *G. niger* was 50%. In *N. melanostomus* 10 out of 15 localities harboured infected fish and the prevalance was 8–31% in most of the other hosts, the infection was found in one locality only.

SITE OF INFECTION: Xenomas were found in the submucosa of the intestine, rarely in the liver and serosa of the internal organs. No description was given.

REMARKS: It is improbable that it was a single species parasitising all the hosts. Of these *N. melanostomus* or *N. fluviatilis,* could have harboured *G. shulmani,* which has been described from the same hosts in the Caspian Sea.

Glugea sp. of Pfeiffer, 1895

Found in *Phoxinus phoxinus* (freshwater) in Germany. No other data available.

Glugea sp. of Roman, 1955

Found in 2 specimens of *Perca fluviatilis* (freshwater) in the Danube river, Rumania. Numerous small "cysts" found on the gills contained oval spores, 2–3 × 4–5.7 μm.

Glugea sp. of Voronin, 1974

Found only once in the skin of *Lota lota* (freshwater) from Lake Vrevo near Leningrad, U.S.S.R. Forms small white xenomas 0.5–0.7 mm, mostly in groups on the fins. Voronin (1974), who gave no further data, considered it as a new species.

B. Genus *Pleistophora* Gurley, 1893, emend. Canning and Nicholas, 1980

Twenty-five named species and a number of unnamed forms attributed to the genus are listed in Table 2.1, grouped according to the type of tissue invaded. Not all are well described and it is certain that not all belong to *Pleistophora: P. peponoides, P. scianae, P. cepedianae* and *P. shiplei* formerly attributed to *Pleistophora* have been transferred to the genus *Glugea* or the collective group *Microsporidium* in the present work. Yet others may be transferred as more evidence becomes available. Two spore sizes have been found in 9 of the named species and even 3 spore sizes have been recorded (Table 2.1). Although the formation of macrospores is given as a generic character their apparent absence in the remaining species does not constitute grounds for transfer to a new genus, since macrospores are rare, may not always be expressed or may not easily be seen. The role of the macrospores is a subject worthy of investigation.

TABLE 2.1   Species of *Pleistophora* Classified According to Tissues Invaded

Muscle	*P. anguillarum* *2	*P. carangoidi* +1
	P. destruens +1	*P. duodecimae* *2
	P. ehrenbaumi *2	*P. gadi* +1
	P. hippoglossoideos *1	*P. hyphessobryconis* *1
	P. ladogensis +1	*P. littoralis* *2
	P. macrospora +1	*P. macrozoarcidis* *2
	P. tahoensis *1	*P. tuberifera* +1
	P. typicalis *2	*P. vermiformis* *1
	Pleistophora sp. of Bond, 1938 +1	
	Pleistophora sp. of Drew, 1909 +1	
	Pleistophora sp. of Ghittino, 1974 +1	
Oocytes	*P. elegans* +2	*P. mirandellae* +2
	P. oolytica *3	*P. ovariae* *1
	P. sulci *1	
Testes	*P. longifilis* *2	
	Pleistophora sp. of Lucký, 1957 *1	
Liver	*Pleistophora* sp. of Woodcock, 1904 +1	
Connective tissue	*P. dallii* +1	
"Visceral tissue"	*P. sauridae* +1	
Mesentery and intestine	*P. acerinae* +1,	*P. priacanthi* +3

*Well described species
+Poorly described species
1,2,3 Number of spores sizes recorded

Pleistophora typicalis Gurley, 1893 (Figs. 1.9, 1.15, 2.47A-I)

SYNONYM: Glugeidé Thélohan, 1892

HOSTS: The original description of Gurley (1893) based on Thélohan's reports (1891–1892) gave only *Myoxocephalus scorpius,* which is the type host. Thélohan (1895) added *Blennius pholis, Taurulus bubalis, Pungitius pungitius.* Shulman (1962) listed *Myoxocephalus quadricornis labradoricus.* These are all marine except the euryhaline *P. pungitius.*

Canning, Hazard and Nicholas (1979) recognised that the infection in blennies was due to an independent species, which they named *P. littoralis* (p. 108). It is possible that parasites from all except the type host are closely related species, incorrectly assigned to *P. typicalis.* Resolution of this problem depends on re-examinations of these hosts.

GEOGRAPHICAL DISTRIBUTION: Atlantic coast of France, off Concarneau, in *M. scorpius, B. pholis* and *T. bubalis* (Thélohan, 1895); coast of Scotland, off Aberdeen in 6 out of 17 *M. scorpius* (Canning and Nicholas, 1980); White and Baltic seas in *M. quadricornis labradoricus* (Shulman and Shulman-Albova, 1953); Rennes, France in *P. pungitius* (Thélohan, 1895).

SITE OF INFECTION: Skeletal muscles. Common in muscles of ventral abdominal wall of *M. scorpius* (Canning and Nicholas, 1980). Shulman and Shulman-Albova (1953) found a species, which they identified as *P. typicalis* in liver and in the walls of the urinary bladder and gall bladder of *M. scorpius.*

SIGNS OF INFECTION AND PATHOLOGY: Opaque white to yellowish-brown threads or spindle shaped loci, visible to the naked eye in the muscles. Each represents an infected muscle bundle, comprised of individually infected muscle fibres. The entire locus may be transformed into a solid mass of close-packed SPOVs containing spores, with some merogonic and sporogonic stages interspersed.

At the ultrastructural level (Canning and Nicholas, 1980), intact myofibrils are seen interwoven between parasite stages and at the edge of the loci, the frayed ends of the myofibrils are in direct contact with the parasites. During the active development of the parasites there is no infiltration by host macrophages and no cyst wall separating the parasites from the muscles.

STRUCTURE AND LIFE CYCLE (based on Canning and Nicholas, 1980; Fig. 2.47A-F)

Merogony: Meronts rounded, 7 μm and larger, with 2 to several nuclei; the plasmalemma invested with an amorphous layer 0.5 μm thick, permeated by channels connecting with a mass of vesicles directly abutting on to the myofibrils. Division by plasmotomy.

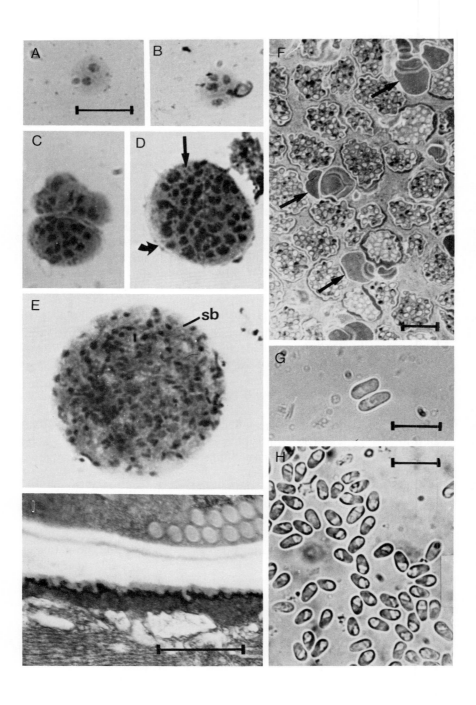

Sporogony: Sporogonial plasmodia, 20–40 μm with numerous nuclei, distinguished from meronts by secretion of a fine electron-dense layer midway through the amorphous wall. This wall is later modified to form a double- or triple-layered envelope enclosing the spores. The mature sporogonial plasmodium withdraws inside the SPOV envelope and progressively segments into uninucleate sporoblasts.

SPOV: 2 types (Fig. 1.9); those containing numerous microspores, commonly occurring, measuring 23 (15–29) μm (fresh); those containing 8 macrospores, rare, measuring 13 (11–14) μm (stained).

Spores (Figs. 1.15, 2.47G,H): Fresh microspores are ovoid, uninucleate, one side slightly more arched, 2.3 × 4.4 μm (2.4–2.7 × 3.7–5.6 μm) with the posterior vacuole shifted sideways, mostly with an oblique anterior border, occupying one third or more of spore length. There are 10–22 coils of the polar tube. Fresh macrospores are elongate, tapering more anteriorly than posteriorly, 3 × 7.5 μm (3–3.3 × 6.5–8.3 μm). Posterior vacuole reaches almost to mid-spore length. Up to 33 coils of the polar tube. Possibly binucleate.

Thélohan (1892) gave 3 μm for the spore measurements and Gurley (1893) gave 1.5–2 μm, neither author mentioning whether the spores were measured fresh or from stained smears or sections. Shulman and Shulman-Albova (1953) gave measurements of 2.6–2.9 × 3.5–5 μm and posterior vacuole 1.7 × 2 μm but this parasite of *M. scorpius* was not from muscle.

Pleistophora acerinae Vaney and Conte, 1901

HOST: *Gymnocephalus cernuus;* and *Perca schrenki* according to Agapova (1966). Freshwater.

GEOGRAPHICAL DISTRIBUTION: France at Lyon (Vaney and Conte, 1901); U.S.S.R.: river basins of the Dnieper, Western Dvina, Yenisei and Barabinskie lakes (Shulman, 1962); Lake Chelkar (Agapova, 1966) and Lake Balkash (Maksimova, 1962) in Kazakhstan.

Fig. 2.47 *Pleistophora typicalis.* A,B, meronts with 3 and 4 nuclei; C, group of meronts after division; D, undivided sporont, curved arrow points at the amorphous SPOV envelope, straight arrow points at the chromophil granules amongst the sporont nuclei; E, sporont in which the cytoplasm has finally separated into uninucleate sporoblasts (sb); F, a mass of parasites within the degenerate muscle fibre, which are mostly microspore SPOVs but a few are sporonts not yet completely divided (arrows); G, two fresh macrospores; H, fresh microspores; I, the envelope of a mature SPOV as seen in the electron microscope. It is separated from the muscle fibre by a loose layer of vacuolar spaces and three layers of different electron density can be resolved. Scale = 10 μm on Fig. A applies also to Figs. B–E; scale on Figs. G and H = 10 μm; scales on Figs. F and I = 20 μm and 0.4 μm, respectively. (Reproduced from Canning and Nicholas, 1980, with permission of authors and publisher.)

SITE OF INFECTION: Mesentery (Vaney and Conte, 1901); mesentery and intestinal walls (Agapova, 1966; Kashkowsky, Razmashkin and Skripchenko, 1974).

SIGNS OF INFECTION AND PATHOLOGY: Whitish elongate mass (up to 3 mm) attached to the mesentery (Vaney and Conte, 1901); or white nodules giving the intestine a greyish white hue (Agapova, 1966); both consist of closely packed SPOVs. Shulman (1962) regarded each parasitic mass (0.8– 3 mm in size) as a possible hypertrophic derivative of a single host cell.

STRUCTURE AND LIFE CYCLE

Sporogony: Thick-walled SPOVs vary in size according to the number of spores they contain (Vaney and Conte, 1901). 16 spores per vesicle (Shulman, 1962).

Spores (fixed?): Ovoidal, 2 × 3 μm, with a posterior vacuole (Vaney and Conte, 1901); oval, 2–2.5 × 3–4 μm, length of the vacuole 1.7–2 μm (Shulman, 1962); 2–2.8 × 3–3.4 μm (Agapova, 1966).

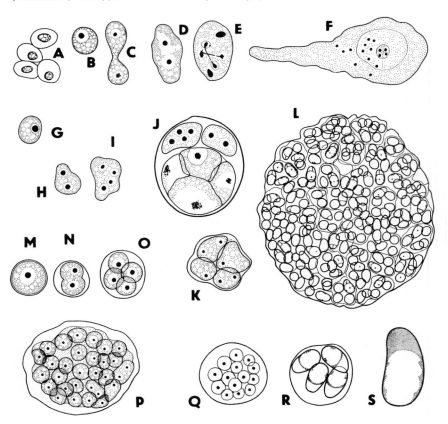

REMARKS: The available descriptions and illustrations of this species are inadequate for proper identification. Spores of *P. acerinae* are of a similar size and shape to those of *Glugea acerinae* (Sect. A, 2, p. 74). This and Shulman's comment on the nature of the parasitic mass suggest that *P. acerinae* and *G. acerinae* may be identical.

Pleistophora anguillarum Hoshina, 1951 (Fig. 2.48)

HOST: *Anguilla japonica*. Euryhaline.
GEOGRAPHICAL DISTRIBUTION: Japan and Taiwan; common in eel culture farms.
SITE OF INFECTION: Skeletal muscles.
SIGNS OF INFECTION AND PATHOLOGY: Small whitish spots beneath skin, sometimes external swellings due to the presence of numerous cysts. The parasite develops in muscle fibres, in direct contact with host cells, but after the breakdown of SPOVs in terminal infections the spores may be found on other organs. In terminal infections the muscle degenerates into a soft mass containing spore-filled vesicles. In infected fishes erythrocyte counts were slightly higher than control counts.
COMMERCIAL IMPORTANCE: Severe infections result in considerable mortality. In less severe cases growth of young eels may be retarded and old eels may be unsuitable for market. A definite threat to eel culture.
STRUCTURE AND LIFE CYCLE: The stages of development described by Hoshina (1951) are illustrated in Fig. 2.48 and described in the legend. The large cyst (Fig. 2.48L) resembles a complete parasite locus enclosing numerous spores within SPOVs while the small cysts (Fig. 2.48K,Q,R) resemble the SPOVs themselves, with the sporogonial plasmodium fragmenting to produce sporoblasts. Kano and Fukui's (1982) paper includes micrographs suggesting that possibly the separate segments of the plasmodium do not develop synchronously. The measurements given by

---

Fig. 2.48 *Pleistophora anguillarum*. Some stages of the life cycle redrawn from Hoshina (1951): A–L represent the typical sequence of developmental stages within the muscles. A, meronts 3-7 × 4-11 μm, in the fresh state; B, stained meront; C, dividing meront; D–F, allegedly, growth of the meront into a large plasmodium. Within this plasmodium, or after transformation into a cyst, sporonts develop, at first uninucleate, then tetranucleate, G to I; J, young trophozoite or cyst containing (as well as two other cells) a group of four cells representing the earliest stage of a sporont; K, a sporont in its cyst; L, large mature cyst in which each sporont gives rise to 4 to 8 macrospores. M–P, another mode of development in the muscle in which the meront produces a thin membrane around itself (M), undergoes successive binary division (N,O), each product allegedly transforming into a sporont, so that a cyst is formed (P); Q, sporoblasts of the microsporous sequence; R, SPOV with macrospores; S, mature spore.

Hoshina (1951) of 117 × 195 μm for the large cysts visible as white spots in the muscle and 10–17 × 13–18 μm for the small cysts would be in accord with this interpretation.

Spores (fresh, Fig. 2.48S): Two types, rare microspores (many per SPOV) and macrospores (4–8 per SPOV), both uninucleate and of similar elongate ovoid shape, very slightly bent to one side. A large posterior vacuole occupies almost two thirds of the spore and is encircled by prominent coils of the polar tube. According to Hashimoto and Takinami (1976), there are 44 coils of the polar tube in macrospores.

TRANSMISSION: Kano and Fukui (1982) infected juvenile eels experimentally by oral administration of spores. Early schizonts, cysts in muscles and whitish lesions on the body surface mainly around the abdomen, were observed 10, 20 and 25 days after infection, respectively. When they immersed eels in spore suspensions, lesions developed all over the body and this led the authors to suggest that direct infection had taken place through the skin. However, their experiment did not preclude infection by ingestion of spores. Infection was reduced to very low levels below 16°C.

TREATMENT: Kano, Okauchi and Fukui (1982) found that fumagillin administered with food at 5 or 50 mg/kg/day for periods of 60 or 20 days, respectively, could control the infections and be used as a preventive measure.

REMARKS: The description of Hoshina (1951) is controversial and very difficult to interpret in terms of the known life cycle of *Pleistophora*. Stages, not mentioned here, which he saw in the gut epithelium may have been early stages of infection or unrelated parasites. Further reference: Hashimoto, Sasaki and Takinami (1979).

Pleistophora carangoidi Narasimhamurti and Sonabai, 1977

HOST: *Carangoides malabaricus*. Marine. Infected fish were 14–16 cm in length.

GEOGRAPHICAL DISTRIBUTION: Indian ocean (unspecified).

SITE OF INFECTION: Skeletal muscles.

STRUCTURE AND LIFE CYCLE: As many as 15 foci of infection can be found in a host. Small foci, 700–800 μm in diameter, contain sporogonial plasmodia, while large foci, 10–15 mm in diameter, contain numerous spore-filled SPOVs 70–90 μm in size.

Spores: Oval 6.8–7.2 × 1.8–3 μm.

REMARKS: No other relevant data and no illustrations were supplied.

Pleistophora dallii Zhukov, 1964 (Fig. 2.46K)

HOST: *Dallia pectoralis*. Freshwater.
GEOGRAPHICAL DISTRIBUTION: Waters near Laurentia bay, Chukotka, U.S.S.R. In 2 out of 20 fish examined.
SITE OF INFECTION: Subcutaneous connective tissue near the base of pectoral and caudal fins.
SIGNS OF INFECTION AND PATHOLOGY: The parasites form tumour-like structures appearing as swellings at the bases of the fins.
STRUCTURE: Ellipsoid spores with slightly tapering anterior poles; 3.9–5.5 × 2.1–2.2 μm (not stated whether fresh or fixed). SPOVs contain numerous spores.
REMARKS: An inadequately described species. The description by Shulman (1962), namely "cysts visible to the naked eye, apparently formed from one hypertrophied cell", suggests that the species might belong to the genus *Glugea*.

Pleistophora destruens Delphy, 1916

HOST: *Mugil auratus*. Marine.
GEOGRAPHICAL DISTRIBUTION: France, the island of Tatihou, near Cherbourg.
SITE OF INFECTION: Skeletal muscles.
SIGNS OF INFECTION AND PATHOLOGY: Lateral curvature (scoliosis) of the hinder part of the body, due to degeneration of invaded muscles.
STRUCTURE AND LIFE CYCLE: SPOVs, with persistent membranes, were present at various stages of development and when mature were compressed to a polygonal form.
   Spores: Ovoid or pyriform, with a large posterior vacuole, 1.5–2.5 × 2.5–3.5 μm.
REMARKS: Not found since Delphy (1916) who supplied no illustration; a redescription is urgently needed.

Pleistophora duodecimae Lom, Gaievskaya and Dyková, 1980 (Figs. 2.23A, 2.46L, 2.50)

HOST: *Coryphaenoides nasutus*. Marine.
GEOGRAPHICAL DISTRIBUTION: Atlantic ocean, 20° 32'02" South and 12° 03'04" East, depth about 500 metres. One out of 15 specimens found infected.
SITE OF INFECTION: Skeletal muscles.

SIGNS OF INFECTION AND PATHOLOGY: The extremely enlarged infected muscle fibres were filled with SPOVs and were encased in an envelope of fibroblasts, compacted internally and looser externally. In some of the fibres, the SPOVs envelopes had broken down. There was a moderate degree of cell infiltration between the infected fibres. The infected fibres were a target for tissue reaction (Sect. IIB,2).

STRUCTURE AND LIFE CYCLE: Merogony was not observed.

Sporogony: Multinucleate sporogonial plasmodia, 18–30 μm in diameter, with nuclei 0.4–0.6 μm in diameter fragment via smaller multinucleate segments into sporoblasts. Nuclear division continues during this fragmentation. SPOVs with substantial envelopes contain from about 50 to 100 spores each. A very small fraction of vesicles are smaller and contain macrospores with not more than 8 spores per vesicle. Sporonts developing into macrospores were distinguished by large nuclei.

Spores: (formol-fixed) (Figs. 2.23A, 2.46L): Microspores 2.7 (2.4–3.7) × 4.3 (4–4.8) μm, are broadly rounded anteriorly and have a single nucleus, 0.3 × 0.8 μm and a large posterior vacuole reaching to about the middle of the spore. Macrospores, 3.3 (3–3.7) × 6.2 (6–7) μm are somewhat more stubby.

Pleistophora ehrenbaumi Reichenow, 1929 (in Doflein and Reichenow, 1927–1929)

HOSTS: *Anarhichas lupus* and *A. minor*. Marine.

GEOGRAPHICAL DISTRIBUTION: North Sea.

SITE OF INFECTION AND PATHOLOGY: Subcutaneous tumour-like swellings up to 13 × 5 cm.

The parasite invades muscle fibres gradually replacing their contents by innumerable SPOVs leaving only a thin peripheral layer of sarcoplasm. At this stage, no host tissue reaction is apparent. The remnants of muscular tissue eventually disintegrate leaving a mass of SPOVs and free spores dispersed between other, still intact, fibres. The tissue debris is now pervaded by host phagocytes which ingest the spores.

COMMERCIAL IMPORTANCE: The host fish are part of commercial catches.

STRUCTURE AND LIFE CYCLE: SPOVs containing 4, 8, 16 or more spores were the only stages observed.

Spores: Roughly ovoid uninucleate, with a large posterior vacuole showing distinct coils of the polar tube. Spore size varies from 3 × 1.5 μm to 7.5 × 3.5 μm.

REMARKS: Reichenow's (1929) illustrations of SPOVs with fixed and stained spores are of no diagnostical value, so that they are not reproduced here. This species needs to be re-examined to clarify the life cycle and the relationship with other apparently similar species (e.g. *P. macro-zoarcidis, P. duodecimae)*. (Other references: Reichenow 1932, Jírovec 1932).

Pleistophora elegans Auerbach, 1910 (Fig. 2.46M)

HOST: a hybrid of *Abramis brama* and *Rutilus rutilus*. Freshwater.
GEOGRAPHICAL DISTRIBUTION: The Rhine (Auerbach, 1910); the basins of rivers emptying into the Black Sea (Shulman, 1962).
SITE OF INFECTION: Oocyte, ovary.
STRUCTURE AND LIFE CYCLE: Meronts multinucleate, present in connective tissue as well as ovary, probably distributed in the blood.
   Spores (fresh): Long and narrow. Microspores and macrospores were reported but only the size of the macrospores 10×4 μm was given. The threads of the polar tube were distinct around the large posterior vacuole. Two nuclei were reported in mature spores.
REMARKS: The original description is unsatisfactory, lacks detail and is based on a single infected specimen. The presence of two nuclei in the spores, if correct, does not accord with the generic description. Auerbach (1910), himself, thought it possible that the species might later be found to be identical to *P. mirandellae* (p. 115). Shulman (1962) gave no details when he recorded the presence of this species in "ovaries of bream and roach" in the Black Sea river basins.

Pleistophora gadi Polyansky, 1955 (Fig. 2.23D)

SYNONYM: *Pleistophora* sp. Young, 1969.
HOST: *Gadus morhua morhua*. Marine.
GEOGRAPHICAL DISTRIBUTION: Barents Sea. Only one infected fingerling found by Polyansky.
SITE OF INFECTION: Skeletal muscles.
STRUCTURE AND LIFE CYCLE: SPOVs of diameter 18–23 μm containing numerous pear-shaped spores $5.4–7.2 \times 2.7–3.6$ μm (in fixed material).
REMARKS: Illustrations in Polyansky (1955), as well as in Polyansky and Kulemina (1963) are inadequate for the differentiation of this species.
   Young (1969) reported an emaciated cod caught off the Essex coast of England, which bore skin lesions up to 3 cm in diameter often surrounded by fibroblasts. The muscle beneath them contained brown tumour-like structures about 1×0.2 mm, filled with SPOVs containing up to 100 spores. The lack of more precise data on the parasite makes its identification with *P. gadi* a matter of conjecture.

P. gadi may perhaps occur in other hosts, too. A sample of muscle tissue of *Gadus virens,* obtained by the author through the courtesy of Dr MacKenzie, contained spores which could belong to this species.

Pleistophora hippoglossoideos Bosanquet, 1910 (Figs. 2.23B, 2.45E, 2.49)

HOSTS: Type host is *Drepanopsetta hippoglossoides* Bosanquet (1910). Also *Hippoglossoides platessoides* and *Solea solea* (Lom, unpublished). All marine.

GEOGRAPHICAL DISTRIBUTION: Common over the entire northern North Sea (Kabata, 1959).

SITE OF INFECTION: Muscles of the fins and of the walls of the visceral cavity.

SIGNS OF INFECTION AND PATHOLOGY: Whitish nodules in the musculature, either small and rounded, 1.2 mm (Bosanquet, 1910) or larger and oblong, 2.5 × 10 mm visible externally on the fish and projecting somewhat into the body cavity (Kabata, 1959; Fig. 2.49). The nodules consist of a honey-combed mass of SPOVs lying in a structureless matrix containing remnants of muscle fibres (Bosanquet, 1910) or in apparently empty spaces (Lom, unpublished).

A slight degree of cellular infiltration may occur at the edge of the nodules (Dyková, unpublished). Kabata (1959) observed that parts of the muscle were compressed and/or displaced by the adjacent nodules, but there is no great damage to the host.

COMMERCIAL IMPORTANCE: Heavy infections with large, striking nodules may render the fish unsightly and cause their rejection as food.

STRUCTURE AND LIFE CYCLE

Merogony: Young meronts are rounded, with single, vesicular nuclei (Bosanquet, 1910).

Sporogony: Multinucleate sporogonial plasmodia fragment into sporoblasts. Even the largest of these plasmodia are smaller than SPOVs containing spores. The vesicles in the type host measure 21 (16–30) μm (Kabata, 1959) or 25–25 μm (Bosanquet, 1910); in other hosts they have similar size range (Lom, unpublished).

Spores (Figs. 2.23B, 2.45E): Elongated-ovoid or pyriform, often more arched on one side. The posterior vacuole does not often reach the mid-point of the spore; its anterior border is either straight or slightly arched and its cytoplasmic margins lining the spore wall are very thin, with the polar tube coils indistinct. There is a single kidney-shaped nucleus and a distinct 0.4 μm large PAS positive polar cap.

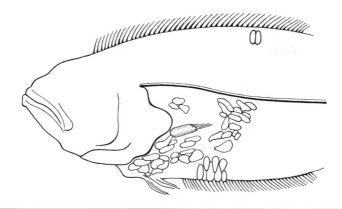

Fig. 2.49 Blind side of long rough dab *(Drepanopsetta hippoglossoides)* showing nodular cysts of *Pleistophora hippoglossoideos* (redrawn from Kabata, 1959).

MEASUREMENTS OF FRESH SPORES FROM *Hippoglossoides platessoides:* 2.7 (2.3–3.2) × 4.8 (4.4–5.3) μm; from *Solea solea:* 2.6 (2.5–2.9) × 4.7 (4.1–5) μm (Lom, unpublished). Spore measurements given by other authors (possibly from fixed material): Bosanquet (1910) gave 2 × 3.5 μm; Kabata (1959) gave 2.2 (1.7–3.2) × 3.7 (2.5–6.1) μm, (including almost round spores, e.g. 2.8 × 2.3 μm and greatly elongated ones, e.g. 6.1 × 2.2 μm, some of them bent in the shape of a fat sausage).

REMARK: Some of Bosanquet's (1910) sketches of the sporont fragments appear to have paired nuclei. As diplokarya are not characteristic of the genus, they may simply be dividing nuclei.

Pleistophora hyphessobryconis Schäperclaus, 1941 (Figs. 1.13, 2.1, 2.2, 2.23C, 2.52–2.55, 2.61–2.63)

HOSTS: Originally described from *Paracheirodon inessi* and *Hemigrammus erythrozonus.* Spontaneous infections occur also in *Hemigrammus ocellifer; Brachydanio rerio* (Opitz, 1942); *Hyphessobrycon flammeus* (Steffens, 1956); *H. callistus callistus; Cheirodon axelrodi; Hasemania nana: Barbus lineatus; Brachydanio nigrofasciatus; Opisthogramma seitzigi* (Steffens, 1962); *Hemigrammus pulcher; Hyphessobrycon rosaceus; Xiphoporus helleri* (van Duijn, 1956); *Hyphessobrycon heterorhabdus* (Sterba, 1956). It is common in the fan-tailed variety of *Carassius auratus auratus* (Lom, unpublished). Richert (1958) claimed to have found it in *Carassius auratus gibelio* and in *Phoxinus phoxinus.* The hosts belong to four families of freshwater fish.

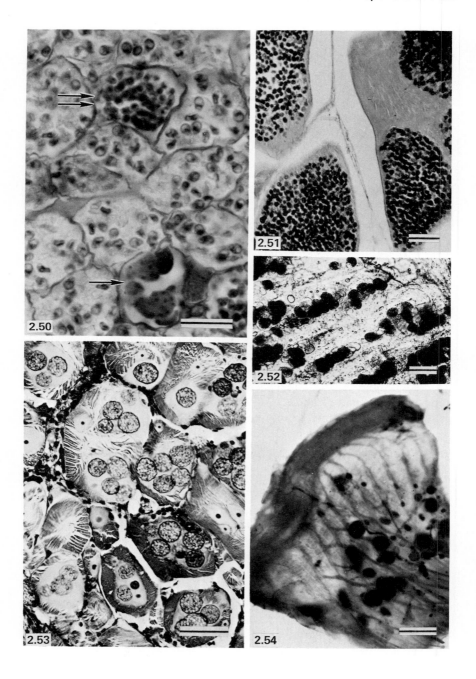

GEOGRAPHICAL DISTRIBUTION: The fish from which the species was originally described were brought from the upper course of the Amazon river in Peru. The parasite is commonly occurring in fish kept in aquaria.

SITE OF INFECTION: Skeletal muscle. In heavy infections the parasite may spread to other body organs and may be found in the connective tissue of the ovaries, intestinal epithelium and skin, and between the renal tubules. Mature spores, loose or in SPOVs released from muscle, pervade the body cavity and may be found in a variety of organs. A predilection site for their concentration is subcutaneous tissue, whence they penetrate the skin; they may completely fill the skin gland cells. Skin scrapings of heavily infected fish will reveal numerous mature SPOVs and free spores.

SIGNS OF INFECTION AND PATHOLOGY: Fading or even total loss of vivid colours, and anomalous behaviour or movements. Muscles pervaded by the parasites (Figs. 2.52, 2.53) appear greyish or greyish-white, regions of heavily infected musculature may appear as white patches under the skin. Heavily infected tetras may be emaciated, with a distorted body (scoliosis, kyphosis or both) due to the heavy muscular damage. Fantailed goldfish may display bristled scales.

Each parasite within the muscle fibre is encircled by what appears in stained sections as a "halo" (Fig. 2.2). It is an area of sarcoplasm replaced (Fig. 2.1) by a mass of disorganized cisternae of smooth endoplasmic reticulum, free ribosomes and myofibrils. With the progress of the infection, whole muscles are destroyed and contain masses of SPOVs and free spores, together with host phagocytes, which ingest the spores. In cases of more limited "nests" of SPOVs the aggregations may be surrounded by host pigment. It is not clear how the infection spreads from muscles to other body organs.

Fig. 2.50 *Pleistophora duodecimae,* showing closely packed SPOVs with spores: arrow points at a sporogonial plasmodium undergoing fragmentation; double arrow points to SPOV containing developing sporoblasts. H & E; scale bar = 20 μm.

Fig. 2.51 *Pleistophora macrozoarcidis,* SPOVs partly filling the muscle fibres. There is no host cell infiltration between the fibres. Heidenhain's; scale bar = 50 μm.

Fig. 2.52–2.54 *Pleistophora hyphessobryconis,* Fig. 2.52 SPOVs located along the muscle bundles of an experimentally infected crucian carp. Fresh squash preparation; scale bar = 50 μm. Fig. 2.53. Transverse section through the muscle bundles of a heavily infected neon tetra. Parasites are encircled with a halo of disintegrated sarcoplasm and there is almost no cell infiltration between the bundles. H & E; scale bar = 50 μm. Fig. 2.54. Part of the fresh body wall of fantail goldfish, with large cyst-like parasite foci visible in the myomeres; scale bar = 3 mm.

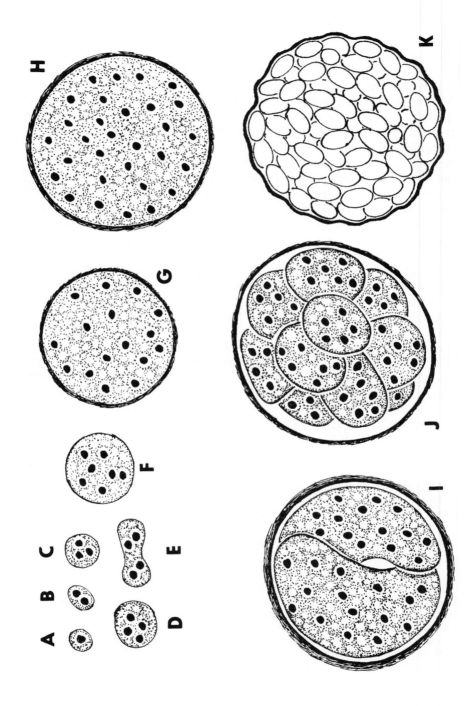

In goldfish kidney, developing SPOVs lie between the tubules and are encased by a double layer of connective tissue cells. Thieme (1954) described atrophy of the testes ("parasitic castration") and liver in heavily infected tetras without specific infection of these organs.

In heavily infected goldfish, phagocytes packed with spores (Fig. 2.63) may concentrate to form granuloma-like formations up to 3 mm in size, in which a thick envelope of connective tissue cells encompasses a mass of living and dead spores released from phagocytes. These formations are situated in the mesentery or on the surface of the viscera.

Encapsulation though rare, is more effective in limiting the spread of infection than phagocytosis. In neon tetras, rounded cysts up to 200 μm in diameter, with a connective tissue wall up to 15 μm thick, are formed around heavily infected muscle fibres (Thieme, 1954; Dyková and Lom, unpublished). In fan-tailed goldfish, the cysts, located predominantly in the body wall muscles, can attain dimensions of up to 2 mm (Fig. 2.54). They contain viable, mature spores with a minimal amount of cell remnants in between. More advanced cysts contain only connective tissue, fat droplets, pigment and dead spores. This suggests that encapsulation serves to isolate and destroy the parasites (Dyková and Lom, unpublished).

Heavy infections are lethal. Thieme (1954) reported the survival of neon tetras for an average of only 14 days from the first appearance of the whitish opacity of the muscle tissue. The fish had displayed anomalous behaviour for an average of 90 days before the appearance of the lesions. The length of the asymptomatic phase was not determined. In contrast, heavily infected goldfish, unable to move or feed normally, with scales bristled and spores oozing from the skin, may survive for several weeks.

COMMERCIAL IMPORTANCE: Pleistophorosis is among the most serious diseases of neon tetras and the other hosts, causing heavy losses particularly in large scale breedings. Unfortunately, the only course of action is total destruction and replacement of the infected stocks.

STRUCTURE AND LIFE CYCLE (Fig. 2.55)

Merogony: Meronts develop within a halo of disintegrated sarcoplasm up to 24 μm across; products of its development later fill the space completely. Oval uninucleate cells 2–4 × 1.5–2 μm in size, giving rise to binucleate meronts 6 to 8 μm in size, which divide by binary fission. Oval meronts may attain 15 μm and have 4 nuclei, and may be difficult to dis-

---

Fig. 2.55 Diagrammatic representation of the life cycle of *Pleistophora hyphessobryconis.* A–D, growth of the meronts; E, division of a meront with four nuclei; F, young sporogonial plasmodium; G,H, growth of the sporogonial plasmodium, with the gradually thickening future SPOV wall; I, plasmodium detached from the thick wall and divided into two segments; J, plasmodium divided into numerous still multinucleate segments; K, SPOV containing spores.

tinguish from young sporonts. Uninucleate meronts are perhaps responsible for dissemination of the infection to other parts of musculature.

Sporogony: The growing spherical sporont can be identified at the four nuclei-stage when a thick layer is laid down at its surface. The nuclei measure 1.5 μm in diameter. The sporogonial plasmodium grows into a sphere 25–35 μm in diameter, then retracts within an SPOV and divides by plasmotomy eventually to produce 20–130 uninucleate sporoblasts, and, later, spores. At this stage the developing SPOV wall is an electron dense layer up to 0.3 μm thick (Lom and Corliss, 1967). It has a smooth inner face, while the outer surface is rugged and bears small projections (Fig. 2.62); a number of small pores penetrate the wall. Later, the wall thins as its inner surface is disrupted by the formation of small blisters (Fig. 2.61) and in mature SPOVs the wall is thin and fragile so that spores are commonly released from them. Within the SPOV, the space between the sporoblasts and the wall is filled by a thin, amorphous substance.

Spores (Fig. 1.13, 2.23C): Ovoid, one side slightly more vaulted. The posterior vacuole occupies more than half the spore length, has an arched anterior border; its cytoplasmic borders are thick and prominently ridged around the large number of polar tube coils—about 34 coils. There is a single nucleus 1.5 × 0.5 μm, located transversely beneath the spore wall. PAS positive polar cap about 0.4 μm in size.

Measurements of fresh spores: 4 × 6 μm (average dimensions)—Lom & Corliss (1967); 3.5 × 6 μm—Schäperclaus (1941); 3–3.5 × 5–6 μm— Steffens (1956, 1962); 2.2 × 5.8–6.1 μm—Thieme (1956). Spores remain viable for at least a year at 4°C.

TRANSMISSION: In live hosts, spores may be excreted with urine or released through the skin. Normally, however, the spores contaminate the water when released from decomposed dead hosts. Schäperclaus (1941) found that even young fry, 3.5 mm long, carried infections and succumbed at the age of 8 days. The mode of infection was not established and transovarial transmission has not been proven. Autoinfection, by the hatching of spores in the host where they are produced, may be possible but also, has not been confirmed.

Experimental infection after peroral administration of spores was successful in 22% of the neon tetras and 63% of one-year-old goldfish (Lom, unpublished). These figures, however, are lower than in spontaneous infections where 100% prevalence is common. *Rutilus rutilus* and *Cyprinus carpio* were refractory. Infections have also been achieved by intramuscular inoculation, as spores will germinate in the inoculated tissue and produce a heavy local infection (Lom, 1969). Stages of the parasite pervaded the inoculated area and sometimes resulted in its complete destruction. Intramuscular inoculation has facilitated the maintenance of the species in goldfish and has been used to infect tench *(Tinca tinca)* and carp *(Cyprinus carpio).*

TREATMENT: The only report of successful treatment is that of Andodi and Frank (1969), who added sodium bicarbonate ($NaHCO_3$), in quantities sufficient to adjust the aquarium water to pH 7.5–8, while introducing ozone at a rate of about 1 mg per hour per 100 l. They claimed that they achieved a 100% cure of moderately advanced cases, and that the treatment was effective in preventing the disease in fish up to 5 months of age. The most widely practised method of prevention requires thorough disinfection of the tanks and use of healthy parents to guarantee infection-free offspring.

REMARKS: *P. hyphessobryconis* is a very suitable model for experimentation, still largely neglected. Reference papers are those of Dyková and Lom (1980), Lom and Corliss (1967), Lom and Vávra (1961), Nigrelli (1953), Porter and Vinall (1956), Reichenbach-Klinke (1952) and Thieme (1952, 1954, 1956).

Pleistophora ladogensis Voronin, 1978 (Figs. 2.23E, 2.56)

SYNONYM: *Pleistophora* sp. of Lopukhina and Strelkov, 1972; *Pleistophora* sp. of Voronin, 1974.

HOSTS: *Lota lota,* freshwater and *Osmerus eperlanus eperlanus.* Euryhaline.

GEOGRAPHICAL DISTRIBUTION: Vrevo ($=L.$ *lota*) and Ladoga ($=O.$ *eperlanus*) Lakes near Leningrad, U.S.S.R. *L. lota* under 25 cm were uninfected but in older fish the infection rate could reach 32–50%. In the 45–50 cm age class, only mature spores were found. The prevalence in *O. eperlanus* was 5%–8% (Voronin, 1981).

SITE OF INFECTION: Skeletal muscles; also in striated muscles of the swim bladder and gill opercula of *L. lota.*

SIGNS OF INFECTION: In *O. eperlanus,* white "cysts" in the musculature, which consisted of fibres filled with parasites, show through the skin.

STRUCTURE AND LIFE CYCLE (Voronin, 1978)

Merogony: Meronts with 2, 4 and 6 nuclei transformed into rounded sporonts.

Sporogony: Sporogonial plasmodia, up to 18 μm in size, with up to 20 nuclei, give rise to thick-walled SPOVs which averaged 43 μm in diameter.

Spores (fresh (Fig. 2.23E): Ovoid, 5.4 (5–5.8) × 2.9 (2.7–3.3) μm in size, with a large posterior vacuole occupying about one half of the spore volume; the anterior border of the vacuole is flat or slightly concave, the

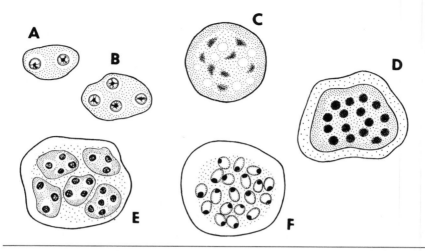

Fig. 2.56   Life cycle of *Pleistophora ladogensis*. A,B, meronts; C, sporogonial plasmodium; D, sporogonial plasmodium detached from the SPOV wall; E, segmentation of the sporogonial plasmodium; F, sporoblasts within the SPOV (redrawn from Voronin, 1978).

sides are very thin and regular. Spores from *O. eperlanus* have slightly more pointed anterior ends than those from *L. lota*.

REMARKS: Voronin (1978) differentiated this species from *P. hyphessobryconis,* but not from other species from freshwater fish. *P. ladogensis* may actually represent two species, one from each host. Voronin (1978) himself suggested that two "races" or subspecies existed because of the different infection rates in the two hosts (e.g. in Vrevo lake where *L. lota* was heavily infected, the infection was absent in *O. eperlanus*) and spore shape differences in the two fish hosts.

Pleistophora littoralis Canning and Nicholas, 1980 (Figs. 1.8, 1.10, 1.11, 1.14, 2.57–2.60)

SYNONYM: *Pleistophora* sp. of Canning, Hazard and Nicholas, 1979.
HOST: *Blennius pholis*. Marine.
GEOGRAPHICAL DISTRIBUTION: Start Point, Devonshire coast, England.
SITE OF INFECTION: Skeletal muscles.
SIGNS OF INFECTION AND PATHOLOGY: Individually infected muscle fibres appear as white streaks measuring 10 × 0.3 mm but when grouped together, infection sites may measure 10 × 2–3 mm. All stages of the parasite abut directly on to intact myofibrils. There is no cyst formation or degeneration of host tissue (Canning, Hazard and Nicholas, 1979).

STRUCTURE AND LIFE CYCLE

Merogony: Uninucleate or binucleate meronts 2.5 μm in diameter and surrounded by a hyaline layer almost 0.3 μm thick, grow and divide by multiple fission and perhaps also by plasmotomy, producing uninucleate cells.

Sporogony: Sporonts differ from meronts by producing up to 200 rounded nuclei. At the ultrastructural level, there are conspicuous stacks of smooth endoplasmic reticulum close to the nuclei. Sporogony is initiated by the deposition of an electron dense secretion at the inner border of the hyaline layer. The secretion extends as digitations into it. This hyaline and electron dense layer persists as the wall of the SPOV around the sporont which produces uninucleate sporoblasts, via segmentation into cytomeres, each with up to 10 nuclei. Two types of SPOVs were seen: macrospore SPOVs measuring 4 × 15 μm (stained) with 8 spores, and microspore SPOVs measuring 20–40 μm (mean 28 μm, stained) with up to 200 spores (Fig. 1.9).

Spores (Figs. 2.57, 2.58): Fresh macrospores with average dimensions of 3.8 × 7.7 μm are elongate ovoid with a large posterior vacuole. The number of coils of the polar tube, shown by electron microscopy, is 33–39, arranged in five ranks (Fig. 1.11).

Microspores, which outnumber the macrospores by several thousands to one, are ovoid to pyriform, with a single nucleus and a posterior vacuole, which reaches almost to midspore length and has a straight, often oblique or slightly vaulted anterior border. The average dimensions of fresh spores are 2.5 × 4.6 μm. The number of coils of the polar tube varies from 10 coils in one or two ranks to 17 coils in two ranks (Fig. 1.10).

REMARKS: Thélohan (1891, 1895) described microsporidian infections from 3 marine fish (including *B. pholis*) and 1 freshwater fish. Thélohan attributed them to the same species which was named *P. typicalis* by Gurley (1893), who selected *Myoxocephalus scorpius* as the type host. Canning, Hazard and Nicholas (1979) pointed out the diversity in data and measurements given by Thélohan from the various hosts and Canning and Nicholas (1980) demonstrated differences between *P. typicalis* in the type host and the parasite of *B. pholis;* they established the latter as an independent taxon which they named *P. littoralis.*

Pleistophora longifilis Schuberg, 1910 (Fig. 2.64)

HOSTS: *Barbus barbus; Barbus capito conocephalus; Rutilus rutilus.* Freshwater.

GEOGRAPHICAL DISTRIBUTION: A single infected *Barbus barbus* was found by Schuberg (1910) in Heidelberg, Germany; and infection of this species

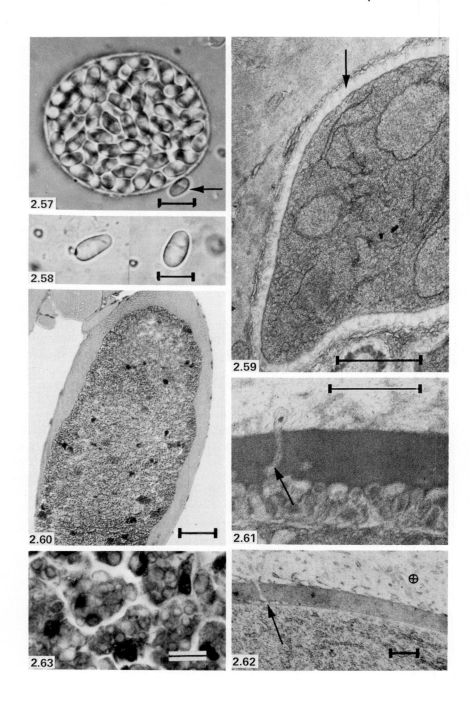

was reported from the basins of Black Sea bound rivers by Shulman (1962); a single infected *B. capito conocephalus* was reported by Osmanov (1971); Walliker (1966) reported infection in *Rutilus rutilus* from England.

SITE OF INFECTION: Testes.

SIGNS OF INFECTION AND PATHOLOGY: Whitish, very slightly bulging spots on the surface of the testis. The parasite develops within epithelial cells of the seminiferous canaliculi. During parasite growth the host cell hypertrophies and its nucleus swells up into an irregularly elongate lobular and large structure. The host cell membrane cannot be seen. If most of the epithelial cells are infected, the lumen of the canaliculi are obliterated. The original structure of the canaliculus epithelium is no longer recognisable, and its site is indicated merely by the surrounding connective tissue envelope. Heavily infected canaliculi fuse together, the connective tissue septa around them giving way and transforming into fibrous strands. The neighbouring tissue is also subject to degeneration. After degeneration of cells, mature SPOVs lie free in cell detritus. The contrast between this and xenoma formation is striking.

STRUCTURE AND LIFE CYCLE

Merogony: Stages have not been recorded.

Sporogony: Uninucleate and binucleate stages grow into multinucleate sporogonial plasmodia with a thick membrane, although in mature SPOVs, the membrane is thinner. Spherical SPOVs, 18–45 μm in diameter, contain as few as 8 but mostly more than 20–30 spores. The vesicle size is related to spore dimorphism, microspore vesicles measuring up to 30 μm and macrospore vesicles up to 45 μm in diameter.

Spores (probably measured fresh): Pyriform 2 × 3 μm (microspores) and 6 × 12 μm (macrospores). There is a single nucleus and a large

---

Fig. 2.57–2.60 *Pleistophora littoralis.* Fig. 2.57. Mature SPOV with microspores, their shape can be seen well from one free spore (arrow). Fig. 2.58. Two macrospores. Fig. 2.59. Multinucleate meront bordering on an intact myocyte; the interface between the two is a hyaline layer permeated by small vesicles (arrow). Fig. 2.60. Muscle fibre, which, except for its periphery, has been replaced by a mass of closely packed SPOVs. Scale bars on Figs. 2.57, 2.58 = 10 μm; 2.59 = 1 μm; 2.60 - 100 μm. (Reproduced from Canning, Hazard and Nicholas, 1980, with permission of authors and publisher.)

Figs. 2.61–2.63 *Pleistophora hyphessobryconis.* Figs. 2.61, 2.62. The SPOV vesicle wall maturation; at all stages, there are narrow channels through the wall (arrows) and around the multinucleate sporogonial plasmodium, the wall is compact (Fig. 2.61). Later as the plasmodium starts to divide, it is subtended by a meshwork surrounding electron dense particles and the space outside the wall is full of filiform or microtubular projections, possibly originating in the SPOV wall (Fig. 2.62). Fig. 2.63. Closely packed macrophages with ingested spores of *P. hyphessobryconis* in a late stage of infection in a goldfish. Figs. 2.61, 2.62, scale = 0.25 μm; 2.63 = 10 μm.

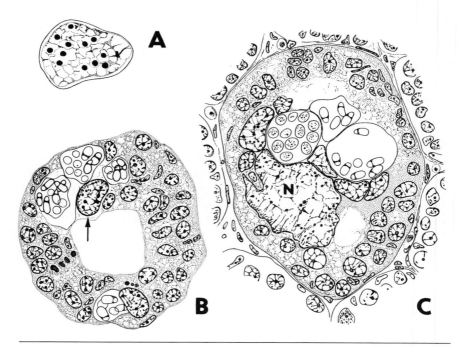

Fig. 2.64 *Pleistophora longifilis* (redrawn from Schuber, 1910): A sporogonial plasmodium; B, transverse section through an infected seminiferous tubule of the testis, in which the lumen is still preserved and the nuclei (arrow) of infected cells are hypertrophied; C, the same but with the lumen of the tubule now completely obliterated and the nuclei (N) of infected cells are enormously hypertrophied.

posterior vacuole, reaching up to half or more of the spore length.

REMARKS: All data taken from Schuber (1910). Though Osmanov (1971) gave no description, it is probable that the data he reported relates to *P. longifilis*. Goreglyad (1962) claimed, without reasonable evidence, that he had found this parasite in the swim bladder of carps. Plehn (1924) mentions the possible occurrence of *P. longifilis* in *Cottus gobio*.

Pleistophora macrospora Cépède, 1906 (Fig. 2.46P)

HOST: *Noemacheilus barbatulus*. Freshwater.

GEOGRAPHICAL DISTRIBUTION: France near Grenoble where only one infected fish was found by Cépède (1906); basins of rivers emptying into Black Sea (Shulman, 1962).

SITE OF INFECTION: Abdominal muscles.

SIGNS OF INFECTION: Intramuscular, yellowish-white ellipsoidal "cysts" up to 3 mm large, distending the abdominal wall.

STRUCTURE AND LIFE CYCLE

Sporogony: Sporonts are spherical or subspherical, 25–30 μm in diameter; mature SPOVs each with a double contoured envelope enclosing a large and variable number of spores (Cépède, 1906).

Spores (Fig. 2.46P): Elongate-ovoid or ovoid, with a large, globular posterior vacuole; fresh spores 4.2 × 8.5 μm (Cépède, 1906); length 8 μm (Léger and Hesse, 1916).

REMARKS: Cépède (1906) and Léger and Hesse (1916) gave a description of the internal structure of the spore, describing polar capsules. This is considered erroneous and has been omitted from this account.

Pleistophora macrozoarcidis Nigrelli, 1946 (Figs. 2.11, 2.12, 2.23F, 2.45F, 2.46N, 2.51)

SYNONYM: *Ichthyosporidium* sp. of Fischtal, 1944; and also of Sandholzer, Nostrand and Young, 1945.

HOST: *Macrozoarces americanus*. Marine.

GEOGRAPHICAL DISTRIBUTION: Common to the western part of the North Atlantic. According to Sandholzer, Nostrand and Young, (1945) the percentage varied in 1943 from 4% to 38% in various localities along the U.S. Atlantic coast. This covers the range given by Fischtal (1944) and Sheehy, Sissenwine and Saila (1974). The prevalence of infection increases with the age and size of the host fish.

SITE OF INFECTION: Skeletal muscles.

SIGNS OF INFECTION AND PATHOLOGY: Small or large tumour-like masses, often up to 8 cm, sometimes recognisable by bulges of the body. A pus-like exudate is often visible when larger tumours are cut. Even small lesions (1–5 mm) are detectable by the candling method (Chap. 5, Sect. I). The early stages of infection appear as minute whitish cylinders, 0.5 mm or more in length, lying along the axis of the muscle fibres. In advanced stages the parasites occupy practically the entire mass of each muscle fibre and only a thin peripheral layer of sarcoplasm is left (Fig. 2.11, 2.51). The groups of infected fibres may be encapsulated together within a common layer of concentrically arranged connective tissue. The centre of such a "pseudocyst" consists of free spores liberated by the breakdown of SPOVs (Dyková and Lom, unpublished). The tissue may assume a brownish appearance, and ultimately, the worst affected muscle fibres are completely hyalinized and destroyed, leaving a granular debris full of spores. These reactive processes follow the pattern described in Sect.

IIB,2. Ulcerative conditions were not observed and spores probably depend on death and decay of the host for release into the environment.

COMMERCIAL IMPORTANCE: During the years 1943 and 1944, ocean pout were very important market fish, and *P. macrozoarcidis* had already represented a threat to the ocean-pout industry (Sandholzer, Nostrand and Young, 1945). Olsen and Merriman (1946) indicated that microsporidian infections were one of the factors that had contributed to the decline of the use of ocean pout for human consumption and its relegation to trash fishery. According to Sheehy, Sissenwine and Saila (1974), the infection impedes sucessful marketing of ocean pout as fish food.

STRUCTURE AND LIFE CYCLE

Merogony: Uninucleate meronts of variable size grow, divide by binary or multiple fission and invade adjacent regions of muscle fibres to continue merogony. A series of nuclear processes prior to sporont formation, culminating in what Nigrelli (1946) interpreted as synkaryon formation, needs verification.

Sporogony: Growth of the sporonts, followed by nuclear and cell divisions, produces a variable number of sporoblasts within resistant SPOVs which measure 5 to 30 μm in diameter.

Spores: Ovoid, 3.5 to 5.5 μm long when fresh (Fig. 2.23F) with a large vacuole at the wider posterior end. Sausage-shaped macrospores, up to 8 μm in length, were rare. Fischtal (1944) gave the spore size as 2–4 × 4–7 μm.

Autoinfection: Nigrelli (1946) found uninucleate cells, reminiscent of meronts, within the mass of spores. He proposed that hatching of mature spores *in situ* might occur and explain the massive infection of large areas of body musculature. Sandholzer, Nostrand and Young (1945) reported that lesions caused by *P. macrozoarcidis* continued to increase in size and number, at temperatures as low as −13°C in muscles prepared as fillets, and that noninfected fillets placed in contact with infected ones could become parasitised. This extraordinary observation, involving survival, transmission and growth of microsporidia in dead tissue, would have to be verified.

REMARKS: Nigrelli's (1946) illustrations of developmental stages, which need verification are not reproduced here as they are hardly of any diagnostic value. The manner in which the parasite is localized in the centre of the muscle fibres and the tumour-like swellings of heavily invaded areas is reminiscent of *P. ehrenbaumi* infections. However, spores that we were able to examine in stained preparations, differed in shape and structure, as did the general aspect of the SPOVs. Sandholzer, Nostrand and Young (1945) suggested that *P. macrozoarcidis* may have been rather recently introduced into *M. americanus,* since a previous thorough study by Clemens (1920) did not reveal any infections in fish from the same localities.

Pleistophora mirandellae Vaney and Conte, 1901

HOST: *Alburnus alburnus.* Freshwater.
GEOGRAPHICAL DISTRIBUTION: The only record, of one infected fish, is from France.
SITE OF INFECTION: Oocytes in the ovary.
SIGNS OF INFECTION AND PATHOLOGY: Host reaction is manifested by destruction of the infected oocyte by phagocytic cells.
STRUCTURE AND LIFE CYCLE: Vegetative stages briefly mentioned as uninucleate "amoeboid bodies" in cytoplasmic cavities within oocytes.

Spores: Dimorphic; microspores ovoid, 4 × 7.5 μm, with a vacuole and a single nucleus; macrospores 6 × 12 μm, with the same structure. Microspores are found in "small", dark SPOVs with resistant envelopes which do not rupture in the host. The authors suggested that microspores serve for the infection of new host fish. Macrospores are found in "large", clear vesicles with less resistant envelopes, which rupture within the host; Vaney and Conte suggested that they serve for autoinfection.
REMARKS: Originally described from *Alburnus mirandellae* Blanchard, now known to be synonymous with *A. alburnus* (L.).

Pleistophora oolytica Weiser, 1949 (Figs. 2.23H, 2.65, 2.66, 2.68)

HOSTS: *Leuciscus cephalus* and *Esox lucius*—Weiser (1949); *Hucho hucho* —Otte (1964); *Rutilus rutilus* (Lom, unpublished). Freshwater.
GEOGRAPHICAL DISTRIBUTION: Rivers Svitava and Říčka near Brno, Czechoslovakia (Weiser); the river Danube, Austria (Otte).
SITE OF INFECTION: Oocytes in the follicles.
SIGNS OF INFECTION AND PATHOLOGY: The infected oocytes are cloudy, white or even degraded into yellowish clots. The parasite develops predominantly within the yolk vacuoles, which are eventually replaced by the parasite (Figs. 2.65, 2.66).

About 10% and 15% of the follicles were invaded in the ovaries of *L. cephalus* and *E. lucius,* respectively.

Infection of the oocyte possibly takes place before the *zona radiata* is formed, as the *zona radiata* remains undamaged while the SPOVs gradually replace the entire contents of the oocyte, "liquefying" its cytoplasm. Finally, phagocytes infiltrate the contents of the follicle which undergo atresia (Weiser, 1949).

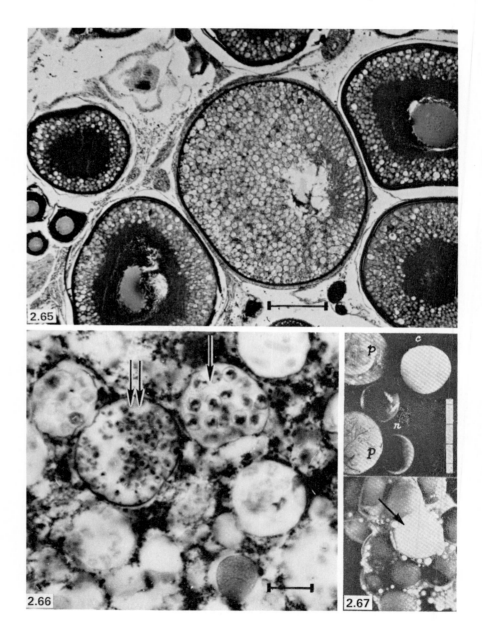

Otte (1964) observed serious damage to the ovary of *Hucho hucho*. Many mature oocytes were infected, in which the yolk granules were liquidized, and the resulting fluid separated the egg membrane from its epithelial cover. Later, the oocytes were destroyed by the parasite's growth. The *zona* was the only thing left around the spores in the overgrown connective tissue.

STRUCTURE AND LIFE CYCLE

Merogony: Oval meronts, 2–3 μm large, with one or two compact nuclei grow and divide, form new meronts with up to 4 nuclei and divide to produce uninucleate meronts again (Fig. 2.68A-C).

Sporogony: Multinucleate sporogonial plasmodia are produced from meronts. The plasmodia then break down progressively within SPOVs into fresh spherical sporoblasts, 2–3 × 3–4 μm (Fig. 2.68D-I).

Spores: Ovoid, with a large posterior vacuole sometimes reaching beyond the mid-line of the spore(Fig. 2.23H). A single nucleus and PAS positive polar cap measuring 1.6 × 0.3 μm in the medium-sized spores (Lom, unpublished).

Weiser (1949) described 3 types of spores: in *L. cephalus* the most frequent type (30–60 spores per SPOV) measured 5.5–6.5 × 3.5 μm; microspores (about 100 per vesicle) measured 3 × 1.5 μm; macrospores (30–40 per vesicle) measured 8.4 × 4.2 μm (Fig. 2.68K-M). In *E. lucius*, the spores measured 5 × 3 μm, 3 × 1.5 μm and 7 × 3–3.5 μm, respectively. Otte (1964) reported the size of spores from *Hucho hucho* to be 1–8 × 1–3 μm.

REMARKS: The infection was found just once in all four host species. Weiser (1949) took the parasites from *L. cephalus and E. lucius* to be conspecific, due to similar development and behaviour in the host cell. Further studies may show if this applies to the parasites from *Hucho hucho* and *Rutilus rutilus*.

Fig. 2.65–2.66 *Pleistophora oolytica* infection in the ovary of *Esox lucius*. Fig. 2.65. An infected oocyte amidst normal ones. H & E; scale bar = 200 μm. Fig. 2.66. Yolk droplets completely replaced by SPOVs containing mature spores; single arrow points at SPOV containing sporoblasts and double arrow indicates sporoblasts which will become microspores. H & E; scale bar = 20 μm.

Fig. 2.67 *Pleistophora sulci* infection in *Acipenser ruthenus*. Top: eggs of *A. ruthenus*– uninfected (n), infected with the coelenterate *Polypodium hydriforme* (P) and with *P. sulci* (c). Bottom: a lobe of the fresh ovary of *A. ruthenus* with one egg infected with *P. sulci* (arrow); scale division = 1 mm, (From Rašín, 1949).

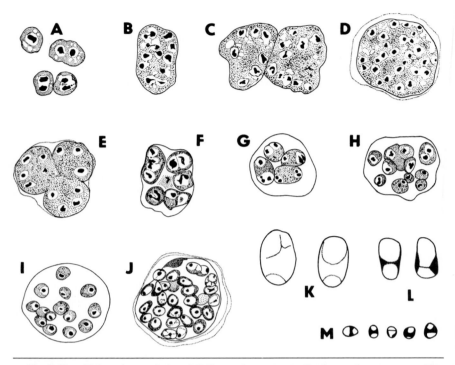

Fig. 2.68 *Pleistophora oolytica:* A,B,C, growing meronts; D, plasmodium as seen within yolk droplets; E,F,G,H, segmentation of the sporogonial plasmodium into sporoblasts; I,J, maturation of the sporoblasts within the SPOV. Spores (drawn to a scale different from A–J): K, macrospores; L, normal spores; M, microspores (redrawn from Weiser, 1949).

Pleistophora ovariae Summerfelt, 1964 (Figs. 2.14, 2.23G)

HOSTS: Females of *Notemigonus chrysoleucas*—Summerfelt (1964); females of *Pimephales promelas*—Nagel and Summerfelt (1977b). Males are refractory. Freshwater.

GEOGRAPHICAL DISTRIBUTION: Recorded in 13 midwest and southern states of the U.S.A. and in California, as a common parasite in shiners especially in those cultured in bait minnow hatcheries. According to Summerfelt and Warner (1970b), the prevalence ranged from 0% (in 4 fish farms out of 49) to 100% with an overall prevalence of 48% in a total of 2759 female fishes examined. In infected fishes, the percentage of space occupied in the ovaries by parasitic lesions ranged from 13 to 65%, with an average of 37%.

SITE OF INFECTION: Developing oocytes in the ovary.

SIGNS OF INFECTION AND PATHOLOGY: The ovary, instead of its normal uniform, greenish or yellowish hue and general translucence, is mottled with scattered white spots or streaks, each representing an amorphous mass of ovarian stroma and spores released from disintegrated oocytes. Summerfelt and Warner (1970a,b) provided a comprehensive survey of host/parasite relations. Atresia of heavily infected ova is common in intermediate or fully mature ova. After disintegration of the ova, spore masses coalesce forming a dense stroma, composed of cellular debris including the *zona radiata,* yolk vacuoles and masses of spores.

The process starts with hyperplasia of the follicular epithelium, after which the *zona radiata* collapses and breaks down. Cellular invasion by follicular phagocytes follows and completes the destruction of the host cell. The ovigerous lamellae of the ovary contain fibroblasts and possibly monocytes, which invade the atretic follicles. Since, also, collagen fibrils are detectable, some degree of fibrosis is present. A dense stroma of connective tissue, present in the infected ovary but absent in the normal ovary, contributes to its increased size. This explains why, in the postspawning season (August, September), the infected ovaries are heavier than healthy ones. In prespawning and spawning fish (May and June), however, destruction of cysts results in normal ovaries being heavier than the infected ones.

No inflammation, or encapsulation of the lesions is induced. However, phagocytes eliminate the spores from the cell debris by digesting them completely, even their walls (Dyková and Lom, 1980; Fig. 2.14). Destruction of the oocytes can result in a pronounced parasitic castration, with fecundity reduced to about 37% of normal in a fish stock, as judged by the average proportion of the ovaries destroyed. The depression of fecundity is actually far greater, since infected oocytes do not mature or do not survive the spawning.

Interestingly, infected fish were generally larger and heavier than uninfected fish probably because reduced egg production allowed nutrients to be used for faster growth (Summerfelt and Warner, 1970b).

RELATION TO HOST AGE: Infections can be acquired very early. Summerfelt and Warner (1970b) found that 30% of 3-month-old *N. chrysoleucas*-fry were already infected. The lowest prevalence was in young-of-the-year and the infection rate increased steadily through the second year (6%) third and fourth years (75%) then dropped to 33 and 35% respectively in age classes 5 and 6. Selective mortality of infected older fish although not observed could have resulted in the fall in prevalence.

Intensity of infection decreased with age and varied with season: it was greatest in May and June. Thus, the maximum number of spores and infected ova occurred during the spawning season of the host.

COMMERCIAL IMPORTANCE: Reduced fecundity of *N. chrysoleucas,* an economically very important bait fish, results in decreased production. *P. ovariae* is regarded as a serious threat to the shiner minnow industry. STRUCTURE AND LIFE CYCLE (based on Summerfelt, 1964; Summerfelt and Warner, 1970a; Wilhelm, 1966)

Merogony: Spherical meronts, up to 8–10 μm, with spherical nuclei 3 μm in size, lie singly or in groups, and divide by binary and multiple fission.

Sporogony: Sporogonial plasmodia become SPOVs measuring 17.8 × 22.8 μm and containing 8, 16, up to 20, but mostly 12 spores.

Spores: Elongate-ovoid, very often slightly bent. The large posterior vacuole occupies ⅔rds of the spore. Its anterior limit is vaulted and its side borders appear crenated, due to the easily visible polar tube threads (Fig. 2.23G). There is a single nucleus, 2 × 1 μm large and a distinct PAS positive polar cap. Electron micrographs show about 26 coils of the polar tube (Lom, unpublished).

RECORDED MEASUREMENTS: 8.4 × 4.2 μm from the original Illinois source (Summerfelt, 1964); 8.5 × 4.6 μm (Lom unpublished); 8.6 × 3.6 μm (Summerfelt and Warner, 1970a). Formalin fixed spores from 22 sources across the U.S.A. showed a variation of mean size from 6.8 × 4.0 to 8.2 × 4.8 μm (Summerfelt and Warner, 1970a).

TRANSMISSION: The parasite remains in the shiner population from one year to the next and there has been no evidence of spontaneous disappearance. Having successfully transmitted the parasite perorally, Summerfelt (1972) also postulated the possiblity of transovarial transmission, by demonstrating the presence of the microsporidian in 5% of 38 blastulas examined. Autoinfection has not been positively proven. Goldfish are refractory to experimental infection (Nagel and Summerfelt, 1977b).

TREATMENT: Nagel and Summerfelt (1977a) tested the efficacy of nitrofurazone as a control method for the infection in golden shiners. The chemical was fed at levels of 1.10, 1.65, 2.20, 3.30, 5.51 and 7.71 g active ingredient/kg feed. The dosage response was non-linear. The maximum drug effect on prevalence was at 2.2 g, while maximum effect on intensity of infection was at 3.3 g. None of the dosages tested produced a complete cure and the net production of shiners was even reduced in one pond, in which the fish were given the 2.2 g treatment. This suggested that high levels of this drug may inhibit gonad development.

Pleistophora priacanthicola He Xiaojie, 1982 (Fig. 2.46O)

SYNONYM: *Pleistophora priacanthusis* Hua and Dong, 1983.
HOSTS: *Priacanthus tayenus* and *P. macrocanthus.* Marine.

GEOGRAPHICAL DISTRIBUTION: South China Sea near Phanjiang or Guandong where the prevalence varied from 36 to 100%.

SITE OF INFECTION: Stomach, pyloric caeca, pericardium, intestinal tract, gonads and less severe infections in most organs of the body.

SIGNS OF INFECTION AND PATHOLOGY: In severe infections, growth is retarded, fish are emaciated and there is loss of lustrous appearance. Macroscopic cysts up to 22 mm in size may fill the body cavity, resulting in pressure atrophy of visceral organs. Cysts consist of a mass of closely packed SPOVs within a connective tissue wall up to 56 μm thick.

COMMERCIAL IMPORTANCE: The infection constitutes a possible threat to the *P. tayenus* landings.

STRUCTURE AND LIFE CYCLE (interpreted from Hua and Dong, 1983)

   Merogony: Not described.

   Sporogony: Sporogonial plasmodia, 13–24 μm in size, divide within SPOVs, via cytomeres, into uninucleate sporoblasts and spores.

   Spores (Fig. 2.46O): Ovoid or elongate-ovoid, anterior end more pointed, with a large posterior vacuole occupying one half or more of the spore length. Macrospores 8.2 (7–10) × 3.8 (3.2–4) μm; medium spores 5.4 (5–6) × 3.1 (3–3.5) μm; microspores 2.9 (2.7–3) × 1.9 (1.6–2) μm (fresh). Fresh SPOVs measure 9.5–63 μm, depending on the spore type, and the number of spores per vesicle is mostly 20–80 (range 6–210).

REMARKS: Amplification of the description of the spores and data on merogony are needed.

Pleistophora sauridae Narasimhamurti and Kalavati, 1972 (Fig. 2.46Q)

HOST: *Saurida tumbil*. Marine.

GEOGRAPHICAL DISTRIBUTION: Indian Ocean at Visakhapatnam, Andra, India where infection was found throughout the year. Not more than 10 "cysts" were found in one host.

SITE OF INFECTION: Smooth muscles.

SIGNS OF INFECTION AND PATHOLOGY: The parasite develops within whitish cysts, 1–2 mm in diameter, attached to the visceral muscles. The outer layer of the cyst wall was described as consisting of host cells. "Less advanced" stages of the microsporidian were found at the periphery of the cyst. No injury to the host tissue was observed.

STRUCTURE AND LIFE CYCLE

   Merogony: Not described.

   Sporogony: Sporogonial plasmodia were rounded; one, which contained 21 nuclei, measured 8 × 10 μm. SPOVs with "several to very numerous" spores have a thin wall, easy to break.

Spores (fresh): Oval, 2–2.2 × 3.6–4.2 μm, with a large posterior vacuole and a single nucleus. The PAS positive polar cap is very conspicuous (Fig. 2.46Q).

REMARKS: There are indications that this incompletely described species might be synonymous with *Microsporidium sauridae* (Sect. L, p. 160). The host, the locality and site of infection were identical and "cyst" formation was similar. The statement that the SPOV membrane is delicate and ruptures easily recalls *Glugea,* where breakdown of the SPOVs in the centre of the cyst is common. The final decision on generic status will only be made possible by a revision of the species.

Pleistophora sulci (Rašín, 1936) Sprague, 1977 (Fig. 2.67)

SYNONYM: *Cocconema sulci* Rašín, 1936.
HOSTS: *Acipenser ruthenus* and *A. guldenstadti.* Freshwater.
GEOGRAPHICAL DISTRIBUTION: Basins of rivers emptying into the Black Sea (= Danube) and Caspian Sea (= Volga, Kura). Prevalence is up to 50%, in some Czechoslovak localities (Rašín, 1949).
SITE OF INFECTION: Oocytes in the ovary.
SIGNS OF INFECTION AND PATHOLOGY: Infected fish-eggs are markedly hypertrophic. The amount of yolk is drastically reduced, and the disappearance of pigment granules from beneath the egg membrane contributes to the milky appearance of the infected eggs (Fig. 2.67). The parasite develops in concentric layers around the nucleus of the fish-egg. Near the centre of the oocyte the SPOVs contain mature or almost mature spores. These are surrounded by a layer of developing vesicles and just beneath the egg membrane, there are nests and ribbons of "small cells" possibly meronts.
STRUCTURE AND LIFE CYCLE
   Merogony: Not described.
   Sporogony: SPOVs, up to 25 μm in diameter, contain a great number of spores. Sporonts and sporoblasts contain diplokarya (Lom, unpublished).
   Spores (fresh): Spherical, average diameter 2.5 μm. Most of their volume is occupied by the posterior vacuole; there are 8 to 9 polar tube coils (Lom, unpublished).
REMARKS: The appearance of SPOVs fully entitled Sprague (1977) to transfer this species from the suppressed genus *Cocconema* to *Pleistophora.* Observations of diplokarya in the sporogony sequence, at variance with the generic diagnosis of *Pleistophora,* make its position within this genus very temporary.

Pleistophora tahoensis Summerfelt and Ebert, 1969

HOST: *Cottus beldingi*. Freshwater.
GEOGRAPHICAL DISTRIBUTION: Lake Tahoe, California, U.S.A. The prevalence was about 6.4%. No significant sex- or age-dependent differences in prevalence were found.
SITE OF INFECTION: Abdominal muscles.
SIGNS OF INFECTION AND PATHOLOGY: Spindle-shaped "cysts", which measure 2–6.5 × 1.9–2.5 mm and may show through the skin, are embedded in the abdominal body wall. The cyst consists of a mass of SPOVs packed closely together and surrounded by a tough connective tissue sheath; the whole structure corresponds to destroyed bundles of muscle fibres fused together. The sheath represents the original interstitial connective tissue or perimysium.

Invasion of muscle fibres results in disintegration of the sarcoplasm and myofibrils; the resulting large parasitic aggregation ("cyst") displaces parts of the muscle and weakens the body wall.
STRUCTURE AND LIFE CYCLE
Merogony: Globular meronts 4–10 μm, with 1–8 compact nuclei.
Sporogony: SPOVs up to 23 μm, 20–48 spores per vesicle.
Spores (fresh): Ovoidal, 6 × 3 μm, large posterior vacuole occupying slightly more than half the body length, with the encircling coils of the polar tube visible; a single nucleus. The values for spore length, when plotted, give a slightly bimodal curve; however, true dimorphism does not exist.
REMARKS: All the above data are extracted from Summerfelt and Ebert (1969). Their micrographs of spores are unclear and are not useful for diagnosis.

Pleistophora tuberifera Gasimagomedov and Issi, 1970 (Fig. 2.69)

HOSTS: *Neogobius kessleri gorlap* (4 out of 23 specimens infected), *N. caspius* (1 found infected), *N. melanostomus affinis* (4 out of 34 specimens infected). Freshwater.
GEOGRAPHICAL DISTRIBUTION: South Caspian Sea near the town of Bekdash, U.S.S.R.
SITE OF INFECTION: Subcutaneous layer of musculature.
SIGNS OF INFECTION AND PATHOLOGY: Groupings of tuberiform cysts up to 0.2 × 0.55 mm on the surface of subcutaneous muscle. Cysts were presumed (sic) to be composed of SPOVs.

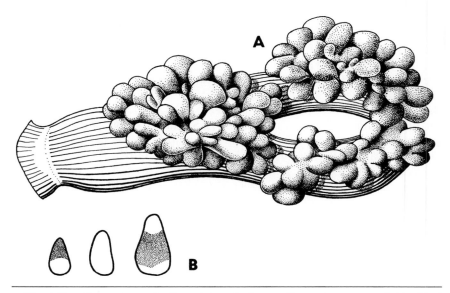

Fig. 2.69   *Pleistophora tuberifera:* A, bunch of grape-like cysts on a muscle bundle of a goby; B, spores embedded in glycerine-gelatine medium (redrawn from Gasimagomedov & Issi, 1970).

STRUCTURE AND LIFE CYCLE
  Merogony: Not described.
  Sporogony: SPOVs 22–32 μm, 20–50 spores per vesicle.
  Spores (Fig. 2.69B): Very variable in shape, either oval with attenuated anterior end or pear-shaped. A posterior vacuole present. Size of glycerin-gelatine preserved spores: 3.6 × 1.8 μm, 4.8–5.4 × 2.4–3 μm (two specimens, respectively) from *N. kessleri gorlap;* 9.6 × 5.4 μm from *N. caspius* (one specimen); 6–7.2 × 3–3.6 μm and 8–8.4 × 4.2 μm (two specimens, respectively) from *N. melanostomus affinis.*
  REMARKS: All above data were taken from the incomplete original description, based on glycerin gelatin preserved material only.

Pleistophora vermiformis Léger, 1905 (Figs. 2.23N, 2.41K)

SYNONYM: *Pleistophora* sp. Otte, 1964.
HOST: *Cottus gobio*. Freshwater.
GEOGRAPHICAL DISTRIBUTION: Small streams in Tourraine, France—several infected specimens (Léger); the river Danube, Austria—1 infected fish (Otte); the Malše river, Czechoslovakia—1 infected specimen (Lom, unpublished).

SITE OF INFECTION: Skeletal muscles.

SIGNS OF INFECTION AND PATHOLOGY: Numerous elongate, vermiform masses, up to 3 or 4 mm long, parallel with the muscle fibres, which may show through the skin (Fig. 2.41K). Infected muscle fibres are dilated and their interior is packed with SPOVs. The infection is never heavy enough to be sufficient for a fatal prognosis.

STRUCTURE AND LIFE CYCLE

Merogony: Not described.

Sporogony: SPOVs 20–60 μm (Otte, 1964).

Spores (Fig. 2.23N): 3–6 × 1–2 μm with an easily visible vacuole (Otte, 1964) ovoid, averaging 4.8 × 3 μm when formol fixed (Lom, unpublished).

REMARKS: The descriptions of Léger (who gave no spore dimensions) and Otte are extremely brief and a thorough redescription is necessary.

Pleistophora sp. of Arthur, Margolis and McDonald, 1982

In the muscles of *Theragra chalcogramma* in the Pacific; a common yet undescribed parasite.

Pleistophora sp. of Blasiola, 1977

In *Gobiodon okinawae*. Marine. Mentioned in a letter to the author; no data supplied.

Pleistophora sp. of Bond, 1937

HOST: *Fundulus heteroclitus*. Euryhaline in Chesapeake Bay, U.S.A.

SITE OF INFECTION: Muscle tissue and outer layer of the spinal cord.

SIGNS OF INFECTION AND PATHOLOGY: A swelling about 4 mm above the body surface. Section of affected muscles revealed a "general oedema . . . and degeneration with much infiltration near the centre of the infected tissue".

STRUCTURE AND LIFE CYCLE: "Small masses" containing 10–18 spores.

Spores: Oval, 1.5–2 × 2.5–3 μm.

REMARKS: Bond (1937) probably mistook host macrophages with ingested spores for "parasitic masses" with "pansporoblastic nuclei" and spores. His sketchy line drawings and description are of no help in diagnosis.

Pleistophora sp. of Drew, 1909

In the skeletal muscles of one specimen of "cod" caught "off Iceland coast".

SIGNS OF INFECTION AND PATHOLOGY: The parasites were located within diffuse pigmented areas of a brownish colour, either in muscle fibres or in the interfibrillar tissue. They caused tissue degeneration, notably in the centre of the infected areas, where also a number of "yellow bodies" (which may have been sites of phagocytosis and melanin deposition) were observed. The cytoplasm of the muscle fibres was replaced by finely granular material and in surrounding regions the interfibrillar spaces were infiltrated.

STRUCTURE AND LIFE CYCLE

Merogony: "Small" meronts, with several nuclei.

Sporogony: SPOVs 30 μm in diameter with "numerous" spores.

Spores: Oval, 3 × 2.5 μm.

REMARKS: All data from Drew (1909); no illustrations are available.

Pleistophora sp. of Dzhalilov, 1966

In *Noemacheilus malapterus longicauda* from the river Vakhsh, Uzbekistan, U.S.S.R. This information—without any detail—is from Osmanov (1971).

Pleistophora sp. of Lucký, 1957, in Lucký and Dyk (1964)

A single cyst was found by Lucký, in 1957 in testis of *Rutilus rutilus,* from the basin of the river Dyje, Czechoslovakia; might very well be *P. longifilis* Schuberg.

Pleistophora sp. of Putz, 1965 in Putz and McLaughlin (1970)

In *Salvelinus fontinalis* (freshwater), U.S.A. It is possibly identical with *Loma salmonae* (p. 132).

Pleistophora sp. of Wellborn, 1966 in Putz and McLaughlin (1970)

In *Dorosoma petenense* (freshwater), U.S.A.

Pleistophora sp. of Woodcock, 1904

SYNONYM: "Sporocyst" of Linton, 1901.

In the liver of *Poronotus triacanthus,* a marine fish caught off the coast of Massachusetts, U.S.A. (Linton, 1901). Globular, 1.5 mm large "cysts" contained short and thick spores, about 2.5 μm long and "a little less" in thickness, with bluntly rounded ends. Spores were aggregated in "globular or oblong clusters'' (possibly SPOVs) up to 20 μm in diameter. Woodcock (1904) assigned this organism to the genus *Pleistophora*.

C. Genus *Loma* Morrison and Sprague, 1981

Morrison and Sprague (1981b) originally proposed as the type species *L. morhua* Morrison and Sprague, 1981. However, *L. morhua* has to be considered a junior synonym of *L. branchialis*.

Loma branchialis (Nemeczek, 1911) Morrison and Sprague, 1981
(Figs. 2.23I, 2.45G, 2.70–2.74, 2.76)

SYNONYMS: *Nosema branchialis* Nemeczek, 1911; *Glugea branchialis* (Nemeczek, 1911) Lom and Laird, 1976; *Loma morhua* Morrison and Sprague, 1981a.

HOSTS: Originally described from *Gadus (Melanogrammus) aeglefinus;* later from *G. callarias* (Bazikalova, 1932; Dogiel, 1936; Fantham, Porter and Richardson, 1941), *G. morhua marisalbi* (Shulman and Shulman-Albova, 1953), *G. morhua kildinensis, Enchelyopus cimbrius* and *Gadus morhua* (Morrison and Sprague, 1981a). Marine.

GEOGRAPHICAL DISTRIBUTION: Boreo-arctic. In the North Atlantic, 3 specimens of *Gadus aeglefinus* out of several thousands (Nemeczek, 1911), 3 out of 5 specimens of the same host (Lom and Laird, 1976) and in 10 out of 91 specimens of *G. morhua* (Morrison and Sprague, 1981b). In the White Sea, 3 infected out of 68 *G. morhua maris-albi* (Shulman and Shulman-Albova, 1953). Also recorded from *G. morhua kildinensis* and *Enchelyopus cimbrius* from the Barents and Baltic Seas (Shulman, 1957).

SITE OF INFECTION: Gills. Although recorded also from the pyloric caeca of *Gadus morhua* (Dogiel, 1936) and the intestinal wall, liver, spleen, kidney, corpus vitreum, brain and medulla oblongata of *Gadus morhua maris-albi* (Shulman and Shulman-Albova, 1953), these exceptional findings should be reinvestigated.

SIGNS OF INFECTION AND PATHOLOGY: Round, whitish xenomas up to 0.5 mm, sometimes even 1.2 mm in the gill filaments (Figs. 2.70, 2.72) and/or pseudobranchs. The plasma membrane of the cell is raised as irregular projections into the thick basement membrane of the pillar system, which together form the xenoma wall up to 1.5 μm thick (Morrison and Sprague, 1981a). By light microscopy it appears as a chromophilic layer (Nemeczek, 1911) or striated band (Lom and Laird, 1976). The single hypertrophic host cell nucleus may be branched or even separated into two pieces.

Xenomas cause distortion and pathological changes in gill lamellae, even

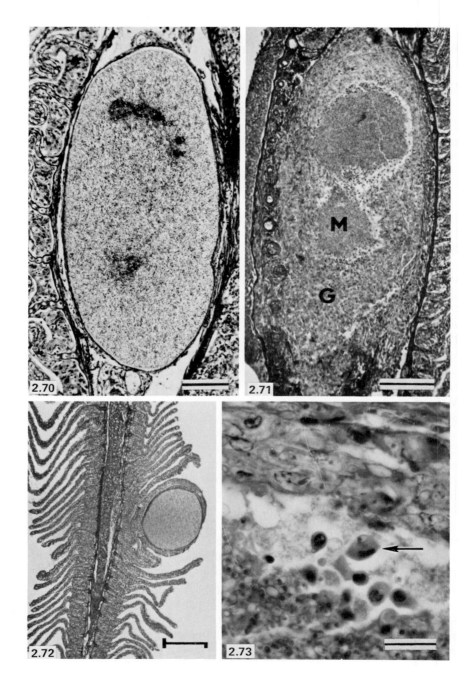

in the adjacent uninfected ones. Kabata (1959) observed disintegration of the connective tissue in the gill filaments, hypertrophy of the filaments and displacement of the blood vessels and of the cartilaginous axis of the filament. Lom and Laird (1976) studied the tissue reaction (Sect. IIB,1) which eliminated the xenoma and restored the gill structure (Figs. 2.71, 2.73).

COMMERCIAL IMPORTANCE: All cases observed have been light infections without any noticeable impairment of the host's health.

STRUCTURE AND LIFE CYCLE (data from Morrison and Sprague, 1981b)

Merogony: Uninucleate meronts have conspicuous perinuclear cisternae of ER (Fig. 2.74) and are surrounded by cisternae of host endoplasmic reticulum.

Sporogony: Sporonts produce a small number of sporoblasts, then spores. The spores lie one to three per SPOV. The SPOV cavity contains fine tubules which extend from the surface of the sporoblasts (Fig. 2.76).

Spores (fresh; Figs. 2.23I, 2.45G): Ovoid to elongate ovoid, 2.3 (2–2.6) × 4.8 (4.1–6.5) μm, from haddock (Lom and Laird, 1976); 3.5 × 6.3 μm from haddock (Nemeczek, 1911); 3.5–4.2 × 5.7–6.6 μm from Atlantic cod (Fantham, Porter and Richardson, 1941); length 5–6 μm from Kildin cod (Dogiel, 1936). The posterior vacuole has a convex, not concave, anterior border and the cytoplasmic side walls are very thin. The oval or crescentic nucleus, 0.5–1 μm, lies transversely in the anterior part of the spore and the PAS positive polar cap is delicate and disc-like (Lom and Laird, 1976). There are 16–17 coils of the polar tube.

REMARKS: Morrison and Sprague (1981b) established a new species *Loma morhua,* for a parasite of the gills of *G. morhua,* although it only differed from *L. branchialis* in having slightly smaller spores. Fixation-induced shrinkage alone may account for the size differences since they measured only fixed spores from *G. morhua.* In spite of stating (1981a) that "possession of smaller spores and occurrence in a different host provide a questionable basis for regarding it as new", they (1981b) identified another finding from haddock as *L. morhua.* In this work *L. morhua* is considered as a junior synonym of *L. branchialis.*

Fig. 2.70–2.73 *Loma branchialis* in the gills of *Gadus aeglefinus:* Fig. 2.70. A spore-filled xenoma located in the central axis of the gill filament. Fig. 2.71. A similar xenoma in the process of destruction by host tissue reaction. Granulation tissue (G) has proliferated into the periphery and is invading the central mass (M) of spores. Fig. 2.72. A grown xenoma located within the epithelium of a gill lamella. Fig. 2.73. Peripheral portion of the granuloma in Fig. 2.71, with remnants of a spore mass attacked by host defence cells; arrow indicates a macrophage with an ingested spore. Scale bar = 100 μm in Figs. 2.70 and 2.71, 200 μm in Fig. 2.72 and 20 μm in Fig. 2.73.

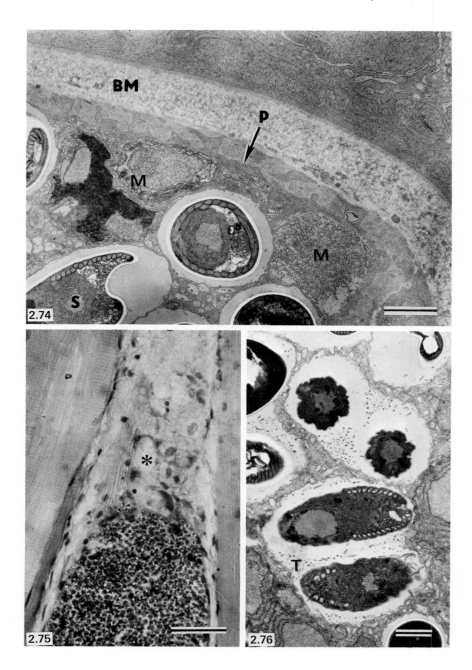

Loma dimorpha Loubès, Maurand, Gasc, de Buron and Barral, 1985

HOSTS: *Gobius niger* and *Zosterisessor ophiocephalus*. Marine.
GEOGRAPHICAL DISTRIBUTION: Étang de Thau, Mediterranean coast of France. In winter time, about ¼ of the host population was infected.
SITE OF INFECTION: Connective tissue of the digestive tract.
SIGNS OF INFECTION AND PATHOLOGY: Round xenomas, 100–300 μm, 1 to about 80 per fish. The plasma membrane of the cell is covered by an amorphous layer pervaded by collagen fibers of host origin.
STRUCTURE AND LIFE CYCLE

Merogony: Uninucleate meronts about 2.5 μm in size.

Sporogony: The SPOV membrane is formed around a multinucleate sporont which grows into a cylindrical or compact multinucleate plasmodium which was seen to give rise to 8 (but supposedly up to 12) sporoblasts in chain- or morula-like arrangement. The SPOV cavity contains tubules similar to those in *L. branchialis.*

Spores (fresh): Oval, averaging 4.5 × 1.8–2 μm, with a large posterior vacuole. Polar tube with 13–15 coils.
REMARKS: All data taken from the original description. Growth of uninucleate sporont within the SPOV into a plasmodium is a feature different from other species of *Loma. Glugea* sp. of Naidenova, 1974, is possibly a pro parte synonym of *L. dimorpha.*

Loma diplodae Bekhti, 1984

HOST: *Diplodus sargus*. Marine.
GEOGRAPHICAL DISTRIBUTION: Brackish lagoons and sea at the Mediterranean coast of France near Montpellier; 8 specimens infected from 1344 examined.

Fig. 2.74 *Loma branchialis:* part of the periphery of the xenoma to show the wall as a highly folded, simple plasmalemma (P) and the subtending basement membrane (BM) of the pillar system, which serves as a capsule around the xenoma; M, meronts; S, single spore in its SPOV; scale bar = 1 μm. (Photograph by Dr. C. Morrison.)

Fig. 2.75 *Tetramicra brevifilum* in the perimysium of *Scopthalmus maximus:* mass of spores is wedged between the myomeres next to necrotic muscle fibre with fragmented myofibrils (asterisk); scale bar = 100 μm. (Photograph by Dr. R. A. Matthews.)

Fig. 2.76 *Loma branchialis:* SPOVs with two immature spores per vesicle and with tubules (T), most abundant during sporulation and fewer as the spores mature; scale bar = 1 μm. (Photograph by Dr. C. Morrison.)

SITE OF INFECTION: Efferent blood vessels of the gill filaments.

SIGNS OF INFECTION AND PATHOLOGY: Xenomas up to 150 μm are invested with a plasmalemma coated with several layers of microfibrils. Xenomas block the blood circulation in the gill filament.

STRUCTURE AND LIFE CYCLE: According to Bekhti (1984), developmental stages are identical with those of *L. salmonae* (p. 133, Remarks).

Spores (fresh): Ovoid, averaging 4.1 × 2.2 μm, polar tube makes 17–18 turns and is attached anteriolaterally.

REMARKS: Differs from *L. salmonae* in host, site of infection and details of spore ultrastructure (Bekhti, 1984).

Loma fontinalis Morrison and Sprague, 1983

SYNONYM: *Microsporidium* sp. Morrison and Sprague, 1981.

HOST: *Salvelinus fontinalis*. Freshwater.

GEOGRAPHICAL DISTRIBUTION: Halifax, Nova Scotia, Canada.

SITE OF INFECTION: Gill lamellae.

SIGNS OF INFECTION AND PATHOLOGY: Xenomas up to 0.5 mm, in the gill lamellae or at their base. Host nucleus at the periphery of the xenoma.

STRUCTURE AND LIFE CYCLE (Morrison and Sprague, 1981c, 1983)

Merogony: Uninucleate meronts.

Sporogony: Elongate sporoblasts, one per SPOV (see diagnosis of *Loma* in Chap. 1, Sect. III, F, p. 14), the cavity of which abounds with thin tubules extending from the sporoblast outer membrane.

Spores: 2.2 (1.6–2.4) × 3.7 (3.1–4.3) μm (fixed). There are 14–15 turns of the polar tube and its basal part is attached anteriolaterally. The posterior vacuole occupies almost two thirds of the spore.

REMARKS: Characters of fresh spores need to be defined to determine whether the parasite is identical with *L. salmonae*.

Loma salmonae (Putz, Hoffman and Dunbar, 1965) Morrison and Sprague, 1981 (Fig. 1.3).

SYNONYMS: *Plistophora salmonae* Putz, Hoffman and Dunbar, 1965; *Plistophora* sp. Wales and Wolf, 1955.

HOST: *Salmo gairdneri, Oncorhynchus nerka, O. masou* (Awakura, Tanaka and Yoshimiz, 1982); *Cottus* sp. (Wales and Wolf, 1955). All freshwater.

GEOGRAPHICAL DISTRIBUTION: North America (California, British Columbia); Hokkaido, Japan, widespread in wild and hatchery-reared salmonids.

SITE OF INFECTION: Secondary lamellae of gills.

SIGNS OF INFECTION AND PATHOLOGY: Whitish xenomas, up to 0.4 mm, seen on gill lamellae, are encased with a thin epithelial layer. The host cell membrane, which forms the xenoma wall, abuts directly on to the basement membrane of the pillar system of the lamella. Mature spores are distributed throughout the xenoma.

In heavy infections, gill filaments are distorted, sometimes fused, due to great epithelial hyperplasia and their tips are badly clubbed. Wales and Wolf (1955) reported a correlation between anaemia and abundance of *L. salmonae*.

COMMERCIAL IMPORTANCE: Hatchery epizootics caused considerable mortalities in Japan and North America. One Californian epizootic resulted in an almost complete loss of 170,000 advanced fingerlings (Putz, 1964).

STRUCTURE AND LIFE CYCLE (interpreted from Morrison and Sprague, 1983, and Bekhti, 1984)

Merogony: Uninucleate meronts develop into elongate plasmodia with at least 5 nuclei (Fig. 1.3).

Sporogony: Elongate sporonts, around which the SPOV cavity forms, give rise to 2–4 sporoblasts in one SPOV. Thin tubules, communicating with the sporoblast membrane, extend into the SPOV cavity. Paramural bodies occur in sporoblasts.

Spores: Pyriform, 4.5 (4.25–5.3) × 2.2 (1.7–2.8) μm (fixed; Putz, Hoffman and Dunbar, 1965); 7.5 × 2.4 μm (fresh; Wales and Wolf, 1955). Morrison and Sprague (1983) reported a single layer of 14–17 turns of the polar tube, which is attached anteriolaterally and that the polaroplast has laminate and vesicular parts. Bekhti (1984) observed 11–12 turns of the polar tube.

REMARKS: Bekhti (1984) showed the existence of multinucleate meronts. She described a sporogony sequence which differed from the account given by Morrison and Sprague (1983). Binucleate cells representing the terminal meronts produce SPOVs. Inside these the cells divide to produce 4 sporoblasts, so that there are 4 spores per SPOV. Unfortunately the morphology and illustration of fresh spores have not been presented which also applies to *L. diplodae* and *L. fontinalis*. An unidentified species of *Loma* found by Dr. A. K. Hauck (1983, personal communication) was said to be the cause of losses of fry of *Oncorhynchus tschawytscha*. The fish suffered subacute systemic infections, the xenomas occurring in cartilage, pseudobranchs, gills, choroid gland, kidney and arteries. Inflammatory and degenerative changes impaired swimming, reduced growth and in addition to direct losses, there were mortalities due to increased predation. The species might have been *Loma salmonae*.

Loma sp. of Bekhti, 1984

In *Tilapia melanopleura,* lake Nohkoué in Bénin, Africa. Xenomas are found at the adductor muscle of the gill filaments. At the ultrastructural level, the plasmalemma of the xenoma is coated with fibers arranged in a reticulated meshwork.

D. Genus *Thelohania* Henneguy in Thélohan, 1892

Thelohania baueri Voronin, 1974 (Figs. 2.41C, 2.77)

HOSTS: *Pungitius pungitius* and *Gasterosteus aculeatus*. Brackish.
GEOGRAPHICAL DISTRIBUTION: The Gulf of Finland near Leningrad, U.S.S.R. The average prevalence was 12.5%
SITE OF INFECTION: Oocytes in the ovaries.
SIGNS OF INFECTION AND PATHOLOGY: The infected eggs are white and slightly swollen. Infection is in the yolk, in oocytes of various sizes and may result in complete replacement of the ovum by the microsporidium. Some of the infected eggs are atretic, being encapsulated and penetrated by connective tissue cells.
STRUCTURE AND LIFE CYCLE (Fig. 2.77)
   Merogony: Rounded binucleate meronts give rise to tetranucleate and later to ribbon-like meronts 21×8 µm in size. The tetranucleate meronts having the shape of a figure of eight with nuclei in pairs and octonucleate ones forming clover-leaf shapes (Fig. 2.77E) as described by Voronin (1974) might have been in fact sporogony stages.
   Sporogony: Early sporonts binucleate, 6–8 µm in diameter, produce tetranucleate ones and finally spores; SPOVs measure 10 µm in diameter. In *G. aculeatus,* 70% of the SPOVs contained 8 spores and the rest had 4, 6, 9–12 and 16 spores per vesicle. In *P. pungitius,* 58% of the vesicles had 9–12 spores and the rest had 6, 8, 16 and more spores per vesicle.
   Spores (fresh, Fig. 2.41C): Ovoid or slightly pyriform measuring 5.4 (4.5–6) $\times$ 2.7 µm. Posterior vacuole often reaches the middle of the spore and has a straight anterior border and thick side walls. One or two nuclei were recorded. Macrospores which were more frequent in *G. aculeatus,* were 6–7.3 µm long.
REMARKS: The life cycle should be reinvestigated with respect to the pos-

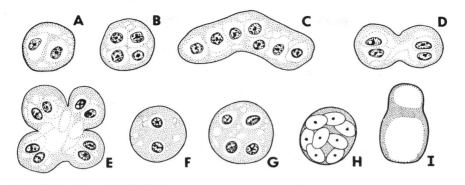

Fig. 2.77 Some stages of the life cycle of *Thelohania baueri:* A–E, meronts and merogony; F,G, sporogonial plasmodium; H, formation of sporoblasts; I, spore (redrawn from Voronin, 1974).

sible occurrence of diplokarya, notably because Voronin (1974) described paired nuclei in some meronts and 2 nuclei in some of the spores, which is not characteristic of *Thelohania*. Sprague (1977) questioned the generic determination, since the pattern of variation in the number of spores in the SPOV is not typical for *Thelohania*. Parasites recorded by Thélohan (1895) and Otte (1964) in the ovaries of sticklebacks and referred to as *Glugea microspora* and *Glugea* sp., respectively, may be identical with *T. baueri*.

Thelohania ovicola (Auerbach, 1910) Kudo, 1924

SYNONYM: *Plistophora ovicola* Auerbach, 1910.
HOST: *Coregonus exiguus bondella*. Freshwater.
GEOGRAPHICAL DISTRIBUTION: Neuchâtel Lake, Switzerland. Several infected fish were found.
SITE OF INFECTION: Ovary.
SIGNS OF INFECTION AND PATHOLOGY: Invaded eggs were whitish and slightly smaller than the intact ones. The parasite developed in the central part of the yolk and its growth resulted in complete destruction of the egg contents which were finally represented only by spores and cell debris.
STRUCTURE AND LIFE CYCLE
Merogony: Not observed.
Sporogony: The youngest sporont observed was tetranucleate, 6 μm diameter; more advanced sporonts were round, 6–10 μm with a distinct membrane. SPOVs, 10–12 μm in diameter, contained 6–8 spores.

Spores (glycerin-preserved): Oval or pyriform, somewhat constricted in the anterior half; 4–6 × 6–8 μm. A rounded posterior vacuole.

REMARKS: A redescription of this poorly described species is needed. Kudo (1924) transferred this species to the genus *Thelohania* mainly on the small number of spores in the SPOVs, as depicted by Auerbach (1910). Sprague (1977) believed that the evidence for generic identification was insufficient.

E. Genus *Heterosporis* Schubert, 1969

This genus is similar in its development to *Pleistophora*. However, because it induces cell hypertrophy (xenoma formation) it is considered as a separate genus. The simple hypertrophy, known in some *Pleistophora* species (e.g. *P. longifilis,* Fig. 2.64) is of a different type; the hypertrophic cell does not form the same kind of envelope around the parasite mass, as in *Heterosporis*. In *Pleistophora* species invading muscle fibers, the infection appears as a diffuse infiltration and the sarcoplasm and nuclei are finally completely destroyed.

Heterosporis finki Schubert, 1969 (Figs. 2.6E, 2.78–2.80)

HOST: *Pterophyllum scalare.* Freshwater.

GEOGRAPHICAL DISTRIBUTION: The infected fish came from a pet shop in Stuttgart, West Germany. A sole finding.

SITE OF INFECTION: Connective tissue around the oesophagus.

SIGNS OF INFECTION AND PATHOLOGY: Nodules on the oesophagus, consisting of xenomas. The host cell originally invaded is probably a connective tissue cell stimulated by the parasite into hypertrophy. The youngest xenomas observed (Schubert's material, Lom, unpublished) were about 30 μm long, with pale cytoplasm and a single oval nucleus, 8 μm long (Fig. 2.78). The cell contained a thin-walled vacuole 10 μm in diameter, containing meronts. At a later stage, the vacuole occupied most of the hypertrophic host cell. The cytoplasm was reduced to a 4–5 μm thick enveloping layer around the central region, delineated by a thick solid wall, and was full of SPOVs. At an advanced stage there were additional nuclei in the cytoplasmic layer which suggested either that nuclear division had occurred or that additional hypertrophic cells had fused with the one originally infected (Fig. 2.79). The xenoma may reach 300 μm in diameter and is coated by a layer of connective tissue. Its organisation is quite different from the xenomas of *Glugea*.

STRUCTURE AND LIFE CYCLE

Merogony: Young meronts are uninucleate rounded cells, measuring 3.5 μm (Lom, unpublished).

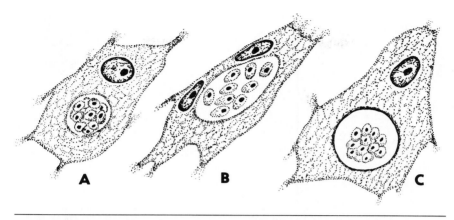

Fig. 2.78 Initial stages of *Heterosporis finki* within hypertrophic connective tissue cells with one (A,C) or two (B) hypertrophic nuclei: parasites, probably meronts, are enclosed within a vacuole.

Sporogony: Sporonts grow into sporogonial plasmodia and mature as SPOVs with 8 macrospores or, less commonly, 16 or more microspores. The SPOV wall is 0.45 μm thick, opaque and highly undulant in the electron microscope (Schubert, 1969b).

Spores (fresh): Ovoid, with a rather tapered anterior end and a large vacuole which occupies the broad posterior half of the spore, and sometimes extends into the anterior half. Macrospores 7–9 × 2–3 μm, with 28–30 coils of the polar tube; microspores 3 × 1.5 μm, with 8 coils of the polar tube.

REMARKS: Schubert (1969b) supplied no light microscope illustration of spores. He mistook the helically arranged polyribosomes in the sporoplasm of the spores for DNA material, claiming that the sporoblasts and spores lacked a true nucleus. The SPOVs have the structure and spore number (8, 16, and more) typical of *Pleistophora*. Although the data on merogony and sporogony are scarce, the development accords well with the pattern of the genus *Pleistophora;* future studies are needed to endorse the validity of the genus *Heterosporis*.

F. Genus *Tetramicra* Matthews and Matthews, 1980

Tetramicra brevifilum Matthews and Matthews, 1980 (Figs. 2.75, 2.81, 2.82, 2.83D–L)

HOST: *Scophtalmus maximus*. Marine.
GEOGRAPHICAL DISTRIBUTION: Coast of Cornwall, U.K. (Matthews and

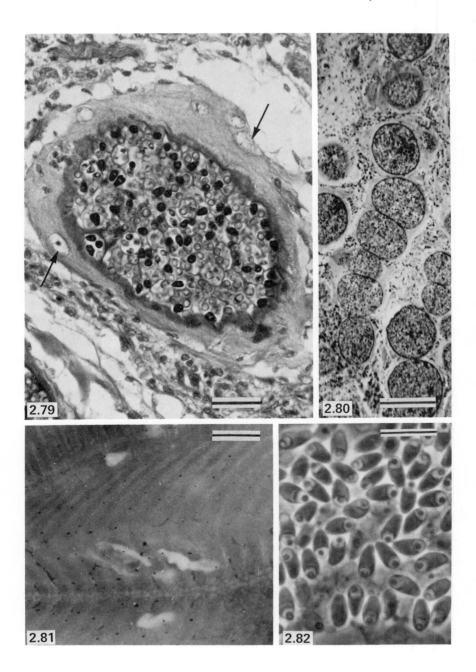

Matthews, 1980) where about 10% of the turbot-of-the-year were infected; Scotland (material sent by Dr. K. McKenzie), U.K.

SITE OF INFECTION: Connective tissue of the skeletal musculature.

SIGNS OF INFECTION AND PATHOLOGY: Whitish nodules visible through the central body wall of young fish (Fig. 2.81), consisting of numerous xenomas up to 200 μm diameter. The anastomosing surface microvilli of the xenomas interlock to form composite formations several millimetres in size. These cysts have spherical or dendritic outlines, depending on the amount of space available in the tissues (Fig. 2.75). The xenomas originate in connective tissue cells. They have a centrally located, hypertrophic nucleus and the developmental stages of the parasite and spores are not stratified within the xenoma.

The pathogenic effects are most noticeable when cysts break down, releasing free spores. Muscle fibres show degenerative signs such as vacuolisation of the sarcoplasm and separation of myofibrils. Leucocyte infiltration involves mainly lymphocytes and macrophages, which ingest and destroy the spores. Encapsulation of spores was not observed.

When infections are heavy a substantial proportion of the body musculature can be inactivated and swimming impaired.

STRUCTURE AND LIFE CYCLE (Fig. 2.83D–L)

Merogony: Subspherical, uninucleate meronts, 3.3 μm in diameter, develop into cylindrical meronts, 3.3 × 7.4 μm with at least 7 nuclei.

Sporogony: Binucleate cells produced at the termination of merogony divide into binucleate sporonts (Fig. 2.83I,J). These in turn grow, undergo nuclear division and produce 4 sporoblasts, measuring 3 × 4 μm, by radial cleavage. The sporoblasts begin their differentiation into spores while still interconnected, the point of attachment becoming the anterior end of the spore.

Spores (fresh): Oval, wider posteriorly, 2 × 4.8 μm. The posterior vacuole, 1.7 μm in diameter, occupies the posterior ⅓ of the spore and contains, unlike all other fish microsporidia, a conspicuous, 1.3 μm large spherical inclusion (Fig. 2.82). The basal part of the polar tube is attached

---

Figs. 2.79–2.80 *Heterosporis finki.* Fig. 2.79. The cyst, with closely packed SPOVs and a heavily stained wall, is located within a hypertrophic, host cell complex with numerous nuclei (arrows); some spores are heavily stained, others not. H & E; scale bar = 20 μm. Fig. 2.80. Section through proliferating connective tissue along the pharyngeal canal, showing numerous cysts of *H. finki.* H & E; scale bar = 100 μm.

Figs. 2.81–2.82 *Tetramicra brevifilum.* Fig. 2.81. Whitish parasite foci in fresh musculature of *Scopthalmus maximus; scale bar* = 2 mm. Fig. 2.82. Cluster of spores, with conspicuous inclusions in the posterior vacuole, as seen by phase contrast; scale bar = 10 μm.

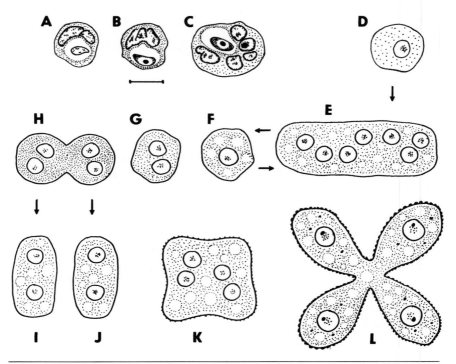

Fig. 2.83   A–C: *Glugea anomala,* early stages of the invasion of a mesenchyme cell, scale bar = 10 μm; A, at day 12 post infection the parasite lies in a vacuole, embraced by the host cell nucleus; B, a more advanced stage of the cell invasion, when the nucleus of the meront has developed a large karyosome and the host cell nucleus has become trilobed; C, the host cell nucleus has separated into a number of peripherally located nuclei. D–L: diagrammatic representation of developmental stages of *Tetramicra brevifilum* (Redrawn from Matthews and Matthews, 1980); D, sporoplasm released from a spore develops into a multinucleate meront (E) which divides into uninucleate meronts (F). These either repeat the cycle or produce a binucleate meront (G) (interpreted as a sporont mother cell by Matthews & Matthews). This gives rise through division stage (H) to two binucleate sporonts (I and J) which in turn each produce a tetranucleate sporont (K) which divides into four sporoblasts (L) which then mature into spores. (A–C redrawn from Weissenberg, 1968.)

anteriolaterally and falls into only about four coils. There is a single nucleus: rarely, binucleate spores occur.

REMARKS: All above data based on Matthews and Matthews (1980).

G. Genus *Microgemma* Ralphs and Matthews, 1986

Microgemma hepaticus Ralphs and Matthews, 1986 (Figs. 1.4, 1.5)

HOST: *Chelon labrosus.* Marine and estuarine.
GEOGRAPHICAL DISTRIBUTION: Coast of Cornwall, England.
SITE OF INFECTION: Liver.
SIGNS OF INFECTION: Xenomas up to 0.5 mm in diameter, more or less spherical, were found in "O" group fish.
STRUCTURE AND LIFE CYCLE

Merogony: Multinucleate meronts, within a cisterna of host endoplasmic reticulum (Fig. 1.4), divide by plasmotomy.

Sporogony: Plasmodia in direct contact with host cell cytoplasm. Sporoblasts produced initially by single exogenous budding, later by multiple budding and fragmentation of the plasmodium (Fig. 1.5).

Spores (fresh): Pyriform, 4.2×2.4 μm in size, uninucleate, polar tube in single rank with 7–10 coils. Posterior vacuole occupies posterior third of spore.

H. Genus *Spraguea* Weissenberg, 1976

Spraguea lophii (Doflein, 1898) Weissenberg, 1976 (Figs. 1.6, 1.7, 2.6C, 2.8, 2.10, 2.13, 2.23K, 2.23L, 2.45I, 2.84, 2.85, 2.87)

SYNONYMS: *Glugea lophii* Doflein, 1898; *Nosema lophii* (Doflein, 1898) Pace, 1908.
HOSTS: *Lophius piscatorius; L. budegassa; L. americanus.* Marine. Probably also in *L. gastrophysus* (Jakowska, 1966), collected off the Brazilian coast.
GEOGRAPHICAL DISTRIBUTION: Common in its hosts along the Mediterranean and Atlantic coasts of Europe; in the North Sea, along the coasts of Britain, Norway and Iceland, at the U.S. Atlantic coast and along the coast of Brazil.

In various landings of *Lophius piscatorius, L. budegassa* and *L. americanus,* the prevalence varied from 32 to 100%. The prevalence was found to increase with age in fish caught off the Icelandic coast, when the in-

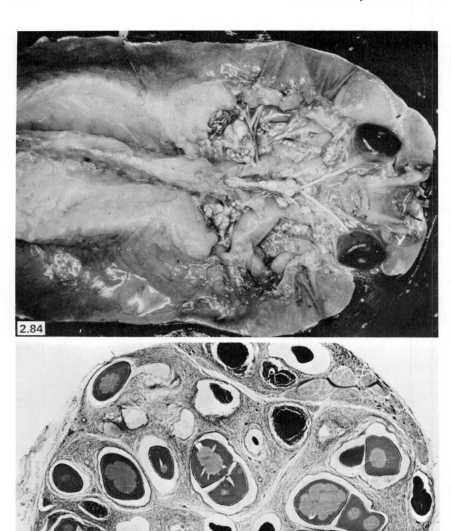

fection ranged from zero in very young fish through 41.2% in middle-sized fish to 46.3% in market-sized fish (Priebe, 1971).

SITE OF INFECTION: Ganglion cells of the central nervous system, mainly of the large extracranial ganglia of the brain and spinal nerves (Fig. 2.84). Essentially, the infection can be found wherever ganglion cells are present, even in the small ganglia along the main nerves, in the walls of the *sinus venosus* of the heart and in ganglia of other viscera (Weissenberg, 1909).

SIGNS OF INFECTION AND PATHOLOGY: The infected ganglia, chiefly in the cerebro-spinal region, are conspicuously enlarged, milky white masses, often with a bunch-of-grapes appearance. These cyst-like structures consist of several to many large xenomas fused together (Fig. 2.85). The first to recognise the true nature of these ''cysts'' was Mrázek (1899).

The sharply delimited parasite mass is restricted to a region of the ganglion cell near the point of exit of the axon where the infection originally started (Weissenberg, 1976) or in the axon itself (Mrázek, 1899). It does not extend into the cell body itself (Fig. 2.6C, 2.87). The entire ganglion cell is enormously hypertrophic, especially the parasitised region, which constitutes the actual cyst measuring 1.5 mm and more. The non-parasitised cytoplasm forms an envelope 10–75 μm thick around the microsporidian mass and lacks both Nissl's granules and neurofibrils. Ultrastructurally it shows little more than small granules, vesicles and mitochondria (Loubès, Maurand and Ormières, 1979). Although these authors observed several host cell nuclei, the ganglion cell nucleus usually remains single but becomes enlarged, lobed and poor in chromatin.

The compact parasitic mass is in no way separated from the rest of the cell. In small xenomas, the mass of parasites appears dark and consists of developmental stages and large, oval spores packed together within a chromophile substance containing neurofilaments. In advanced xenomas, the parasitic mass has, within the dark region, a pale region harbouring a different type of developmental stages and slender curved spores (Fig. 2.85). This pale area is sharply delimited from the outer region and may represent a large spore-filled vacuole (Loubès, Maurand and Ormières, 1979).

Figs. 2.84–2.85 *Spraguea lophii* infection in *Lophius americanus*. Fig. 2.84. A 48-cm-long fish, dissected to expose the brain, and masses of xenomas in ganglia of the trigeminal and facial nerves and in the first dorso-spinal ganglia (Photograph by Dr. S. Jakowska). Fig. 2.85. Xenomas in a cross section of a spinal ganglion; the clear halos around the xenomas may result from shrinkage during fixation; the xenomas have a darker periphery (containing oval spores) and pale core (containing slender spores); some xenomas are in the process of destruction by host cell defences; scale bar = 0.5 μm.

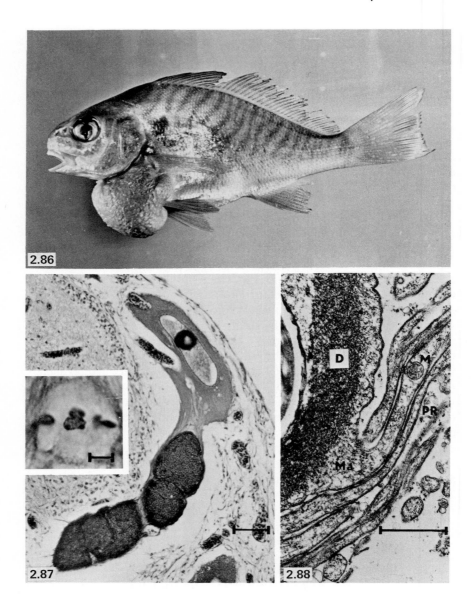

Considerable increase in connective tissue brings about separation of infected ganglion cells and nerve bundles and there is marked proliferation of glial cells. The body of the hypertrophic ganglion cell is ultimately severed from the region containing the parasites and the cell is thus functionally eliminated. Uninfected ganglion cells may perish from pressure atrophy induced by the huge xenomas and as a result of glial cell reaction when satellite glial cells destroy the ganglion cells.

Mature xenomas are ultimately destroyed by host tissue reactions similar to those against other microsporidian xenomas (Sect. IIB,1). The whole spore mass turns into a granuloma and sometimes spore masses are encapsulated by connective tissue (Fig. 2.10). In some cases, spores can be found within the nerves (Fig. 2.8). Complete repair of the infected region of the ganglion is not achieved: the ganglion turns into a tumour-like mass, consisting of destroyed ganglion cells, granulomas and hyperplastic connective tissue. Nevertheless the vital functions of the host are not noticeably impaired and infection is life-long. Tumours in the brain are responsible for only a very slight increase in the intracranial pressure within the rather voluminous skull.

COMMERCIAL IMPORTANCE: Recent studies indicate that the parasite has no deleterious effect on its host.

STRUCTURE AND LIFE CYCLE (based on Loubès, Maurand and Ormières, 1979 unless stated otherwise): The initial stage of infection is unknown. Small uninucleate cells inside vacuoles up to 8 μm in diameter (Fig. 2.87) may possibly be early developmental stages (Lom, unpublished).

DEVELOPMENT IN THE PERIPHERAL ZONE OF THE XENOMA ("Nosemoides" type) (Fig. 1.7)

Fig. 2.86 Spot, *Leiostomus xanthurus,* infected with *Ichthyosporidium giganteum:* the ventral bulge is caused by a compact parasitic mass, 20 × 20 mm wide (Photograph by Dr. F. J. Schwartz).

Fig. 2.87 Part of a transverse section of the medulla oblongata of *Lophius americanus,* showing a hypertrophic neurone harbouring *Spraguea lophii:* the two parasitic masses lie in a region of the cell distinct from that containing the enlarged host cell nucleus; H & E; scale bar = 100 μm. The inset shows a vacuole in the neurone cytoplasm, containing uni- and binucleate parasites, which may represent initial stages of the life cycle; H & E; scale bar = 5 μm.

Fig. 2.88 Portion of the xenoma wall of *Ichthyosporidium giganteum* with microvillus-like processes (PR): M–mitochondria, D–dense granular zone, Ma–cytoplasmic matrix; scale bar = 0.5 μm. (Reproduced from Sprague and Vernick, 1968, with permission of authors and publishers.)

Merogony: Rarely observed, meronts with isolated nuclei have copious concentric filaments around the nuclei.

Sporogony: Multinucleate sporogonial plasmodia with isolated nuclei cleave radially into several sporoblasts.

Spores (fresh) (Fig. 2.23L): Ovoid, uninucleate, 4.2 × 2.5 μm, rather asymmetric and with one side more vaulted. Posterior vacuole is globular or with a rather straight anterior limit. There is a distinct PAS positive polar cap (Lom, unpublished), a polaroplast encircled by an electron dense zone and there are 5–6 coils of the polar tube. The external surface has fine ridges (Lom and Weiser, 1972). Sometimes there are macrospores of double size. Weidner (1972) and Loubès, Maurand and Ormières (1979) provided other details on the spore ultrastructure.

DEVELOPMENT IN THE CENTRAL REGION ("Nosema" *type*) (Fig. 1.6)

Merogony: Meronts have one to several diplokarya.

Sporogony: Diplokaryotic sporonts produce 2 sporoblasts by binary fission.

Spores Fresh: (Fig. 2.23K, 2.45I): Curved, bean-like, rarely straight, 3.7 × 1.4 μm, posterior end is broader. Posterior vacuole occupies ⅓ to ½ of the spore length, has a rounded anterior border and almost no cytoplasmic side-walls (Lom, unpublished). There is a diplokaryon and 3–4 coils of the polar tube.

TRANSMISSION: Weissenberg (1976) speculated that the spores reach the anglerfish within some prey fish and that the sporoplasm released from the spore may migrate from the gut along the nerve axons into the ganglion cells. He failed (1911a) to infect *Lophius piscatorius* and *Gobius paganellus* by feeding them parts of infected ganglia.

REMARKS: Although the life cycle of *S. lophii* is still incompletely known, Loubès, Maurand and Ormières (1979) threw some light on the puzzle of the spore dimorphism. Rather than assuming that the anglerfish are infected by 2 simultaneously occurring different species, they suggested that *S. lophii* might be a dimorphic species in which binucleate spores ("Nosema" type) are produced in an asexual phase of the life cycle and the oval, uninucleate spores ("Nosemoides" type) in a sexual phase. They reached this conclusion by comparison with other dimorphic species of microsporidia and because the two developmental series occurred consistently within the same cysts.

I. Genus *Ichthyosporidium* Caullery and Mesnil, 1905

The recognition of this genus as a microsporidium and not a lower fungus resulted from studies of Sprague (1965, 1966); the fungi once supposed to belong to this genus are properly accommodated within the genus *Ichthyophonus*.

Ichthyosporidium giganteum (Thélohan, 1895) Swarczewsky, 1914
(Figs. 2.23J, 2.45H, 2.86, 2.89–2.93)

SYNONYMS: *Glugea gigantea* Thélohan, 1895; *Nosema giganteum* (Thélohan, 1895); *Ichthyosporidium phymogenes* Caullery and Mesnil, 1905; *Plistophora labrorum* Le Danois, 1910; *Plistophora gigantea* (Thélohan, 1895) Swellengrebel, 1911; *Thelohania gigantea* (Thélohan, 1895) Caullery, 1953; *Ichthyosporidium* sp. Schwartz, 1963.

HOSTS: *Crenilabrus melops* recorded by Thélohan, 1895; Caullery and Mesnil, 1905; Le Danois, 1910; Swellengrebel, 1911; Mercier, 1921. *Crenilabrus ocellatus* recorded by Swarczewsky, (1914). *Leiostomus xanthurus* recorded by Schwartz (1963). Marine.

GEOGRAPHICAL DISTRIBUTION: In *C. melops,* along the Atlantic coast of France and along the coast of Holland; in *C. ocellatus* in the Black Sea at Sevastopol, U.S.S.R. In *L. xanthurus,* the prevalence in 2 landings in Chesapeake Bay (Atlantic Coast, U.S.A.) was 25 and 13% (Schwartz, 1963).

SITE OF INFECTION: Subcutaneous connective tissue (all authors); fat tissue (Sprague, 1969); liver (Le Danois, 1910).

SIGNS OF INFECTION AND PATHOLOGY: A large, ventrally-located swelling (Fig. 2.86) which may extend from head to anal fin. It consists of hyperplastic host tissue, pervaded by xenoparasitic complexes of two types: small multilocular cysts (Fig. 2.89) made up of fibrous capsules, about 0.1–0.2 mm in size; and huge thick-walled xenomas up to 4 mm in size.

Around the small cystic forms (Dyková and Lom, 1980) there is a chronic inflammatory reaction (Fig. 2.89), characterised by diffuse influx of histiocytes, lymphocytes and of solitary multinucleate cells. The concentration of infiltrating cells is so high that it is impossible to recognise the character of the original tissue, whether it was subcutaneous connective tissue or fat.

The tissue reaction to the large lobed xenomas, which are filled with mature spores (Fig. 2.90), is a proliferative inflammatory response characterised by the presence of epitheloid cells and histiocytes. Bordering the xenoma wall, there is a layer of epitheloid cells in a palisade-like arrangement, their long axes being oriented perpendicularly to the xenoma wall. Ultimately, the xenoma wall may disappear and the proliferating granulation tissue grows into the mass of spores.

STRUCTURE AND LIFE CYCLE: The development of *Ichthyosporidium* and

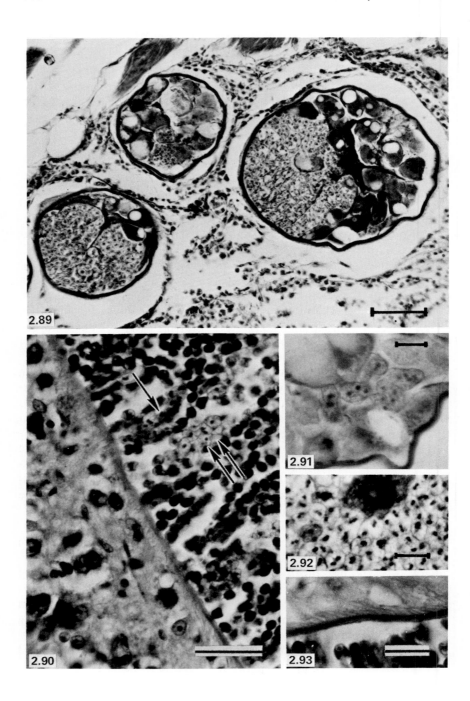

its relation to the formation of the xenoma complex are less well understood than in other vertebrate microsporidia. The following account is based on the material from *L. xanthurus* studied by Sprague (1969), Sprague and Hussey (1980) and Lom (unpublished), rather than on observations on the parasite in the original host as described by Swellengrebel (1912) and Swarczewsky (1914).

Cystic forms: These contain mainly meronts, the earliest stages are found in tissue, which Sprague and Hussey (1980) presumed to be a connective tissue reticulum, with numerous fibroblasts. These stages are tiny meronts measuring 1.5 μm, which divide by binary fission. The hypertrophic cytoplasm of several infected fibroblasts coalesces to form syncytial rounded bodies up to 20 μm. The peripheral zone of the infected cytoplasm later transforms into a fibrous capsule, walling off a parasitised island. A "syncytial" xenoma reults.

The fibrous capsules, first of irregular shape, later always globular, contain masses of parasites and clump-like remnants of host cell nuclei within altered host cell cytoplasm. The capsule becomes subdivided into compartments, some of which may contain degenerate host cell cytoplasm and nuclei, while others contain meront stages, singly or in pairs. Meronts have different appearances in separate compartments and it has not been determined whether they have single or diplokaryon nuclei. Some nuclei are up to 1 μm large, and some are dot-like. Around the groups of meronts, small vacuoles appear forming a honeycomb. The honeycomb and inner compartments later disappear, according to Sprague and Hussey (1980) leaving a mass of late stage meronts with diplokarya. Only exceptionally do the cystic stages include mature spores.

Mature xenomas: Contain mainly sporogonic stages and spores. These xenomas are lobose, possibly as a result of compression against one other. The way in which they arise from the small cystic xenomas is still enigmatic. According to Sprague and Vernick (1974), the wall has the ultra-

---

Figs. 2.89–2.93 *Ichthyosporidium giganteum.* Fig. 2.89. Encapsulated cystic stages containing compartments, within which there are cells of quite different appearance; the capsules are surrounded by tissue (the origin of which cannot be determined) showing chronic inflammatory reaction; H & E; scale bar = 50 μm. Fig. 2.90. Peripheral part of a large xenoma of *I. giganteum*, filled mostly with mature spores; between them there are some moniliform meronts (arrow) and clusters of sporonts (double arrow); nothing is left of the host cell cytoplasm; the outer face of the xenoma wall is surrounded by epithelial cells and, further away, by histiocytes; H & E; scale bar = 20 μm. Fig. 2.91. Small multinucleate plasmodia near the wall of the cystic form; H & E; scale bar = 10 μm. Fig. 2.92. Encapsulated stage with compartments containing parasitic cells apparently with diplokarya; H & E; scale bar = 10 μm. Fig. 2.93. The wall of a huge xenoma; its outer face (top) bears numerous microvillus-like projections; H & E; scale bar = 10 μm.

structure of a flattened cell, 1.0–1.5 μm thick, without any nuclei. It has an inner fibro-granular layer with no distinct inner boundary. On the outer surface there are microvillus-like projections (Fig. 2.88) also visible by light microscopy. These are also present on the surface of some of the small cystic forms.

In sporogony the nuclei are certainly diplokaryotic. Nuclear size increases, by an undetermined process, from their tiny size in the meronts, up to 2 μm in sporonts. Moniliform aggregations of sporonts, perpendicular to the xenoma wall, constrict into stages with considerably smaller nuclei. These in turn divide into sporoblasts with nuclei, which have reverted to the small size of those in meronts. Sprague and Hussey (1980) suggested that the fluctuation in nuclear size might indicate karyogamy followed by meiosis. Clusters of sporogonial stages can sometimes also be seen in the centre of the xenoma.

Spores (all in fixed state; Figs. 2.23J, 2.45H): From *C. melops* ovoid with a pointed anterior end, 5 × 7–8 μm (Thélohan, 1895); 4–5 × 5–6 μm (Swellengrebel, 1912); from *C. ocellatus:* ovoid, 3 × 5 μm; "macrospores", 4 × 7 μm (Swarczewsky, 1914); from *L. xanthurus:* ovoid, 4 × 6 μm (Schwartz, 1963).

Sprague and Vernick (1968b, 1974) found 27 and 25–27 coils of the polar tube in spores from *C. melops* and *L. xanthurus,* respectively.

TRANSMISSION: Thousands of simultaneously developing multilocular cysts led Sprague and Hussey (1980) to presume that autoinfection took place. Experimental infections of *C. ocellatus* by the oral route failed (Swarczewsky, 1914).

REMARKS: The early development of *Ichthyosporidium* is still unknown. The course of development as interpreted by Sprague and Hussey (1980) with multi-locular cysts arising as syncytial xenomas, and their transition into large xenomas is so strange that confirmation of those events is needed.

As far as it can be seen, parasites from all hosts (Sprague, 1977) are identical. In view of their wide distribution, reinvestigation should not be difficult.

Ichthyosporidium hertwigi Swarczewsky, 1914 (Fig. 2.94)

HOST: *Crenilabrus tinca*. Marine.
GEOGRAPHICAL DISTRIBUTION: Black Sea at Sevastopol, U.S.S.R.
SITE OF INFECTION: Connective tissue of the gills.
SIGNS OF INFECTION AND PATHOLOGY: Cystic structures on gill filaments up to 4 mm in size, containing rounded compartmentalized cysts 60–100

μm large. The parasite inflicts serious injury on the gill tissue of the host.
STRUCTURE AND LIFE CYCLE: Many features in common with *I. giganteum*. Early stages were uninucleate amoeboid stages, measuring 5–8 μm, located in groups within a nest of hypertrophic connective tissue cells, the whole being encased by layers of concentrically arranged fibroblasts and located within the overgrown connective tissue in the gills. The presumed life cycle is given in Fig. 2.94. Attempts at experimental infection were unsuccessful.

Spores: Measure 4.5–6 μm, larger than those of *I. giganteum*.

REMARKS: The life cycle is not understood and the specific independence of *I. hertwigi* needs to be established. The species has not been found since Swarczewsky's original and only description.

The arrangement of connective tissue cells and fibroblasts around the parasite is reminiscent of host defence reaction.

J. Genus *Mrazekia* Léger and Hesse, 1922

Mrazekia piscicola Cépède, 1924 (Fig. 2.46T)

HOST: *Odontogadus merlangus*. Marine.
GEOGRAPHICAL DISTRIBUTION: One fish from an unknown locality was dissected in France: it has not been found since.
SITE OF INFECTION: Pyloric caeca.
STRUCTURE AND LIFE CYCLE

Merogony and Sporogony: Present but not described.

Spores: Cylindrical, tapering posteriorly, measuring 20 × 6 μm and with a longitudinally striated surface. The manubrium extends almost the entire length of the spore, before tapering into a fine, spirally coiled filament. The caudal projection at the posterior end of the spore equals the length of the spore.

K. Genus *Encephalitozoon* Levaditi, Nicolau and Schoen, 1923

The only record of a representative of this genus from fish is that of Jensen, Moser and Heckmann (1979) who found a rare infection by an *Encephalitozoon* sp. in skeletal muscle of California lizardfish, *Synodus lucioceps*. They attributed the parasite to this genus because there were uninucleate spores with corrugated exospores and 4.5 coils of the polar tube. No further details were given.

REMARKS: Information on the life cycle and type of cell invaded are urgently needed, as well as confirmation of the generic status of this parasite.

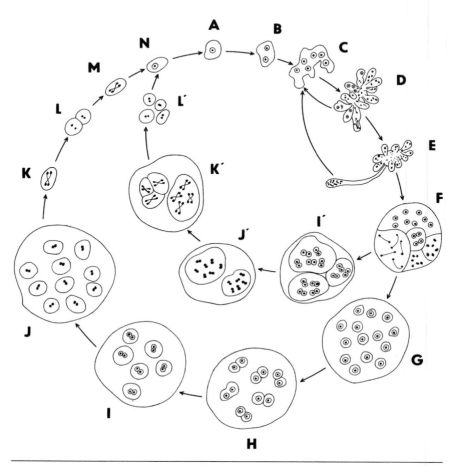

Fig. 2.94 Diagrammatic representation of the developmental cycle of *Ichthyosporidium hertwigi:* uninucleate amoeboid stages (A) (perhaps sporoplasms hatched from spores) grow into lobose plasmodia (B,C) which multiply by pinching off parts of the plasmodium (D,E); such lobose plasmodia encyst (F) while nuclear division continues; after some time, these divisions stop and further development can take place in two ways. In some cases, the plasmodium splits into numerous gametes (G) which fuse ("copulate") (H,I) to produce cells with two nuclei (J); each of these cells divide into two sporoblasts after meiosis (K,L,M); thereafter the sporoblast matures in one spore (N). In other cases, the paired arrangement of nuclei occurs within the plasmodium without the formation of gametes (I') and is followed by similar nuclear processes (J',K',L') as in I to M; again only after completion of meiosis are the sporoblasts produced (L') which mature into spores (N) (redrawn from Swarczewsky, 1914).

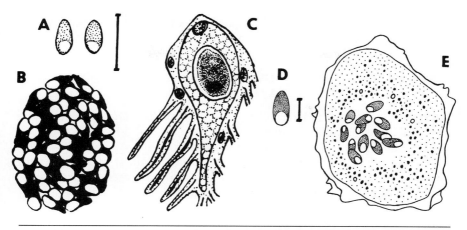

Fig. 2.95 A–C: *Microsporidium bengalis;* A, fresh spores, scale bar = 5 μm; B, part of the spore mass found in the "cyst"; C, transverse section through the gills showing "cyst" (redrawn from Weiser *et al.,* 1981). D,E: *Nosema marionis;* D, a fresh spore, scale bar = 5 μm; E, a young trophozoite of *Ceratomyxa coris* infected by *N. marionis* (redrawn from Thélohan, 1892).

L. Collective Group *Microsporidium* Balbiani, 1884

Microsporidium bengalis (Weiser, Kalavati and Sandeep, 1981) comb. nov. (Fig. 2.95A-C)

SYNONYM: *Nosema bengalis* Weiser, Kalavati and Sandeep, 1981
HOST: *Nemipterus japonicus*. Marine.
GEOGRAPHICAL DISTRIBUTION: Coast at Visakhapatnam, Gulf of Bengal, India. 24 out of 163 fish were infected.
SITE OF INFECTION: Gills.
SIGNS OF INFECTION AND PATHOLOGY: Minute spherical grey cysts, 0.5–0.8 mm; cyst walls, said to be of nuclear origin, were 2–3 μm thick.
STRUCTURE AND LIFE CYCLE
   Merogony: Not described.
   Sporogony: Kidney-shaped or oval, binucleate sporonts and oval sporoblasts were observed. Spores were encountered in groups, but no SPOVs were observed.
   Spores (fresh): Ovoid or ellipsoidal, 2 × 3 μm (pyriform when fixed). Posterior end with a large vacuole. Two "minute spherical nuclei".
REMARKS: The authors identified the species as a *Nosema,* perhaps on the basis of the binucleate spores. On the scanty evidence presented it seems preferable to allot the species temporarily to the collective group

Microsporidium. To date, no microsporidia invading fish are unequivocally identified as belonging to the genus *Nosema*.

Microsporidium cotti (Chatton and Courier, 1923) comb.n. (Figs. 2.6D, 2.96A,B)

SYNONYMS: *Nosema cotti* Chatton and Courier, 1923; *Glugea cotti* (Chatton and Courier, 1923) Sprague, 1977.
HOST: *Taurulus* (= *Cottus*) *bubalis*. Marine.
GEOGRAPHICAL DISTRIBUTION: French Atlantic coast at Roscoff; the prevalence of infection was 5%. Shulman (1957) reported its occurrence in the eastern part of the Baltic Sea but did not name the host.
SITE OF INFECTION: Testis.
SIGNS OF INFECTION AND PATHOLOGY: The lesions are white bodies up to 0.7 mm. The microsporidium develops within a xenoma in the testicular tissue. The hypertrophic host cells apparently float freely inside a small pocket filled with a serose fluid. A fully developed xenoma is invested

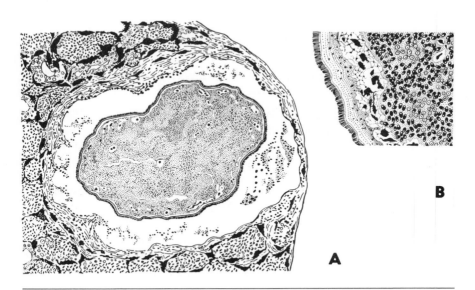

Fig. 2.96 *Microsporidium cotti:* A, the xenoparasitic complex, formed within the testis of *Cottus bubalis,* "floats" in a cavity formed by necrosis of the testicular tissue; a heavy pigment deposit is seen in the testicular interstices. B, section through the periphery of the xenoma showing the brush-like microvilli at the surface and an ectoplasmic layer of host cytoplasm subtended by a layer of host nuclei; the interior is filled with mature spores and islets of developmental stages.

with a brush border, 10–15 μm thick with a basal layer of granulation, subtended by a zone of clear cytoplasm, 10–20 μm thick. Beneath it, there is a layer containing fragments of the host nucleus, which anastomose to form a basket covering the central region. The latter is filled with developmental stages and mature spores of the parasite (Fig. 2.96).

Chatton and Courier (1923) suggested that the xenoma started with a cluster of cells of mesenchyme origin, pervaded by the microsporidium and that the cluster was later encapsulated by a layer of fibroblasts. The cells in the cluster were thought to fuse into one large mass, then to develop the brush border. They thought that the xenoparasitic complex secreted a fluid which digested the fibroblast barrier and left a pocket filled by the fluid and a liquefied capsule of fibrocytes.

STRUCTURE AND LIFE CYCLE: The only comment made on the life cycle is that each sporoblast gave rise to a single spore.

Spores (not stated if fresh): Ovoid, 8–10 μm long.

REMARKS: The illustration of a cluster of infected cells which Chatton and Courier (1923) thought was an early stage of development suggests rather an encapsulated mass of phagocytes packed with ingested spores, as would result from a typical cellular infiltration. This view is supported by the fact that cells in the early stages of infection would hardly be packed with mature spores. Also the uptake of spores into cells of the fibroblast capsule as depicted, is reminiscent of uptake by fixed macrophages.

Sprague (1977) transferred this species to the genus *Glugea* on the basis of the structure of the mature cyst. Even if one assumes the usual unicellular origin of the xenoma instead of the reported multicellular origin the parasite differs from *Glugea* and is unique in its possession of a brush border and situation in a fluid-filled lacuna. These features, if confirmed, might warrant the establishment of a new genus but, in the absence of information on the development of the parasite itself, a definition would be incomplete and we have thus made use of the collective genus *Microsporidium*.

Microsporidium girardini (Lutz and Splendore, 1903) Sprague, 1977

SYNONYM: *Nosema girardini* Lutz and Splendore, 1903.

HOST: *Girardinus caudimaculatus*. Freshwater. Collected near Sao Paolo, Brazil.

SITE OF INFECTION: Skin, muscles, serosa and mucosa of the intestine.

STRUCTURE

Spores (not stated if fresh): Pyriform, 2–2.5 × 1–1.5 μm.

Reidentification with certainty is very unlikely; it can be regarded as a *nomen nudum*.

Microsporidium ovoideum (Thélohan, 1895) Sprague, 1977 (Figs. 2.41D, 2.97)

SYNONYMS: *Glugea ovoidea* Thélohan, 1895; *Nosema ovoideum* (Thélohan, 1895) Labbé, 1899.

HOSTS: *Motella tricirrata* recorded by Thélohan (1895); *Cepola rubescens* recorded by Raabe (1936). Reimer and Jessen (1974) claimed to have found this species in *Merluccius hubbsi*. J. Martinez (personal communication, 1974) found a parasite which is possibly identical in *M. gayi*.

GEOGRAPHICAL DISTRIBUTION: Roscoff, Brittany coast of France in *M. tricirrata;* common in *C. rubescens* at Banyuls, in *M. barbatus* at Marseilles and Monaco, Mediterranean coast of France; in *M. hubbsi* off the Patagonian coast and in *M. gayi* on the coast of Peru.

SITE OF INFECTION: Liver.

SIGNS OF INFECTION AND PATHOLOGY (based essentially on Raabe, 1936): The parasite is found in white cyst-like spots, 1–1.5 mm in diameter, on the surface and deeper in the liver. Martinez (1974, personal communication) observed whitish spots in the liver, up to 0.45 mm, packed with spores.

The parasite develops within slightly enlarged liver cells, in which the nucleus is flattened and pushed aside to adhere to the cell membrane (Fig. 2.97). Doflein (1898) mentioned two nuclei in the host cell. Ultimately, the host cell is completely filled with spores (up to 30 in number).

In the liver, there were also haemorrhagic spots, which were actually necrotic foci of disintegrated liver cells, erythrocytes and infiltrated lymphocytes. Infected liver cells which contained vegetative stages or spores and macrophages packed with ingested spores were also present in the foci. The infection, which could be so extensive, that it reminded Raabe

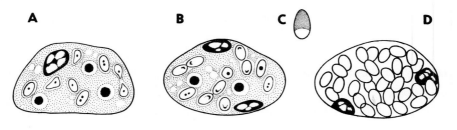

Fig. 2.97 Liver cells from a cyst infected with *Microsporidium ovoideum*: A, binucleate and uninucleate meronts, amoeboid or ellipsoid in shape, and three "globular corpuscles"; B, first division of binucleate meronts; the host cell has two nuclei; C, mature spore (after Thélohan, 1895); D, binucleate host cell, filled with mature spores (redrawn from Raabe, 1936).

(1936) of toxic degeneration of liver, was responsible for the death of infected fish which perished without any external symptoms. Raabe did not elaborate on whether necrotic foci and cysts represented two separate types of host reaction to the infection or two phases of one reaction process.

STRUCTURE AND LIFE CYCLE (based on Raabe, 1936)

Merogony: The earliest stages are uninucleate amoeboid cells, 1.2–2 μm, which divide by binary fission, sometimes forming chains of 2–3 individuals. As a prelude to sporogony, meronts transform into elliptical cells, measuring 2.5 μm, with 2 nuclei. From the description it is not clear whether the nuclei are in diplokaryon arrangement or isolated.

Sporogony: Binucleate stages divide into two sporonts each with a nucleus consisting of several granules. Each sporont produces 2 uninucleate sporoblasts, which transform into spores.

Spores (probably fixed, Fig. 2.41D): Ovoid, 1.5 × 2.5 μm, with a large rounded posterior vacuole (Thélohan, 1895; Raabe, 1936).

REMARKS: Raabe, on account of the presence of white spots on the infected liver and of the identical spore size, identified the species, perhaps correctly, with Thélohan's *G. ovoidea*. However, he reassigned it to the genus *Nosema*. In practice, it is neither a *Glugea* (absence of multinucleate stages and of host cell hypertrophy), nor a *Nosema* (no diplokarya in the sporoblast). Sprague (1977) pointing out the inadequacy of the existing data assigned this species (or possibly 2 species, Thélohan's and Raabe's) into the collective group *Microsporidium*. It seems that *M. ovoideum* is a fairly frequent parasite and could be reexamined in the near future.

Microsporidium peponoides (Schulman, 1962) Sprague, 1977

SYNONYM: *Pleistophora peponoides* Schulman, 1962.

HOST: *Percottus glehni*. Freshwater.

GEOGRAPHICAL DISTRIBUTION: Amur river basin, U.S.S.R.; 2 out of 15 fish examined were infected (Shulman, 1962).

SITE OF INFECTION: Subcutaneous connective tissue beneath the skin (Shulman, 1962) or fins (Vinichenko, Zaika, Timofeev, Shtein and Shulman, 1971).

SIGNS OF INFECTION: Spherical "cysts" up to 1 mm in diameter. Shulman (1962) considered the cysts to have been derived from a single host cell.

STRUCTURE AND LIFE CYCLE: The only information is that there are SPOVs containing 8 spores.

Spores (probably fixed): Anterior pole highly tapered, resembling a gourd or flask. 3.6–4 × 2–2.3 μm.

REMARKS: In spite of the explicit statement of 8 spores being formed within each SPOV, Sprague (1977) transferred the species to the collective group *Microsporidium* on the basis that "the production of cell hypertrophy tumours *(sic!)* containing octosporous pansporoblasts is not characteristic of *Pleistophora,* or any other established genus". He did not consider the genus *Thelohania* and we prefer to leave it in the collective group until future reexamination provides more information.

Microsporidium pseudotumefaciens (Pflugfelder, 1952) comb.n.
(Fig. 2.98)

SYNONYM: *Glugea pseudotumefaciens* Pflugfelder, 1952.
HOSTS: Aquarium fishes of the genera *Xiphophorus, Lebistes, Brachydanio* and *Gambusia*. Named species were *Platypoecilus maculatus* var. *pulchra, Molliensia sphenops, Colisa lalia*. Freshwater.
GEOGRAPHICAL DISTRIBUTION: From aquaria in Stuttgart (Germany); quite common in Pflugfelder's fish stocks.
SITE OF INFECTION: Ovarian follicles and other organs, including liver, kidney, spleen and body cavity.
SIGNS OF INFECTION AND PATHOLOGY: Parasites are located in cells of "residual bodies" ("Restkörper"), consisting of concentrically arranged cell layers. In the ovary, parasites lie within the "mesenchymal" cells.

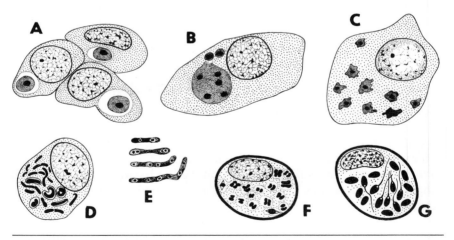

Fig. 2.98   Life cycle of *Microsporidium pseudotumefaciens:* A, amoeboid "germs" (possibly meronts) in host cells of the "residual body"; B, merogony; C, young meronts dispersed throughout the host cell cytoplasm; D, the parasites assume rod-like shapes; E, chains of parasites (cylindrical meronts); F, formation of binucleate "pansporoblasts"; G, mature spores in a host cell with a thickened surface membrane. The spores have partially extruded polar filaments (redrawn from Pflugfelder, 1952).

As a result of intense growth of the parasites, the host cells often disintegrate into a detritus found in the centre of residual bodies.

The host reacts by encapsulating the cells which harbour cylindrical meronts but not those containing sporogonic stages. The disease may cause considerable losses in the fish stocks.

STRUCTURE AND LIFE CYCLE (interpreted from Pflugfelder, 1952)

Merogony: Early amoeboid or rounded meronts are located in cells of the residual bodies within a vacuole (Fig. 2.98A), which disappears when the meronts become large and multinucleate (Fig. 2.98B). Large meronts divide into about 10 amoeboid cells (Fig. 2.98C) which transform into cylindrical cells with up to 6 nuclei (Figs. 2.98D,E). These cells divide by binary fission.

Sporogony: Cylindrical meronts produce cells which round off as sporonts and divide into two sporoblasts (Fig. 2.98F). Meanwhile the host cell becomes invested by a thick membrane.

Spores: No data on size or structure were given.

Some nodules were subcutaneous and spores from those were released from open skin ulcers.

REMARKS: Pflugfelder's (1952) is the only finding of the parasite; his statement that most of the stages were discernible only with difficulty and his picture of stages transforming from amoeboid to rod-like structure makes it doubtful whether all the stages described were really microsporidia. The life cycle does not accord with our knowledge of *Glugea*. We are transferring it provisionally to the collective group *Microsporidium*. The parasite may not even belong to the phylum Microspora.

Microsporidium rhabdophilia Modin, 1981 (Fig. 2.99)

HOSTS: *Oncorhynchus tschawytscha* (type host), *O. kisutch, Salmo gairdneri gairdneri* and strains of domesticated rainbow trout, *Salmo gairdneri irideus*. Freshwater.

Fig. 2.99 *Microsporidium rhabdophilia:* group of spores (redrawn from Modin, 1981).

GEOGRAPHICAL DISTRIBUTION: Arcata, California, U.S.A.

SITE OF INFECTION: Nucleus of rodlet cells in skin, gills and intestine.

STRUCTURE AND LIFE CYCLE

Spores: Observed in clusters of 16 (possibly these were SPOVs). Fresh spores were subcylindrical, slightly curved, 2.9 (2.6–3.5) × 1.1 (0.8–1.2) μm and had a posterior vacuole which occupied ⅓ of the spore length. Fixed spores, in histological sections, appeared spherical with a diameter of 1.48 μm.

REMARKS: Fixation-induced change in spore shape and insufficient description make a reexamination essential.

Microsporidium sauridae (Narasimhamurti and Kalavati, 1972) Sprague, 1977 (Fig. 2.46S)

SYNONYM: *Nosema sauridae* Narasimhamurti and Kalavati, 1972

HOST: *Saurida tumbil.* Marine.

GEOGRAPHICAL DISTRIBUTION: Found throughout the year in fish collected at Visakhapatnam, Andhra, India.

SITE OF INFECTION: Visceral muscles.

SIGNS OF INFECTION AND PATHOLOGY: Whitish cysts, 1–2 mm in diameter, were attached to visceral muscles and were easy to dislodge. The cyst, possibly a xenoma, had a wall with an outer, muscular and an inner, connective tissue layer. The light infection caused no apparent injury to the host.

STRUCTURE AND LIFE CYCLE

Merogony: Early stages were found within connective tissue cells. Advanced stages were said to be intercellular which probably meant that they lay in the lumen of the cyst.

Sporogony: Uninucleate sporonts in the lumen of the cyst were pyriform, 1.6 × 2 μm, lay in long chains and gave rise to a single spore.

Spores (fresh, Fig. 2.46S): Pyriform, 1.8–2 × 2.3–3.8 μm in size, allegedly without a clear posterior vacuole. A single, laterally located nucleus and a small, dot-like PAS positive polar cap.

REMARKS: The presence of a single nucleus in sporonts and spores and the formation of a large "cyst" is justification for Sprague's (1977) exclusion of the parasite from the genus *Nosema* and its transfer to the collective group *Microsporidium*. It is impossible to assign it to any established genus. The original Fig. 2 H, a line-drawing of a section through a mature "cyst" is reminiscent of a *Glugea*-type xenoma, in the process of destruction by tissue reaction of the host.

Microsporidium sciaenae (Johnston and Bankroft, 1919) comb.n.

SYNONYM: *Plistophora sciaenae* Johnston and Bankroft, 1919.
HOST: *Sciaena australis*. Freshwater.
GEOGRAPHICAL DISTRIBUTION: Brisbane river, Central Queensland, Australia.
SITE OF INFECTION: Connective tissue covering the ovary.
STRUCTURE AND LIFE CYCLE: Myriads of pyriform spores, 3–5 × 2–3 μm in size, filled cysts on the ovary. Cyst diameter was approximately 0.8 mm, as calculated from the published micrograph. The cyst protruded into the ovary from the surface layer of connective tissue.
REMARK: As distinct SPOVs were neither illustrated nor described, the species does not accord with the genus *Pleistophora*. It may belong to *Glugea,* but in the absence of positive data and since *Glugea* spp. also have SPOVs, the species is transferred to the collective group *Microsporidium*.

Microsporidium seriolae Egusa, 1982 (Figs. 2.101–2.103)

SYNONYM: *Pleistophora* sp. Ghittino, 1974
HOST: *Seriola quinqueradiata*. Marine.
GEOGRAPHICAL DISTRIBUTION: Common in Japanese cultures of this host.
SITE OF INFECTION: Lateral skeletal muscles.
SIGNS OF INFECTION AND PATHOLOGY: Several small or large depressions

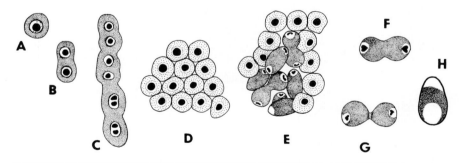

Fig. 2.100 *Microsporidium takedai:* some developmental stages of the life cycle. A, an early stage; B,C, development into a cylindrical meront; D, sporonts produced by the division of a meront; E, a group of sporonts in which some of the cells undergo a final division to produce the sporoblasts; F,G, sporont division; H, mature spore (redrawn from Awakura, 1974).

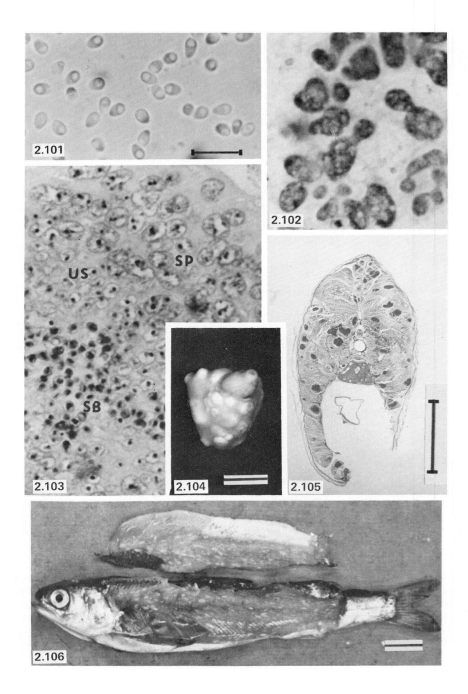

on the lateral body surface. When dissected, the depressed area revealed large, macroscopic masses, lobate in cross section and elongate in longitudinal section. These masses represented regions of disintegrated muscle tissue and consisted of a cheese-like matrix, in which there were developmental stages and spores, bounded by a fibrous capsule. Heavy infections produced the "Beko" disease of *S. quinqueradiata*.

COMMERCIAL IMPORTANCE: The disease endangers cultures of this important commercial fish in Japan.

STRUCTURE AND LIFE CYCLE

Merogony: Not described.

Sporogony: (Figs. 2.102, 2.103): multinucleate sporogonial plasmodia produce sporoblasts by multiple fission, sporogony vacuoles, SPOVs and diplokarya were reportedly absent.

Spores (fresh, Fig. 2.101): Ovoid, sometimes slightly bent and pyriform, 3.3 (2.9–3.7) × 2.2 (1.9–2.4) μm, with a spherical posterior vacuole which does not reach to the mid-spore position. There are 6 coils of the polar tube.

REMARKS: The stated absence of SPOVs excludes this parasite from the genus *Pleistophora* but this character should be reexamined.

Microsporidium takedai (Awakura, 1974) comb.n. (Figs. 2.100, 2.104–2.106)

SYNONYMS: *Plistophora* sp. Takeda, 1933, *Plistophora* sp. Awakura, Kurahashi and Matsumoto, 1966; *Glugea takedai* Awakura, 1974; *Nosema takedai* Miki and Awakura, 1977.

HOSTS: *Salmo gairdneri irideus, Oncorhynchus masou, O. keta, O. nerka* var. *adonis, O. tschawytscha, O. gorbuscha, Salvelinus leucomaenis, S. malura malma*. All freshwater.

Figs. 2.101–2.103 *Microsporidium seriolae.* Fig. 2.101. Fresh spores. Fig. 2.102. Giemsa stained sporogonial plasmodia from a dry smear. Fig. 2.103. Section through a part of a cyst showing some sporogony stages; SP, sporogonial plasmodia; US, uninucleate stages; SB, sporoblasts; scale bar = 10 μm in Figs. 2.101–2.103. (Reproduced from Egusa, 1982, with permission of author and publisher.)

Figs. 2.104–2.106 *Microsporidium takedai.* Fig. 2.104. Hypertrophy of the heart of *Oncorhynchus nerka* var. *adonis,* with opaque parasitic nodules. Fig. 2.105. Cross section of infected *Salmo gairdneri irideus,* with cysts of the microsporidium appearing as black spots scattered in the muscle tissue. Heidenhain's haematoxylin. Fig. 2.106. Spindle shaped, opaque white cysts in the trunk muscles of *O. nerka* var. *adonis;* scale bar = 4 mm in Figs. 2.105 and 2.106 and 1 cm in Fig. 2.107.

Geographical distribution: Chitose River, Lake Tokito Nusua and Lake Akkan, Hokkaido Island, Japan (Awakura, 1978). In *S. gairdneri irideus,* the prevalence was 70–100%; in *O. masou, O. keta* and *S. leucomaenis,* it was 87, 95 and 86%, respectively (Awakura, 1974).

Site of infection: In chronic cases only heart muscle was affected but in acute cases it was also in skeletal muscles and fin, masticatory, eye, throat and gullet muscles.

Signs of infection and pathology: Cyst-like bodies in the trunk musculature, which sometimes showed through the skin. In heart muscle cysts were globular (Fig. 2.104), less than 2 mm in size. In skeletal muscles (Fig. 2.105) they were spindle-shaped (Fig. 2.106), 2–3 × 3–6 mm in size.

The type of cell which the parasite infected was not clear. The parasite multiplies within the cysts, in a cytoplasmic mass which lacks host cell nuclei, xenoma wall or any other distinct boundary, so that Miki and Awakura (1977) were under the impression that the parasites were extracellular.

In chronic cases with a low mortality, cystic formation in the heart caused hypertrophy and deformation. The tissue response included infiltration of inflammatory cells, phagocytosis, hyperplasia of the connective tissue and inflammatory oedema, which resulted in destruction of parasites, whole cysts and formation of granulomas.

Acute cases caused high mortality and were characterized by a massive occurrence of cysts in the trunk musculature; in *O. keta* fry, there were up to 130 cysts per gram of muscle. The tissue response, similar to that observed in the heart muscles, was noticeable in older fish.

In both chronic and acute cases there was an increase of protein in the serum and decrease in the albumin/globulin ratio which led Awakura (1974) to suppose that there was production of antibodies.

Commercial importance: The disease resulted in heavy losses of cultured salmonids, notably of rainbow trout.

Structure and life cycle (based on Miki and Awakura, 1977)

Merogony: Uni-, bi- and tetranucleate meronts produce multinucleate cylindrical meronts, all located at the periphery of the cyst (Fig. 2.100A–C).

Sporogony: Sporonts transform directly into sporoblasts and spores. There are no SPOVs and spores are in close contact with the host cell cytoplasm (Fig. 2.100D–G).

Spores (fresh, Fig. 2.100H): Ellipsoidal to ovoid, 3.4 (2.8–3.9) × 2 (1.7–2.3) µm. The polar tube is attached subapically, anteriolaterally and makes 4 coils.

Transmission: Awakura, Kurahashi and Matsumoto (1966) believed that other water organisms were the source of infection for the fish. Awakura

(1974) thought that he had demonstrated the role of planktonic organisms and pearl-mussel glochidia in the transmission. However, the organisms that he found in rotifers and glochidia do not appear to be microsporidia and the role of intermediate hosts has yet to be proven.

Direct transmission of infection was achieved by oral administration of the spores and by exposure of fish to water contaminated with spores (Awakura, 1974).

COURSE OF INFECTION: In rainbow trout yearlings, meronts appeared in the heart muscle 10 days after peroral infection. Inflammatory cells appeared on day 11. Spores were detected on day 13, and phagocytes appeared on day 17. Cyst-like bodies in the trunk musculature appeared on day 24 and the number of phagocytes dwindled to a minimum on day 30. In fingerlings, developmental stages appeared by day 7, but in 2-year-old trout developmental stages reached lower levels of intensity and were of shorter duration, while the phagocytic response was more pronounced. The fishes which survived the infection acquired "a remarkable immunity" to reinfection, lasting as long as one year.

Temperature plays a pivotal role. At 18°C the parasites grew well. Transfer of infected hosts to colder water, at 8°C, stopped the growth and the microsporidia were restricted to the heart muscle.

TREATMENT: The survival rate of infected fish, treated with sulfadimethoxine, sulfamonothoxine, nitrofurasone and amprolium, was increased but merogony was prevented only by amprolium administered daily during a period of up to 48 days up to the amount of 0.06% of body weight. However, at this level amprolium is toxic and actually reduces the survival rate of treated fish. It is, therefore, of no practical value (Awakura and Kurahashi, 1967).

REMARKS: The absence of SPOVs and lack of a true wall to the cysts precludes the genus *Pleistophora* and *Glugea* for this species and it cannot be assigned to any other genus known to invade fish. Future studies of this excellent experimental model should include an investigation of its taxonomic position.

Microsporidium valamugili (Kalavati and Lakshminarayana, 1982) n.comb. (Fig. 2.46R)

SYNONYM: *Nosema valamugili* Kalavati and Lakshminarayana, 1982.
HOST: *Valamugil* sp. Estuarine.
GEOGRAPHICAL DISTRIBUTION: Estuaries at Visakhapatnam and Bheemunipatmam, India.
SITE OF INFECTION: Intestinal wall.
SIGNS OF INFECTION: White cysts in the outer epithelial layer, which ap-

peared to represent hypertrophic host cells with the nucleus pushed aside. The spore mass was found in a vacuole inside the host cell.

STRUCTURE AND LIFE CYCLE

Merogony: The early meronts 3.4–4.2 μm, mature meronts ribbon-like, 3.2–4.6 × 16.5–18 μm with up to 16 paired nuclei. All merogony stages have diplokarya.

Sporogony: Early sporonts oval or bean-shaped, 5.2–6.4 × 9–10.5 μm with one or 2 diplokarya. No details of division were given.

Spores (Fig. 2.46R): Pyriform, with a pointed anterior and rounded posterior end, 3–3.2 × 5.4–6.2 μm (fresh). A large posterior vacuole, spherical dot-like PAS positive polar cap and 2 minute spherical nuclei side by side were also described.

REMARK: Cylindrical meronts and xenoma-like cysts suggest *Glugea;* however, presumably the authors considered that this species belonged to *Nosema* because of the diplokarya but there is no certainty that this genus occurs in fish.

Microsporidium sp. of de Kinkelin, 1980

This parasite was found in the side of the spinal cord of *Brachydanio rerio* (freshwater) obtained at Thiverval-Grignon, France. It caused deformities of the body—lordosis, spinal curvatures, impaired swimming and about 15% mortality. The only information on the parasite is that fixed spores are found in clusters and in stained sections measured 1.5–2 × 3–4 μm.

Microsporidium sp. of Gaievskaya and Kovaleva, 1975

Forms white cysts in muscles of *Micromessistius poutassou* (marine), collected from the Ireland shelf.

The prevalence was 12% in the years 1972–1974. Infected fish were unusable as food.

Microsporidium sp. of Gasimagomedov and Issi, 1970.

In the intestine and kidney of *Vimba vimba persae* (euryhaline) from the Kirov Bay of the Caspian Sea, U.S.S.R. One out of 15 specimens found infected. Spores were elongate ovoid, 2.4–2.5 × 5–6 μm.

Microsporidium sp. of Herman and Putz, 1970

SYNONYM: *Microsporidia* gen. sp. incert. of Herman and Putz, 1970
HOST: *Ictalurus punctatus* (freshwater) from Maryland, U.S.A.

SITE OF INFECTION: Heart muscle and intestinal submucosa.

SIGNS OF INFECTION: Xenomas, 110–220 μm, with a single, centrally located and hypertrophic nucleus, up to 30 μm large. The wall of the xenoma was extremely thin (0.2 μm).

STRUCTURE AND LIFE CYCLE: The developmental stages observed were rounded and occasionally cylindrical plasmodia, with 1–9 nuclei which reached 6–7 μm in diameter. These were dispersed throughout the xenoma. SPOVs were discernible.

REMARKS: Herman and Putz (1970) mistook the multinucleate plasmodia, which may be merogonial or sporogonial plasmodia, for SPOVs and concluded that the parasite might belong to the genus *Thelohania* or *Pleistophora*. However, the description of the xenoma, reminiscent of the type of xenoma found in *Glugea acerinae,* indicates that it could be a species of *Glugea*. The species is left within the collective group *Microsporidium,* as alloted by Sprague (1977).

Microsporidium sp. of Jones, 1979

Located around the pericardial cavity and brain nerve trunks of *Trachurus declivis* (marine) collected at Nelson beach, New Zealand. Cysts contain binucleate spores, measuring 1.8 × 4.7 μm. Jones (1979) supposed that the heavily infected mackerel washed ashore had succumbed to the infection.

Microsporidium sp. (Marchant and Schiffman, 1946) comb.n. (Fig. 2.23M)

SYNONYMS: *Nosema* sp. of Bazikalova in Polyansky, 1955; *Glugea* sp. of Marchant and Schiffman, 1946.

HOST: *Mallotus villosus*. Marine.

GEOGRAPHICAL DISTRIBUTION: Off the coasts of Newfoundland (Marchant and Schiffman, 1946; Templeman, 1948), the prevalence being 25% according to Vávra and Undeen (1979). Also from the Barents Sea (Polyansky, 1955).

SITE OF INFECTION: Body cavity, peritoneal epithelium, sometimes also ovary (Marchant and Schiffman, 1946) and adipose tissue surrounding the external blood vessels of the intestine (Vávra and Undeen, 1979).

SIGNS OF INFECTION: Ovoid, white, cyst-like xenomas up to 2 mm in length (Marchant and Schiffman, 1946). Unlike that of *Glugea anomala,* the xenoma lacks a laminar wall. According to Vávra and Undeen (1979) the surface of the hypertrophic cells seems to be a cell membrane coated with fibrils, which can be interpreted as possibly of collagen.

STRUCTURE AND LIFE CYCLE
Merogony: Not described.
Sporogony: Sporogonial plasmodium cleaves into sporonts by multiple fission, producing SPOVs with 16–30 spores per vesicle (Vávra and Undeen, 1979).
Spores (fresh, Fig. 2.23M): Elongate ovoid, sometimes slightly curved, 3.6 × 7.4 μm, with a large posterior vacuole. The PAS positive polar cap is located subapically (Vávra and Undeen, 1979). Marchant and Schiffman (1946) gave spore measurements as 3 × 6–6.7 μm.
REMARKS: This species differs in peripheral organisation of the xenoma from *Glugea,* but other features, including the SPOVs suggest this genus. Polyansky (1955) mentioned a *"Nosema"* from *M. villosus,* which may be the same species, although there are no morphological data to compare.

Microsporidium sp. of Noble and Collard, 1970

In the muscles beneath the dorsal side of the peritoneum of *Lycodopsis pacifica* (marine) collected at the Californian coast, U.S.A.; one infected specimen found. Forms relatively large cysts surrounded by a considerable proliferation of the connective tissue.

Microsporidium sp. of Plehn, 1924

In the ovary of *Salmo trutta trutta,* and *Salmo salar* (euryhaline) in southeastern Norway. Signs of infection were small dots in the ova, which turned completely white. The majority of ova were infected. Invaded eggs died off or gave rise to fry with an infected yolk sac.

Plehn (1924) considered it to be *Glugea anomala* or a closely-related species. There is a possibility that this is *Glugea truttae* Loubès, Maurand and Walzer (1981).

Microsporidium sp. of Raabe, 1935 (Fig. 2.107)

HOST: *Pleuronectes flesus.* Marine.
GEOGRAPHICAL DISTRIBUTION: Baltic Sea coast at Hel, Poland.
SITE OF INFECTION: Skin lesions caused by *Lymphocystis.*
STRUCTURE AND LIFE CYCLE
Merogony: Meronts rounded often with pseuopodia, give rise by binary fission to binucleate stages and later to multinucleate plasmodia, which divide by multiple fission. Later, ellipsoid, uninucleate stages were produced in chains.
Sporogony: Initiated by the ellipsoid cells which grew into plasmodia,

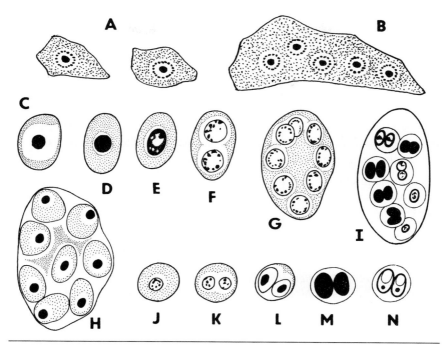

Fig. 2.107 Some stages of the life cycle of *Microsporidium* sp. of Raabe, 1935 from *Pleuronectes flesus:* A, amoeboid meronts; B, multinucleate plasmodia: C, meront with clear cytoplasm; D, meront with dark cytoplasm; E,F,G, nuclear division in the meront; H, formation of the eight sporoblast mother cells; I, transformation of each mother cell in two sporoblasts and into spores; J–N, separate stages of this transformation; M, two sporoblasts; N, two mature spores.

up to 20 × 12 μm, which transformed into stages like SPOVs which contained up to 8 "sporonts". These, which might correspond to sporoblast mother cells, were liberated from the vesicle envelope and underwent one more division to produce 2 spores per "sporont". Strangely enough Raabe (1935) only exceptionally found completely mature spores.

Spores (not stated if fixed): Elliptical, about 3.3 × 2 μm (calculated by the present author from the figures).

REMARKS: *Lymphocystis* disease is a viral disease causing tumour-like hypertrophy of epidermal cells, and affects various marine fishes. Accordingly Raabe's (1935) assumption that the microsporidium was the aetiological agent of cell hypertrophy in the lesions was probably erroneous. His description was evidently based on a concurrent microsporidian invasion of *Lymphocystis*-infected cells and his data must be taken with extreme caution.

Microsporidium sp. of Reimer, 1975

In "cysts" in the stomach wall of *Paralepis elongata* (marine). Personal communication of Dr. H. Reimer (1975).

Spores (Fixed and stained): Ovoid, average dimensions 3.1 × 1.5 μm, rarely up to 5 × 2 μm, with a large posterior vacuole.

In addition to these poorly described species listed as *Microsporidium* sp., findings of microsporidian parasites were briefly recorded from the following hosts (see also pages 18 and 20):

1. In the intestine of *Atherina mochon pontica* n. *caspia* from the Caspian Sea near Bagdesh, U.S.S.R. Two out of 15 specimens were infected (Gasimagomedov & Issi, 1970).

2. In the liver of one out of 24 specimens of *Blicca bjoerkna* from the Caspian Sea, Agrakhan Bay, U.S.S.R. (Gasimagomedov and Issi, 1970)

3. In *Scorpaena porcus* caught in the Adriatic Sea at Split, Yugoslavia (Jírovec 1938, unpublished)

4. In the liver of *Bairdiella chrysura* and in the gall bladder of *Gobiosoma bosci* collected at the Gulf of Mexico coast at Ocean Springs, Mississippi, U.S.A. (A. Lawler 1978, unpublished)

5. In the stomach wall of *Paralepis elongata* from the Atlantic Ocean (H. Reimer 1975, unpublished).

IV. MICROSPORIDIAN SPECIES ENCOUNTERED IN FISH HOSTS AS HYPERPARASITES OF MYXOSPOREA

For the sake of completeness two species of *Nosema* are listed, which are hyperparasites in the cytoplasm of Myxosporea infecting fish, since they could be mistaken for an infection of the fish host.

Nosema marionis (Thélohan, 1895) Labbé, 1899 (Fig. 2.95D,E)

SYNONYM: *Glugea marionis* Thélohan, 1895
HOST: Trophozoites of the myxosporean, *Ceratomyxa coris* Géorgevitch, living in the gall bladder of *Coris julis,* a marine fish.
GEOGRAPHICAL DISTRIBUTION: Mediterranean coast of France.
STRUCTURE AND LIFE CYCLE: Vegetative stages, according to Stempell (1919), resemble those of *N. bombycis.*
Spores (not stated if fresh): Elongate-ovoidal, 3 × 8 μm, with conspicuous posterior vacuole (Thélohan, 1895).

Nosema notabilis Kudo, 1939 (Fig. 2.41F)

HOST: Trophozoites of the myxosporean, *Ortholinea polymorpha* (Davis) living in the urinary bladder of *Opsanus beta* and *O. tau,* both marine.
GEOGRAPHICAL DISTRIBUTION: Atlantic coast of the U.S.A.
STRUCTURE AND LIFE CYCLE: Binucleate meronts divide repeatedly into 2 uninucleate forms, which give rise to binucleate fusiform stages, these later producing 2 binucleate sporoblasts.

Spores (fresh): Ovoid to ellipsoid, average 3.3×2 μm
REMARKS: Detailed description in Kudo (1944).

3. The Microsporidia of Amphibia and Reptiles

I. INTRODUCTION

Almost all the descriptions of microsporidia from amphibia and reptiles came during the early years of microsporidian study, at the end of the 19th century and beginning of the 20th. Many of the stages, particularly those described as gametes of 2 types, were clearly seen in the light of emerging information on other parasitic Protozoa and have not subsequently been confirmed. Total credance cannot be given to these early accounts and the original observations must be re-interpreted according to our knowledge of the group today.

All but two of the species described from these poikilothermic vertebrates produce spores in large groups within SPOVs and these can be attributed to the genus *Pleistophora*. Some of these species were originally placed in the genus *Glugea* but the absence of xenomas, formed by hypertrophy of host cells, as well as differences in the morphology of the meronts and division processes of the sporonts, preclude this genus. They parasitise muscle fibres and closely resemble, in their development and relationships with host cells, other species of the genus *Pleistophora* which infect fish. They are highly pathogenic in some hosts, the wasting of muscles leading to emaciation and death. Whether all are distinct species is difficult to tell without re-examination of material from different hosts and conducting cross infectivity tests. One species, *Pleistophora danilewskyi*, has been described from both amphibian and reptilian host species and, if this is correct, the broad host range suggests that parasites, named as distinct species from other amphibian and reptilian hosts, may be identical with *P. danilewskyi*.

Apart from the *Pleistophora* species, there are species of indeterminate genus from the connective tissue of *Triturus vulgaris* (Amphibia) and oocytes of *Rana pipiens* (Amphibia). A species infecting the gut epithelium of *Podarcis muralis* (Reptilia) belongs to the genus *Encephalitozoon* and is one of the few known representatives of this genus, of which the type species, *E. cuniculi* occurs in mammals.

II. DESCRIPTIONS OF THE SPECIES INFECTING AMPHIBIA AND REPTILES

A. *Pleistophora danilewskyi* Pfeiffer, 1895 (Figs. 3.1, 3.2)

SYNONYMS: Mikrosporidien Danilewsky, 1891; Mikrosporidien Pfeiffer, 1891; *Glugea danilewskyi* Pfeiffer, 1895, *Plistophora danilewskyi* (Pfeiffer, 1895) Labbé, 1899; *Plistophora danilewskyi* Guyénot, Naville and Ponse, 1925; *Microsporidium danilewskyi* (Pfeiffer, 1895) Canning 1976.

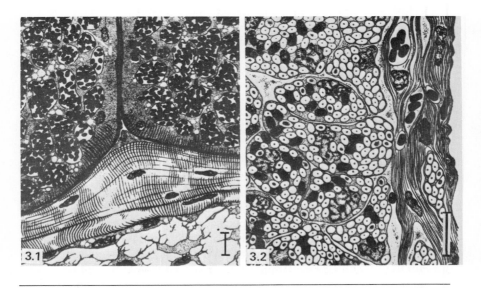

Fig. 3.1 *Pleistophora danilewskyi:* parasitised muscle fibres of grass snake *Natrix natrix,* showing numerous SPOVs lying directly in a matrix of disorganised myofibrils adjacent to normal muscle fibres. Scale bar = 10 μm. (Reproduced from Guyénot and Naville, 1922a.)

Fig. 3.2 *Pleistophora danilewskyi:* macrophages full of spores in the connective tissue of grass snake. Scale bar = 10 μm. (Reproduced from Guyénot and Naville, 1922a.)

Now that the genera *Pleistophora* and *Glugea* have been redescribed at the ultrastructural level, it is clear that this parasite does not have the essential characters of the genus *Glugea*. It has many features in common with the species of *Pleistophora* inhibiting the muscle of fish and it is now possible to assign it to this genus with reasonable certainty, in agreement with Labbé (1899).

HOSTS AND LOCALITIES: Frogs, lizards and turtles, according to Danilewsky (1891), in Poland. The turtles and frogs were further identified as *Emys orbicularis (= E. lutaria = Cistudo europaea)* and *Rana temporaria* by Pfeiffer (1891, 1895a) from material sent to him by Danilewsky.

Probably also *Lacerta* sp. and *Chalcides chalcides (= Seps chalcides = Chalcides tridactylus)* according to Labbé (1899).

Natrix natrix (= Tropidonotus natrix) according to Debaisieux (1919b) in Belgium and Guyénot and Naville (1922a) in Italy.

LESIONS: Whitish, spindle-shaped foci, a few mm long in the striated muscles, always elongated in the direction of the long axis of the muscle fibres. Danilewsky (1891) found them especially common in the posterior extremities ("hinteren Extremitaten") by which he probably meant hind legs. In grass snakes they were found in the intercostal and spinal muscles and even in tendons and intervertebral ligaments.

SPOVs and other stages lie directly in the cytoplasm of the muscle cell, in contact with degenerate myofibrils which occupy the cell periphery and the spaces between SPOVs (Fig. 3.1) (Debaisieux, 1919b; Guyénot and Naville, 1922a). A zone of connective tissue surrounding spore masses was correctly interpreted as a reactional zone by Debaisieux. Guyénot and Naville considered that development in connective tissue was distinct from that in the muscle cysts but the cells which they described, containing amoeboid stages and spores, can now be recognized as phagocytic cells with ingested parasites (Fig. 3.2).

STRUCTURE AND LIFE CYCLE

Merogony and sporogony: Debaisieux (1919b) and Guyénot and Naville (1922a) proposed life cycles which involved sexual phenomena of autogamy (Debaisieux) and true fusion of gametes (Guyénot and Naville). In the light of present knowledge of microsporidian development these interpretations can be discounted but there is sufficient evidence, from their illustrations and descriptions, to conclude that there are uninucleate and plasmodial merogonic stages, interspersed with SPOVs. The plasmodia may have up to 24 nuclei before division.

Sporogonial plasmodia in SPOVs appear to divide via multinucleate segments into sporoblasts but a direct multiple fission into uninucleate sporoblasts cannot be excluded. The binucleate stages (thought to be autogamous copulae by Debaisieux) could be sporoblast mother cells which divide by binary fission into sporoblasts or, more likely, are simply the final stages of sequential divisions of the plasmodium (as thought by Guyénot and Naville). The number of sporoblasts varies between 16 and 60.

Guyénot and Naville described macronuclear and micronuclear sequences of sporogony which, they postulated, represented sexual and parthenogenetic types of spore production respectively. They are probably only stages in a single sporogonic sequence, in which the nuclei become more compact as sporulation progresses.

Spores: Debaisieux was probably correct in believing that the spores have a single nucleus, rather than Guyénot and Naville who thought that the original nucleus divided into two during spore maturation. Spores are ovoid or pyriform with distinct anterior and posterior vacuoles. The anterior vacuole may represent the polaroplast and is probably only revealed by staining. Measurements of 2–4 μm (Pfeiffer, 1891) and 3–4 μm with macrospores measuring 6–7 μm (Debaisieux) were probably from sectioned material. Guyénot and Naville distinguished, on spore size, those which were derived from micronuclear (3 μm) and macronuclear sequences (4 μm). They gave the polar tube length as 50–70 μm.

Species Related or Identical to *P. danilewskyi*

1. *Pleistophora heteroica* (Moniez, 1887) Labbé, 1899

Seen by Vlacovich (1866) in a snake which was probably the black whip snake, *Coluber viridiflavus carbonarius*. It was recorded from many tissues as forming "cysts" containing 10, 20 or many spores which measured 6–7 × 3–3.5 μm. Pfeiffer (1895a) evidently thought this species was identical with *P. danilewskyi*, as he listed *C. carbonarius* as a host for *P. danilewskyi*. The fact that *P. heteroica* parasitises many tissues and has larger spores casts some doubt on this, but measurements may not have been made accurately in 1866.

2. *Pleistophora encyclometrae* (Guyénot and Naville, 1924) Canning, 1976

P. encyclometrae was established for a hyperparasite of trematodes, *Encyclometra bolognensis* and *Telorchis ercolanii*, inhabiting the stomach and intestine respectively of grass snakes, *Natrix natrix*. The snakes were themselves infected with *P. danilewskyi* and the authors found that there were no greater biometric differences between *P. danilewskyi* and the trematode parasite than between muscle and connective tissue forms of *P. danilewskyi*. Even in the face of similarities in development and spore structure and co-existence with *P. danilewskyi*, they chose to give it separate identity, naming it *Glugea encyclometrae*, a name which was later changed to *Nosema (Plistophora) encyclometrae* by Guyénot, Naville and Ponse (1925), indicating uncertainty of its generic status. A summary of the chief findings on this parasite has been given by Canning (1976a).

3. *Pleistophora ghigii* (Guyénot and Naville, 1924)

P. ghigii was first found by Guyénot, Naville and Ponse (1922) in plerocercoid larvae *(Plerocercoides pancerii* (= *Diphyllobothrium erinacei) and in connective tissue of the host snakes, Natrix natrix,* around the encysted cestodes. The microsporidium was later found (Guyénot and Naville, 1924) also in adult *Telorchis ercolanii* (= *Cercorchis ercolanii)* and in muscles of the snakes and was named *Glugea ghigii*. Guyénot, Naville and Ponse (1925), uncertain of its systematic position, used the name *Nosema (Plistophora) ghigii*. Although the spore size was given as 2 × 2.5 μm, circumstantial evidence suggests that, as with *P. encyclometrae,* the trematodes had become infected with *P. danilewskyi* from the snakes. A summary of the characters of this species has been given by Canning (1976a).

4. *Pleistophora myotrophica* (Canning, Elkan and Trigg, 1964) Canning, 1976 (Figs. 3.3–3.9)

SYNONYMS: *Plistophora* sp. Elkan, 1963; *Plistophora* sp. Canning and Elkan, 1963: *Plistophora myotrophica* Canning, Elkan and Trigg, 1964.
HOST AND LOCALITY: *Bufo bufo* in England.
LESIONS: Fusiform spaces, carved in the striated musculature of all regions of the body (Fig. 3.3), are packed with meronts, sporonts and SPOVs. Stages are in direct contact with the myofibrils, which are frayed at the edges of the lesions (Figs. 3.4).

The infections are slowly progressive: many of the toads collected in the locality of Hemel Hempstead, England and housed in crowded laboratory conditions, became emaciated and died over a period of about 2 years or more.

In late infections, phagocytes infiltrate into the lesions, break up the masses of spores and ingest them. The appearance of these infiltrating cells is similar to the cells said by Guyénot and Naville (1922a) to be connective tissue cells parasitised by *P. danilewskyi* (Sect. II, A). Regeneration of muscle in *P. myotrophica* infections takes place but does not keep pace with destruction.

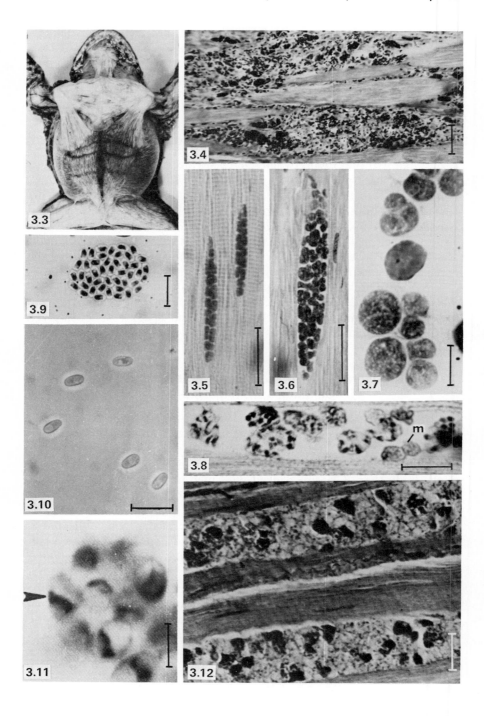

MORPHOLOGY AND DEVELOPMENT

Merogony: Meront stages with 1 to about 8 nuclei, which divide by binary or multiple fission or plasmotomy (Figs. 3.5, 3.6). Larger stages which divide by plasmotomy and give rise to clumps of parasites (Fig. 3.7) are also part of the merogonic cycle, these culminate in sporogonial plasmodia.

Sporogony: Sporogonial plasmodia with isolated nuclei (Fig. 3.8) give rise to groups of up to 100 sporoblasts and spores (Fig. 3.9). It is not known whether the division process is by multiple fission or by a series of sequential divisions. The SPOV was not observed around the spores by light microscopy but is almost certainly present.

Spores: Uninucleate, measuring 3.5–6.7 × 2.3 μm (fresh), 3–4.5 × 2–2.7 μm in smears and 3.2–4.5 × 1.9–2.6 μm in sections. Anterior and posterior vacuoles visible in fixed material. The polar tube measures up to 220 μm.

Fig. 3.3 *Pleistophora myotrophica: Bufo bufo* adult ♂: abdominal, pectoral and femoral muscles with white streaks indicating sites of infection. Life size. (Reproduced from Canning *et al.,* 1964, with permission of the publisher.)

Fig. 3.4 *Pleistophora myotrophica:* heavy infection in the femoral muscle of *B. bufo* (male): Heidenhain's haematoxylin–Bismark Brown. Scale = 100 μm. (Photograph by Dr. E. Elkan.)

Figs. 3.5, 3.6 *Pleistophora myotrophica:* sections of muscle of *B. bufo* showing fusiform sites of infection of increasing size, containing mainly meronts dividing by plasmotomy: Heidenhain's haematoxylin. Scale bars = 100 μm. (Photographs by Dr. E. Elkan.)

Fig. 3.7 *Pleistophora myotrophica:* Giemsa-stained smear from muscle of *B. bufo* showing multinucleate meronts dividing by plasmotomy. Scale bar = 10 μm. (Photograph by Dr. E. Elkan.)

Fig. 3.8 *Pleistophora myotrophica:* section of muscle of *B. bufo* showing meronts (m) but mainly sporogonial plasmodia. Heidenhain's haematoxylin. Scale bar = 10 μm. (Photograph by Dr. E. Elkan.)

Fig. 3.9 *Pleistophora myotrophica:* Giemsa stained group of spores; the spores remain clumped but an SPOV is not visible. Scale bar = 10 μm. (Photograph by Dr. E. Elkan.)

Fig. 3.10 Fresh spores resembling those of *P. myotrophica* from the striated musculature of Western Chorus frogs *Pseudacris triseriata.* Scale bar = 10 μm. (Photograph from material supplied to the authors by Dr. J. Vanderbergh.)

Fig. 3.11 *Pleistophora* sp: SPOV containing spores (arrowhead) from muscle of tuatara, *Sphenodon punctatus.* H. & E.; Scale bar = 10 μm. (Reproduced from Liu and King, 1971, with permission of the authors and publisher.)

Fig. 3.12 *Pleistophora* sp.: section of heavily infected muscle of tuatara, *Sphenodon punctatus.* H. & E.; Scale bar = 100 μm. (Reproduced from Liu and King, 1971, with permission of the authors and publisher.)

TRANSMISSION: Adult toads but not tadpoles were easily infected when fed on infected toad muscle. *Rana temporaria* was almost refractory to infection as only 1 out of 12 frogs similarly fed showed any infection and that was limited to a few foci in the tongue muscle.

NOTES: The authors considered this parasite to be distinct from *P. danilewskyi*, principally because cyst walls were said to enclose the foci of *P. danilewskyi* in the muscle and because *R. temporaria*, a listed host of *P. danilewskyi*, was found to be almost refractory to infection with *P. myotrophica*: only 1 out of 12 frogs, force fed with infected *B. bufo* muscle, became lightly infected, in the tongue only (Canning, Elkan and Trigg, 1964). Although *R. temporaria* has been recorded as one of the hosts of *P. danilewskyi*, there remains some doubt about the identity of the frog, because Danilewsky had not himself identified the hosts of the material, which he had collected and sent to Pfeiffer. Allowing for inaccuracies in the earlier observations and for the uncertainties surrounding the early host identifications, there now seems less justification for separating *P. myotrophica* from *P. danilewskyi*, especially as the spore measurements from sections are similar. *P. danilewskyi* possibly has a wide host range among amphibia and reptiles.

5. *Pleistophora* sp. of Liu and King, 1971 (Figs. 3.11, 3.12)

HOST AND LOCALITY: Two tuataras, *Sphenodon punctatus*, captured on Stephens Island, New Zealand and transported to the Bronx Zoo, U.S.A., died 12 and 13 months after capture.

LESIONS: On necropsy, the tongue and skeletal muscles were white and fragile and showed lesions packed with microsporidia in "cysts" containing 100 or more spores (Fig. 3.12). There had been extensive infiltration of inflammatory cells and replacement of diseased muscle by granulation tissue. Death was attributed to infection.

MORPHOLOGY AND DEVELOPMENT

Merogony and Sporogony: Fusiform bodies 18–21 × 7–11 μm, containing uninucleate sperical bodies were considered to be schizonts but could have been early sporonts. Sporoblasts and immature spores are organized in groups.

Spores: Mature spores (Fig. 3.11), pyriform with anterior and posterior vacuoles measure 3.96±0.31 × 2.22±0.29 μm in stained sections. A membrane was not seen around the groups of spores but this might have been difficult to detect in sectioned post-mortem material. The general appearance is typical of *Pleistophora* SPOVs in muscle.

Although the authors postulated that the infections had been acquired during sojourn in the zoo, the progressive and chronic nature of this type of infection could have meant that the infections were acquired in the natural habitat, possibly with death somewhat hastened by the stress of capture.

The authors did not assign the parasite to a known species, though they were aware of the records of *P. danilewskyi* and *P. myotrophica* in snakes and toads. The spore size accords with that of *P. danilewskyi* and *S. punctatus* probably represents another host for this species.

6. *Pleistophora* sp. of Vanderberg and Canning (unpublished observations) (Fig. 3.9)

A microsporidium producing numerous spores in SPOVs has been observed in the striated muscle of Western Chorus frogs, *Pseudacris triseriata* in Ontario, Canada. The parasite has been transmitted to *B. bufo* and *Bufo americanus,* in which the muscle infections closely resemble those of *P. myotrophica.* Spores are elongate ovoid, slightly broader at one end (Fig. 3.9). They measure 4.9 × 2.8 μm fresh. This parasite may be conspecific with *P. danilewskyi.*

B. *Pleistophora bufonis* (King, 1907)

SYNONYMS: *Bertramia bufonis* King, 1907; probably *Plistophora bufonis* Guyénot and Ponse, 1926.

HOSTS AND LOCALITIES: *Bufo lentiginosus* in U.S.A. (King) and *Bufo bufo* in Switzerland (Guyénot and Ponse). In both cases in Bidder's organ, a rudimentary ovary in male toads.

LESIONS: Both Bidder's organs of one toad almost destroyed by the parasite (Guyénot and Ponse, 1926).

MORPHOLOGY AND DEVELOPMENT

Merogony and Sporogony: Stages dividing by binary fission and stages corresponding to sporogonial plasmodia, with numerous isolated nuclei, were described by King (1907). The complex life cycle involving macrospore and microspore development reported by Guyénot and Ponse (1926) is an unlikely interpretation.

Spores: Oval measuring 3 × 1.5 μm (King, 1907). The distinction between microspores measuring 2.4–3.5 μm and macrospores measuring 4–5.2 μm according to the number produced in the SPOVs (Guyénot and Ponse, 1926) (up to 64 spores in microspore SPOVs and 4, 8, or 16 spores in macrospore SPOVs) more likely represents a normal distribution.

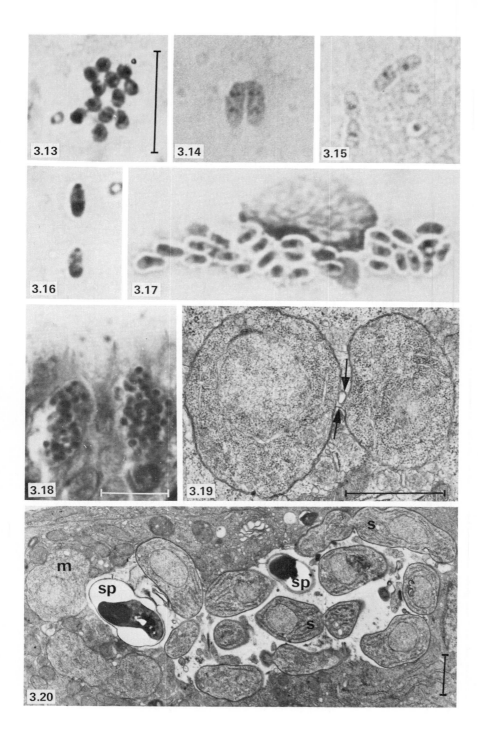

3.13

3.14

3.15

3.16

3.17

3.18

3.19

3.20

C. *Encephalitozoon lacertae* Canning, 1981 (Figs. 3.13–3.20)

SYNONYMS: "Microsporidian" Canning and Landau, 1971; *Microsporidium* sp. Canning and Landau, 1971.

HOST: *Podarcis muralis,* in France.

SITE OF INFECTION: Epithelial cells of intestine, especially the large intestine, the parasites forming into aggregates, within a vacuole lying between the host cell nucleus and the mucus goblet (Fig. 3.18).

MORPHOLOGY AND DEVELOPMENT: The parasite was described at light- and electron microscope levels by Canning (1981) and assigned to the genus *Encephalitozoon.*

Merogony: Meronts with one or two isolated nuclei divide by binary fission, often giving rise to short chains of uninucleate daughters, each measuring 2.1 × 1.9 μm (Fig. 3.13). Ultrastructurally the meronts are cells with a simple plasma membrane, within a close-fitting vacuolar

---

Figs. 3.13–3.20   Light and electron micrographs of *Encephalitozoon lacertae* from intestinal epithelium of wall lizard, *Podarcis muralis.*

Fig. 3.13   Group of meronts adhering in short chains. (Reproduced from Canning, 1981, with permission of the publisher.)

Fig. 3.14   Two binucleate sporonts. (Reproduced from Canning, 1981, with permission of the publisher.)

Fig. 3.15   Division of sporonts into sporoblasts. (Reproduced from Canning, 1981, with permission of the publisher.)

Fig. 3.16   Two sporoblasts; they are characteristically pyriform and show the polar sac as a densely staining dot anteriorly. (Reproduced from Canning, 1981, with permission of the publisher.)

Fig. 3.17   Group of mature spores released from an epithelial cell; the spores are slightly curved and show a single nucleus. (Reproduced from Canning, 1971, with permission of the publisher.) (Figs. 3.13–3.17 Giemsa-stained smears. Scale bar on Fig. 3.13 = 10 μm refers to the five figures.)

Fig. 3.18   Adjacent intestinal epithelial cells showing parasitophorous vacuoles containing mainly spores. Giemsa stained section. Scale bar = 10 μm.

Fig. 3.19   Electron micrograph of two meronts immediately after division. The parasitophorous vacuole membrane fits tightly around the meronts and is itself constricted at the point where they have separated (arrows) so that the meronts become isolated in their own vacuoles when division is complete. Scale bar = 1 μm. (Reproduced from Canning, 1971, with permission of the publisher.)

Fig. 3.20   Parasitophorous vacuole containing sporonts(s) dividing into sporoblasts, and spores (sp); a meront (m) lies in its own vacuole, which is just merging with the large vacuole. Scale bar = 1 μm. (Reproduced from Canning, 1981, with permission of the publisher.)

membrane. The vacuolar membrane constricts around the parasites (Fig. 3.19) as they divide giving several vacuoles in each cell.

Sporogony: Elongate sporonts become binucleate (Fig. 3.14) and divide into two sporoblasts (Fig. 3.15). Ultrastructurally the sporonts are seen in a large vacuole formed by merging of several merogonic vacuoles (Fig. 3.20). The sporont is thickened by addition of a surface coat and the amount of endoplasmic reticulum increases. Sporonts measure 5.0×1.5 μm and sporoblasts, which are piriform and show a densely-stained polar granule, measure 3.4×1.2 μm (Fig. 3.16).

Spores: Elongate ellipsoid, slightly curved and have a single, central nucleus (Fig. 3.17). They measure 3.5×1.5 μm (fresh) and 2.7×1.2 μm (stained). The electron microscope revealed up to 7 coils of the polar tube.

D. *Microsporidium tritoni* (Weiser, 1960) (Fig. 3.21)

SYNONYM: *Nosema tritoni* Weiser, 1960
HOST AND LOCALITY: *Triturus vulgaris* in Czechoslovakia.
LESION: A small whitish elevation, just anterior to the anus beneath the skin in the connective tissue.

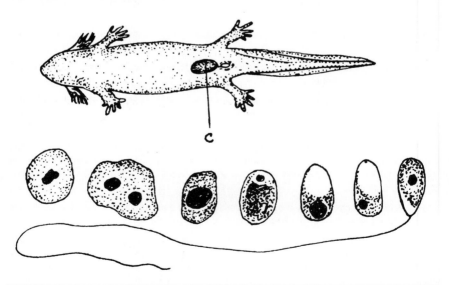

Fig. 3.21 *Microsporidium tritoni:* drawing of ventral surface of common newt, *Triturus vulgaris* showing localisation of cyst(c) containing the parasites; meronts, sporogonic stages and a spore with extruded polar tube are also shown. (Reproduced from Weiser, 1959, with permission of author and publisher.)

MORPHOLOGY AND DEVELOPMENT: Nuclei are isolated. Uninucleate and binucleate amoeboid schizonts and uninucleate ovoid sporoblasts and spores are illustrated. Spores measure 4 × 2 μm and the extruded filament about 43 μm.

COMMENT: As diplokaryon stages are lacking, this parasite cannot belong to the genus *Nosema*. A species of *Encephalitozoon* (Sect. C) has since been described from a lizard and the stages of *M. tritoni* recall this genus.

E. *Microsporidium schuetzi* n.sp. (Figs. 3.22–3.25)

SYNONYM: "Microsporidia" Schuetz, Selman and Samson, 1978
HOST AND LOCALITY: *Rana pipiens* from Vermont, U.S.A.
LESIONS: Enlarged oocytes, pale brown to whitish grey, with homogeneous pigmentation compared with normal asymmetric pigmentation. Infected oocytes are devoid of cortical granules, contain fewer yolk platelets but more lipid droplets and bear surface microvilli at a stage when retraction would normally have occurred (Fig. 3.25). Developmental stages of the microsporidium are distributed at the periphery and spores are concentrated in the centre.

MORPHOLOGY AND DEVELOPMENT: All stages are in direct contact with host cell cytoplasm.

Merogony: Diplokaryotic stages with a simple plasma membrane (Fig. 3.22).

Sporogony: Plasmodia with several diplokarya acquire a surface coat, composed of two distinct layers of differing electron density (Fig. 3.24). Large vesicles in the cytoplasm appear to be responsible for the secretion of the surface coat layers. Division of the plasmodium gives several sporoblasts, which mature into spores.

Spores: The number of nuclei in the spores and spore size in unknown. The exospore, derived from the two surface coats, is a strikingly thick layer. The number of coils of the polar tube cannot be determined with accuracy from the figures but may be only 3 or 4 (Fig. 3.23).

NOTES: Infected oocytes show abnormal ion concentrations, particularly elevated potassium and depressed calcium and they continue to incorporate the yolk protein precursor, vitellogenin, after ovulation. The ultrastructural alterations to the oocyte indicate that invasion occurs at an early stage of differentiation.

The generic status of the parasite cannot be determined on the evidence provided by Schuetz, Selman and Samson (1978). Few of the microsporidia which develop in direct contact with the host cell cytoplasm (no SPOVs) have diplokaryotic nuclei, as were seen in the meronts and sporonts of *M. schuetzi*. Of these genera some have diplokaryotic spores (e.g. *No-*

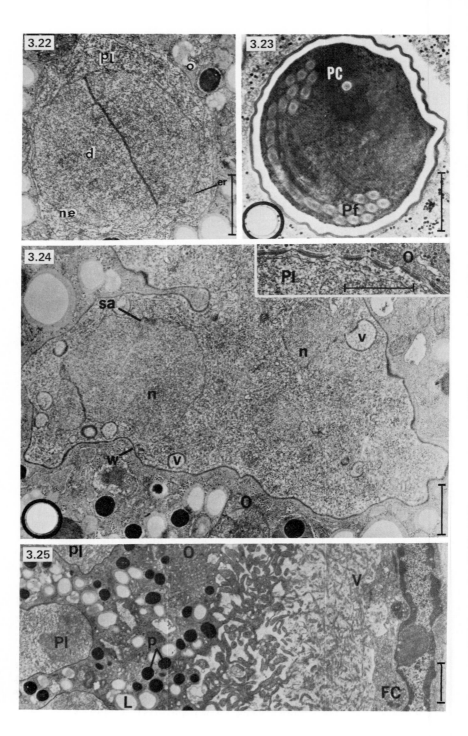

sema, Octosporea) others have unikaryotic spores (e.g. *Enterocytozoon, Perezia)*. As the number of nuclei in the spores of *M. schuetzi* is not known, nor indeed the number of sporoblasts derived from a sporont, the genus is indeterminated and the parasite has here been placed in the collective genus *Microsporidium*.

F. Ambiguous Form

Guyénot and Naville (1922b) while studying the development of *Myxobolus ranae* in *Rana temporaria* found that one plasmodium of the myxosporidium had engulfed a cell, which they believed was of host origin, containing ovoid spores, 4.0 μm long, of a microsporidium. The illustration is not typical of microsporidian spores and this must be considered a doubtful record from amphibia.

Fig. 3.22–3.25 *Microsporidium schuetzi* electron micrographs of stages in oocytes of *Rana pipiens*.

Fig. 3.22 Meront showing a single diplokaryon (d) in an oocyte (o); the parasite cytoplasm (pl) is surrounded by a simple unit membrane; ne = nuclear envelope. Scale bar = 1 μm.

Fig. 3.23 Spore showing coils of the polar tube (Pf) and a section of the rectilinear region of the polar tube (PC) in a dense region which may be the polaroplast. Scale bar = 0.5 μm.

Fig. 3.24 Sporogonial plasmodium with several profiles of the diplokarya (n), one of which shows spindle microtubules and a centriolar plaque (sa); vesicles (v) with electron dense borders may fuse with the surface membrane and give rise to the surface coat (w). Inset: detail of the plasmodium (Pl) and its surface coat, of two layers of different electron density, in contact with the oocyte (O). Scale bar = 1 μm.

Fig. 3.25 Cortical cytoplasm of oocyte (O) containing several sporogonial plasmodia (Pl). The cell shows numerous pigment granules (p) and is abnormal in having fewer lipid granules (L) and having persisting microvilli (V). FC = an attenuated follicle cell. Scale bar = 1 μm. (Reproduced from Schuetz *et al.*, 1978, with permission of authors and publisher.)

4. The Microsporidia of Birds and Mammals

I. INTRODUCTION

Although few species of microsporidia have been recorded from homoeothermic vertebrates, one of them, *Encephalitozoon cuniculi* Levaditi, Nicolau and Schoen, 1923, is common in rodents and lagomorphs and has a wide geographical distribution. It is transmitted orally, especially under crowded conditions, and transplacentally, and, unless specifically eliminated, it is likely to be present in laboratory animal units and among animals bred intensively for economic reasons. In many cases infections are mild and, passing undetected, they may be an important source of error in experiments using laboratory animals. Otherwise there is an acute phase at the onset of infection followed, in survivors, by a chronic phase when the parasite is localized in the brain and other tissues.

Encephalitozoon has also been reported from carnivores and primates and, although slight morphological differences exist between some of these and the rodent parasites, it is likely that the same species is involved. However, the aetiological agent of an encephalitis in one human patient (Matsubayashi, Koike, Mikata, Takei, and Hagiwara, 1959) has been given specific status, probably mistakenly, as *Encephalitozoon matsubayashii* Sprague, 1977. *Thelohania apodemi* Doby, Jeannes and Rault, 1963, another species localizing in the brain of its rodent host *Apodemus sylvaticus,* is also a natural parasite of mammals.

Other microsporidia in mammals may simply be opportunistic parasites, manifesting themselves under conditions of immunological incompetence or stress in the host. These species are difficult to identify and have mostly been placed in the collective group *Microsporidium* in the present work.

The origins of these infections is unknown and, although it was suggested that ingested arthropods may have been the source of a species in the jejunal epithelium of a *Callicebus* monkey (Seibold and Fussell, 1973), it is unlikely that this was so. When microsporidia of invertebrate origin have been grown in tissue cultures, none has survived at mammalian blood temperatures. If these infections are of mammalian origin, the medical importance, especially under conditions of immunodepression, must not be underestimated. This is emphasized by the fatal nature of these illnesses and by recent reports of seropositivity for *E. cuniculi* in patients with severe neurological disorders.

Whether or not microsporidia are widespread in human populations will no doubt be revealed in the near future, now that good serological tests are available for diagnosis.

II. DESCRIPTIONS OF THE SPECIES INFECTING BIRDS

Prior to 1975, the only suggestions that microsporidia could infect birds came from experiments in which tissues from sparrows (Kyo, 1958) and parrots and pigeons (Werner and Pierzynski, 1962) were inoculated into mice. As the recipient mice may have been carrying latent infections of *E. cuniculi,* there is no certainty that the spores, subsequently isolated from the mice, were derived from the birds.

Four well-documented reports of microsporidiosis in birds have all been from lovebirds (Kemp and Kluge, 1975; Lowenstein and Petrak, 1978; Novilla and Kwapien, 1978; Branstetter and Knipe, 1982). The only tissues involved were liver, kidney and intestine (only liver was stated by Branstetter and Knipe). The genus is probably, but not certainly, *Encephalitozoon* but it is not clear whether the same species was involved in all cases; possibly not, because the number of coils of the polar tube differs. The absence of parasites from the brain and minor morphological characters indicate that the species differ from *Encephalitozoon cuniculi* in mammals. The microsporidia were thought to have been the direct cause of death in at least some cases. Lowenstein and Petrak (1978) record in an addendum that several cases of "encephalitozoonosis" were diagnosed in neurologically sick pheasants and a duck from a children's zoo, and in yet another lovebird, *Agapornis roseicollis,* with a history of diarrhoea and wasting.

Encephalitozoon sp. of Kemp and Kluge, 1975 (Figs. 4.2, 4.3)

HOSTS AND LOCALITY: Blue-masked lovebirds, *Agapornis personata,* in captivity in the U.S.A.

LESIONS: In hepatocytes and epithelial cells of kidney tubules, bile duct and intestine (Fig. 4.2). Some focal necrosis in the liver but inflammatory reactions were generally minimal or absent. Caused the death of 10 out of 14 lovebirds.

MORPHOLOGY: Only round or elliptical spores, measuring 1–2 μm were described. At the ultrastructural level the exospore, endospore, 3–4 coils of the polar tube and membranes of the polaroplast were seen (Fig. 4.3).

NOTES: The isolated nuclei, disporoblastic sporogony and development in parasitophorous vacuoles, which are the diagnostic features of the genus *Encephalitozoon,* were not mentioned.

Encephalitozoon sp. of Lowenstein and Petrak, 1978 (Fig. 4.1)

HOSTS AND LOCALITY: 2 peach-faced lovebirds, *Agapornis roseicollis,* in captivity in the U.S.A.

LESIONS AND SIGNS: Focal necrotising hepatitis, nephritis and enteritis with mild lymphocytic inflammation at the periphery of some foci. Organisms were seen within vacuoles in renal epithelial cells and tubule lumina, in hepatocytes, bile duct epithelium and intestinal epithelium. No other organs were infected. The birds died after an emaciating, diarrhoeic illness of several months duration.

MORPHOLOGY: Membrane-lined vacuoles contained clusters of sporoblasts and spores (Fig. 4.1). Stages were uninucleate and spores, measuring 1.5 × 0.75 μm showed 6–8 coils of the polar tube.

NOTES: The parasitophorous vacuole and isolated nuclei suggest the genus *Encephalitozoon* but the number of coils of the polar tube differentiate this species from that seen by Kemp and Kluge (1975).

Encephalitozoon sp. of Novilla and Kwapien, 1978 (Fig. 4.5, 4.6)

HOSTS AND LOCALITY: 1 pied peach-faced lovebird, *Agapornis roseicollis,* in captivity in the U.S.A.

LESIONS AND SIGNS: Organisms in renal tubules (Fig. 4.6) and small intestine, without inflammatory reaction. Heavy infection in hepatocytes accompanied by focal necrosis (Fig. 4.5). The bird was emaciated and died 2 days after exhibiting anorexia, weakness and closed eyes.

MORPHOLOGY: Spores 1.3 × 1.7 μm with PAS positive polar cap.

NOTES: A parasitophorous vacuole was not mentioned but "dense cyst-like clusters" of organisms were mentioned and illustrated. The number of coils of the polar tube and number of nuclei was not mentioned, rendering comparison with the other reports impossible.

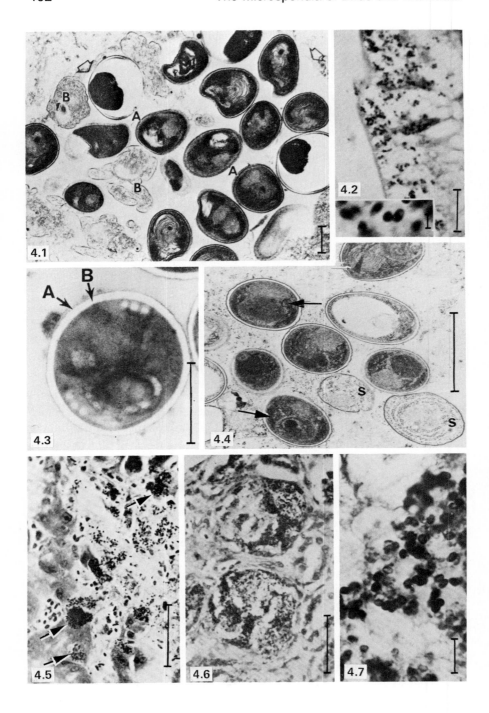

Encephalitozoon sp. of Branstetter and Knipe, 1982 (Figs. 4.4, 4.7)

HOST: 1 peach-faced lovebird, *Agapornis roseicollis,* in captivity in the U.S.A.

LESIONS: Hepatic lesions only were mentioned (Fig. 4.7).

MORPHOLOGY: The only characters given were an irregular wavy contour of the exospore and 6 or less coils of the polar tube (Fig. 4.4).

III. DESCRIPTIONS OF THE SPECIES INFECTING MAMMALS

A. *Encephalitozoon cuniculi* Levaditi, Nicolau and Schoen, 1923
(Figs. 4.8–4.27)

1. HISTORY: *Encephalitozoon cuniculi* is the most extensively studied of all microsporidia. The events leading to its discovery, naming and classification and early reports of its pathogenicity are outlined chronologically

---

Fig. 4.1 *Encephalitozoon* sp. from *Agapornis roseicollis:* electron micrograph of a parasitophorous vacuole containing sporoblasts (B) and spores (A). Scale bar = 1 μm. (Reproduced from Lowenstein and Petrak, 1980, with permission of the authors and publisher.)

Fig. 4.2 *Encephalitozoon* sp. from blue-masked lovebird, *Agapornis personata:* section of ileal epithelial cells showing spores dispersed rather than clumped in vacuoles. Inset: spores showing posterior vacuole. Iron haematoxylin. Scale bar = 10 μm; insert = 1 μm. (Reproduced from Kemp and Kluge, 1975, with permission of the authors and publisher.)

Fig. 4.3 *Encephalitozoon* sp. from *A. personata:* electron micrograph of spore showing exospore (A), endospore (B) and about 3 coils of the polar tube. Scale bar = 1 μm. (Reproduced from Kemp and Kluge, 1975, with permission of the authors and publisher.)

Fig. 4.4 *Encephalitozoon* sp. from *A. roseicollis:* electron micrograph of sporoblasts (s) and spores in a necrotic hepatocyte; arrows indicate the polar tube; there is no certainty that these stages lie in a parasitophorous vacuole. Scale = 1 μm. (Reproduced from Branstetter and Knipe, 1982, with permission of the authors and publisher.)

Fig. 4.5 *Encephalitozoon* sp. from *A. roseicollis:* section of necrotic lesion in liver, showing clusters of spores (arrows) within hepatocytes. McCallum-Goodpasture stain. Scale bar = 5 μm (Reproduced from Novilla and Kwapien, 1978, with permission of the authors and publisher.)

Fig. 4.6 *Encephalitozoon* sp. from *A. roseicollis:* section of renal medulla, showing spores within tubule cells. McCallum-Goodpasture stain. Scale bar = 5 μm. (Reproduced from Novilla and Kwapien, 1978, with permission of the authors and publisher.)

Fig. 4.7 *Encephalitozoon* sp. from *A. roseicollis:* section of liver with lesions, containing spores, apparently dispersed, not in vacuoles. Gomori's methenamine silver stain. Scale bar = 10 μm. (Reproduced from Branstetter and Knipe, 1982, with permission of the authors and publisher.)

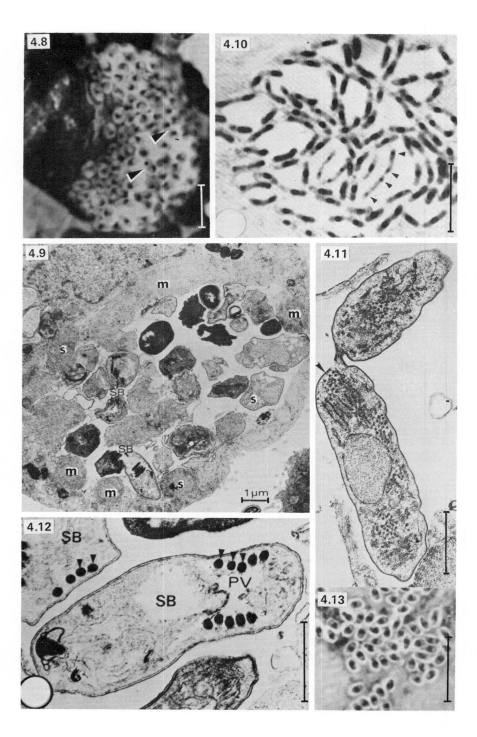

below. These early papers are interesting because the spontaneous infections of *E. cuniculi* in the laboratory animals, which were being used in investigations of human diseases, confused the authors into wrongly implicating the microsporidian parasites in the aetiology of the human diseases. Organisms found in dogs, and named *Encephalitozoon rabiei* and *Encephalitozoon negrii* were purported to cause rabies. These organisms can probably be accepted today as microsporidia but their association with rabies has long since been disproved. Other organisms named and attributed to *Encephalitozoon* were either too poorly described to be identifiable today or have been reclassified by subsequent investigation.

Prior to 1922: Encephalitis and nephritis, of unknown aetiology, was known in rabbits (e.g. Ophuls, 1908; Bull, 1917; Bell and Hartzell, 1919; Twort and Archer, 1922).

1922: Wright and Craighead demonstrated Gram positive microorganisms measuring "never more than 4 × 1.5 μm" in the brain, spinal cord and kidney of rabbits, which were being used in experiments on transmission of poliomyelitis. The rabbits suffered paralytic disease, unrelated to the poliomyelitis, and high mortality.

Fig. 4.8 Tissue culture of *Encephalitozoon cuniculi* mouse isolate in Madin-Darby Canine Kidney (MDCK) cells. Parasitophorous vacuole containing meronts (arrowheads) and sporogonic stages. Giemsa stain. Scale bar = 10 μm. (Photograph by Mrs. W. Hollister.)

Fig. 4.9 Electron micrographs of *E. cuniculi* fox isolate in MDCK cell: meronts (m) lie at the periphery of the parasitophorous vacuole. Sporogonic stages, including sporonts (s) sporoblasts (SB) and spores lie in the centre. Scale bar = 1 μm. (Reproduced from Mohn *et al.*, 1981, with permission of the authors and publisher.)

Fig. 4.10 Tissue culture of *E. cuniculi* mouse isolate in rabbit choroid plexus (RCP) cells. Parasitophorous vacuole containing mainly binucleate meronts; one tetranucleate stage is present (arrowheads). Giemsa stain. Scale bar = 10 μm. (Reproduced from Vávra *et al.*, 1972, with permission of the authors and publisher.)

Fig. 4.11 Electron micrograph of *E. cuniculi* rabbit isolate in RCP cells: sporont, dividing into 2 uninucleate sporoblasts: arrowhead points to surface coat: the endomembrane system is well developed. Scale bar = 1 μm. (Reproduced from Pakes *et al.*, 1975, with permission of the authors and publisher.)

Fig. 4.12 Electron micrograph of *E. cuniculi* fox isolate in MDCK cells: sporoblasts (SB) developing into spores, showing 4-5 coils of the polar tube (arrowheads) and posterior vacuole (PV). The anterior end of the polar tube and developing polar sac are visible at one pole. Scale bar = 1 μm. (Reproduced from Mohn *et al.*, 1981, with permission of authors and publisher.)

Fig. 4.13 Spores of *E. cuniculi* from ruptured mouse peritoneal macrophage: shape and single nucleus are visible. Giemsa stain. Scale bar = 10 μm. (Reproduced from Lainson *et al.*, 1964, with permission of the authors and publisher.)

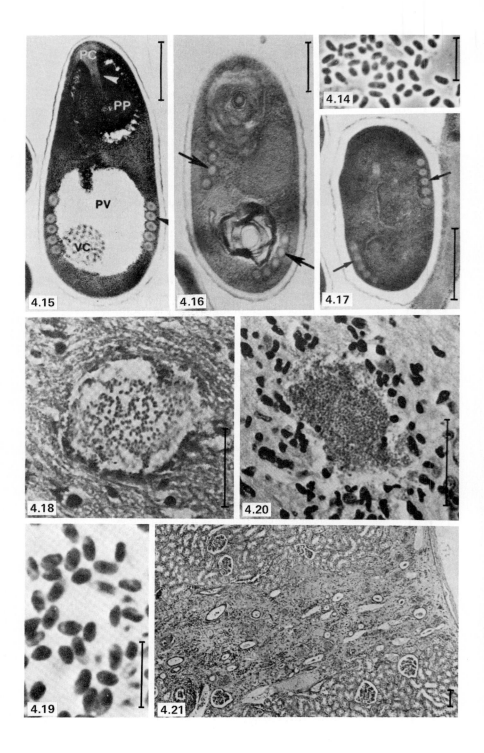

4.15 — PC, PP, PV, VC

4.16

4.17

4.14

4.18

4.20

4.19

4.21

1923, 1924: Doerr and Zdansky (1923a,b) Levaditi and Nicolau (1923) and Levaditi, Nicolau and Schoen (1923a–c, 1924a–d) published a series of papers on lesions and organisms in the brain of rabbits. Doerr and Zdansky found the organisms (1923a) and claimed priority for the discovery (1923b). First the lesions were found (Levaditi and Nicolau, 1923), then the organisms, which were named *Encephalitozoon cuniculi* (Levaditi, Nicolau and Schoen, 1923a). The same authors suggested that they were microsporidia (1923b). In their later papers Levaditi, Nicolau and Schoen, (a) recorded the presence of organisms in kidney as well as brain of rabbits (1923c), (b) proposed that transmission occurred via the urine, found that mice could be infected intraperitoneally and suggested that the Negri bodies in neurones of rabies-infected animals might be very small microsporidia (1924a), (c) found that spores were present in peritoneal macrophages and Kupffer cells of mice before they localised in the brain (1924b), (d) transmitted the parasite to a dog and to rats but not to a monkey *Macacus cynomolgus* (1924c) and (e) studied routes of infection (1924d).

Fig. 4.14   Fresh spores of *E. cuniculi* mouse isolate harvested from MDCK cells. Scale bar = 10 μm.

Fig. 4.15   Electron micrograph of E. cuniculi rabbit isolate from RCP cell: spore showing the polar tube (white arrowhead) originating in the polar sac (PC), passing through the polaroplast (PP), and coils (black arrowhead) around the posterior vacuole (PV); a vesicular complex (VC) lies in the vacuole. Scale bar = 0.5 μm. (Reproduced from Pakes *et al.,* 1975, with permission of the authors and publisher.)

Fig. 4.16   Electron micrograph of *E. cuniculi* isolated from a squirrel monkey, *Saimiri sciureus:* spore showing 4.5 coils of the polar tube (arrows) and polaroplast around the anterior part of the tube. Scale bar = 0.5 μm (approximately). (Reproduced from Anver *et al.,* 1972, with permission of the authors and publisher.)

Fig. 4.17   Electron micrograph of *E. cuniculi* dog isolate in tissue culture in canine embryo fibroblasts: spore showing 5 coils of the polar tube (arrows). Scale bar = 1 μm (approximately). (Reproduced from Shadduck *et al.,* 1978, with permission of the authors and publisher.)

Fig. 4.18   Section of brain of a suricate, *Suricata suricatta* showing large intact aggregate of spores without necrosis of surrounding tissue. Ziehl-Nielsen stain. Scale bar = 50 μm. (Reproduced from Vávra *et al.,* 1971, with permission of the authors and publisher.)

Fig. 4.19   Enlargement of spores from Fig. 4.18. Ziehl-Nielsen stain. Scale bar = 10 μm. (Reproduced from Vávra *et al.,* 1971, with permission of the authors and publisher.)

Fig. 4.20   Section of brain of rabbit with large aggregate of *E. cuniculi:* after rupture of host cell and release of spores there is a tissue reaction. H. & E. Scale bar = 50 μm. (Reproduced from Koller, 1969, with permission of the author and publisher.)

Fig. 4.21   Section of rabbit kidney infected with *E. cuniculi,* showing cortical degeneration and developing fibrosis. H. & E. Scale bar = 100 μm. (Reproduced from Koller, 1969 with permission of the author and publisher.)

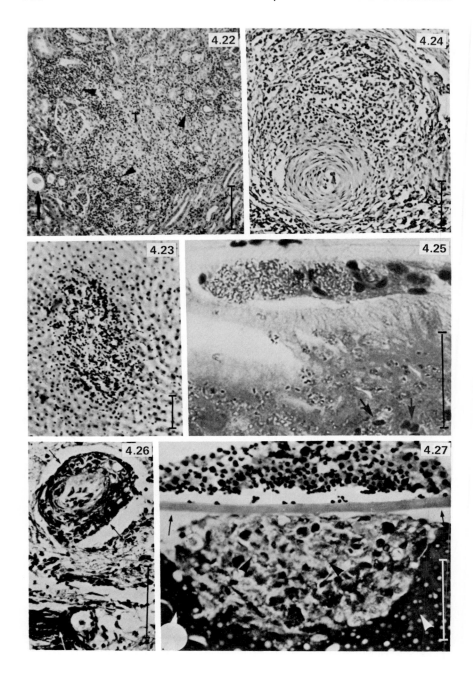

In the last paper they proposed, incorrectly according to the present International Rules of Zoological Nomenclature, that the parasite should be named *Encephalitozoon cuniculi* nov. spec. (Wright and Craighead), in acknowledgement that Wright and Craighead were the first workers to have seen the parasite—and presumably to refute Doerr and Zdansky's claim to the discovery.

1924–1927: E. cuniculi was found by several other authors in rabbits and mice (e.g. Goodpasture, 1924; Cowdry and Nicholson, 1924a,b; Oliver, 1924; Cameron and Maitland, 1924; Smith and Florence, 1925; Bender, 1925).

Levaditi, Nicolau and Schoen (1924e, 1926) and Manouelian and Viala (1924, 1927), investigating the aetiology of rabies and encephalitis in dogs, found organisms in the neurones and salivary glands which resembled *E. cuniculi* and proposed a causal relationship between them and the diseases. They named them *Glugea lyssae* Levaditi, Nicolau and Schoen, 1924 (1924e), *Glugea rabiei* Levaditi, Nicolau and Schoen, 1926, *Encephalitozoon rabiei* Manouelian and Viala, 1924 and *Encephalitozoon negrii* Manouelian and Viala, 1927. All of these might have been *E. cuniculi*. We know today that the diseases are of viral origin.

Fig. 4.22   Section of kidney showing interstitial nephritis caused by *E. cuniculi* in rabbits, 8 weeks after infection: there are numerous lymphocytes (arrowheads) in the interstitium, in a focus of dilated (arrow) and collapsed (T) cortical tubules. H. & E. Scale bar = 10 μm. (Reproduced from Shadduck *et al.,* 1979, with permission of the authors and publisher.)

Fig. 4.23   Section of liver of rabbit showing granuloma formation after infection with *E. cuniculi.* H. & E. Scale bar = 10 μm. (Reproduced from Cox *et al.,* 1979, with permission of the authors and publisher.)

Fig. 4.24   Fibrous thickening of hepatic arterial wall and massive perivascular accumulation of inflammatory cells in blue fox infected with *E. cuniculi.* H. & E. Scale bar = 10 μm. (Reproduced from Nordstoga, 1972, with permission of the author and publisher.)

Fig. 4.25   Ocular lesion in rabbit infected with *E. cuniculi:* equatorial region of lens; the spores are within the lens epithelium and free in the adjacent cataractous cortical tissue; arrows point to degenerate epithelial nuclei. H. & E. Scale bar = 5 μm. (Reproduced from Ashton *et al.,* 1976, with permission of the authors and publisher.)

Fig. 4.26   Ocular lesion in blue fox infected with *E. cuniculi:* retina close to papilla, showing artery with characteristic lesions (arrows); white arrow points to choroid at end of papilla. H. & E. Scale bar = 10 μm. (Reproduced from Arnesen and Nordstoga, 1977, with permission of the authors and publisher.)

Fig. 4.27   Ocular lesion in blue fox infected with *E. cuniculi:* anterior surface of lens with lens capsule (arrows); subcapsular destruction of lens fibres, with accumulation of macrophages and clusters of spores (black arrowheads); white arrowheads point to damaged lens material. H. & E. Scale bar = 10 μm. (Reproduced from Arnesen and Nordstoga, 1977, with permission of the authors and publisher.)

The microsporidian nature of *E. cuniculi* was not accepted (Anigstein, 1925; Wenyon, 1926) and *E. cuniculi* passed into the literature as an organism of uncertain affinity.

1927–1935: Three fatal diseases in man were diagnosed as due to *Encephalitozoon*. Torres (1927a,b,c) described *Encephalitozoon chagasi* from a female baby, who died in Brazil two days after birth, having suffered general muscular contraction and convulsions. The spores measured 3.5 × 1.5 μm. The type material has been lost (according to Weiser, 1964) and no confirmation can be obtained of the nature of this organism.

Coulon (1929) gave the name *Encephalitozoon brumpti* to an organism recovered from the cerebro-spiral fluid of a 17-year-old youth who had died in Corsica after a short illness, involving high fever, coma and acute meningitis. The spores varied in size from 1.25 × 2 μm to 4 × 6 μm. Barker (1974) examined the type material and established that it was not a microsporidium. Thus *E. brumpti* is not a synonym of *E. cuniculi*.

Wolf and Cowen (1937) gave the name *Encephalitozoon hominis* to a parasite causing lesions in the brain and spinal cord of a 4-week-old infant who had died in the U.S.A. They later reconsidered the diagnosis as toxoplasmosis (Wolf, Cowen and Page, 1939). *E. hominis* is therefore a junior synonym of *Toxoplasma gondii*.

1927–1958: No advances were made in understanding the nature of the organism but numerous records, substantiated and unsubstantiated, of *Encephalitozoon* infections in different animal species were reported (Table 4.1).

1959: Matsubayashi, Koike, Mikata, Takei and Hagiwara (1959) recovered spores of *E. cuniculi* from the urine of a child suffering a severe convulsive illness.

1962: Nelson extruded the polar tubes from spores derived from peritoneal macrophages of mice. The parasites were not named but were clearly *E. cuniculi*. The polar tube measured 20–25 μm and the spores 2 μm in length. Thus, 39 years after Levaditi, Nicolau and Schoen (1923a) had named the parasite and had suggested that it belonged to the microsporidia the first proof of the microsporidian nature of *E. cuniculi* was provided.

1964: Lainson, Garnham, Killick-Kendrick and Bird (1964) gave a well illustrated account of the development of *E. cuniculi* and provided ultrastructural evidence of the polar tube. These authors and Weiser (1964) transferred the species to the genus *Nosema*, as they considered that *Encephalitozoon* was a junior synonym.

1970: Cali differentiated the genera *Encephalitozoon* and *Nosema* on the basis of isolated nuclei in the former and diplokarya in the latter and

TABLE 4.1 **Principal Records of *Encephalitozoon cuniculi* from Mammals since 1922**

Authors	Locality	Method of diagnosis[a]
Rabbits (laboratory and domestic)		
Wright and Craighead, 1922	U.S.A.	H(P)
Doerr and Zdansky, 1923	Sweden	H(P)
Levaditi, Nicolau and Schoen, 1923, 1924	Sweden, U.S.A., U.K., France	H(P)
McCartney, 1924	U.S.A.	H(P)
Goodpasture, 1924	U.S.A.	H(P)
Da Fano, 1924	U.S.A.	H(P)
Smith and Florence, 1925	U.S.A.	H(P)
Bender, 1925	U.S.A.	H(P)
Ray and Raghavachari, 1941	India	H(P)
Robinson, 1954	U.S.A.	H(L)
Malherbe and Munday, 1958	South Africa	H(P)
Reddy, 1963	India	H(P)
Howell and Edington, 1968	U.K.	H(P)
Koller, 1969	U.S.A.	H(P)
Shadduck, 1969	U.S.A.	H(P); TC
Flatt and Jackson, 1970	U.S.A.	H(P)
Pattison, Clegg and Duncan, 1971	U.K.	H(P)
Goodman and Garner, 1972	U.S.A.	H(P)
Hunt, King and Foster, 1972	U.S.A.	H(P)
Pakes, Shadduck and Olsen, 1972	U.S.A.	ST; H(P)
Cox, Walden and Nairn, 1972	Australia	IFAT
Chalupský, Vávra and Bedrník, 1973, 1979a,b	Czechoslovakia	IFAT; CIA
Armstrong, Ke, Breinig and Ople, 1973	U.S.A.	TC
Jackson, Solorzano and Middleton, 1973	U.S.A.	IFAT
Testoni, 1974	Australia	H(P)
Cox and Pye, 1975	Australia, U.S.A., U.K.	IFAT; H(P)
Casaubon, 1975	Mexico	H(P)
Ashton, Cook and Clegg, 1976	U.K.	H(P)
Cox, 1977	Australia	IFAT
Waller, 1977	Sweden	CIA; IFAT; ST; H(P)
Cox and Gallichio, 1977	Australia	H(P); IFAT; TC
Cox, Gallichio, Pye and Walden, 1977	Australia	IFAT; H(P)
Pye and Cox, 1977	Australia	IFAT; TC
Cox and Gallichio, 1978	Australia	IFAT; H(P)
Kellett and Bywater, 1978	Switzerland	CIA
Bywater and Kellett, 1978	Switzerland	CIA
Gannon, 1978	U.K.	IFAT; IPT
Shadduck and Geroulo, 1979	U.S.A.	MAT

(*Continued*)

TABLE 4.1 *Continued*

Authors	Locality	Method of diagnosis[a]
Owen and Gannon, 1980	U.K.	IPT
Gannon, 1980	U.K.	IPT; IFAT
Waller, Lyngset, Elvander and Morein, 1980	Norway	CIA (post mortem)
Lyngset, 1980	Norway	H(P)
Julini and Pellegrino, 1981–82	Italy	H(P)
Waller and Bergquist, 1982	Sweden	CIA
Singh, Kane, Mackinlay, Quaki, Ho, Ho and Lim, 1982	U.K., Singapore	IFAT
Hong, Songfong, Yu Dengzi, Si Fongwen and Xu Zhaihai, 1983	China	H(P)
Berkin and Kahraman, 1983	Turkey	H(P)
Waller, Uggla and Bergquist, 1983	Sweden	CIA; IFAT
Julini, 1983	Italy	H(P)
Rabbits (wild)		
Wilson, 1979b	U.K.	CIA
Cottontail rabbits (*Sylvilagus* sp.) (wild)		
Jungherr, 1955	U.S.A.	H(P)
Mice (laboratory)		
Levaditi, Nicolau and Schoen, 1924d	France	H(P)
Cowdry and Nicholson, 1924a,b	U.S.A.	H(P)
Anigstein, 1925	France	H(P)
Twort and Twort, 1932	U.K.	H(P)
Perrin, 1943	U.S.A.	H(P)
Lepine and Sautter, 1949	France	H(P)
Garnham and Roe, 1953	U.K.	H(P)
Morris, McCown and Blount, 1956	U.S.A.	H(P)
Guli and Parisio, 1958	Italy	H(P)
Malherbe and Munday, 1958	South Africa	H(P)
Kyo, 1958	Japan	H(P)
Koike, Kyo and Iino, 1960	Japan	SP
Nelson, 1962, 1967	U.S.A.	H(P)
Innes Zeman, Frenkel and Borner, 1962	U.S.A.	H(P)
Sureau, 1962	Madagascar	H(P)
Werner and Pierzynski, 1962	Germany	H(P)
Ruiz, 1964	Germany	H(P)
Weiser, 1965	Czechoslovakia	H(P)
Jordan and Mirick, 1965a,b	U.S.A.	H(L)
Arison, Cassaro and Pruss, 1966	U.S.A.	H(P)
Akao, 1969	Japan	SP; H(P)
Kaneda, 1969	Japan	SP; TC

TABLE 4.1 *Continued*

Authors	Locality	Method of diagnosis[a]
Bismanis, 1970	Canada	SP; TC
Alvarez, 1971	Mexico	H(P)
Chumakov, Viting, Konosh and Ashmarina, 1970	U.S.S.R.	SP; TC
Vávra, Bedrník and Činátl, 1972	Czechoslovakia	SP; TC
Huldt and Waller, 1974	Sweden	H(P); TC
Somvanshi, Iyer, Gupta and Matanay, 1977	India	H(P)
Gannon, 1980	U.K.	IPT; IFAT
Kellett and Bywater, 1980	Switzerland	CIA; IFAT
Rats (laboratory)		
Cowdry and Nicholson, 1924a	U.S.A.	H(L)
Gordon, 1940	U.K.	H(P)
Perrin, 1943	U.S.A.	H(P)
Lainson, 1954	U.K.	H(P)
Lainson, Garnham, Killick-Kendrick and Bird, 1964	U.K.	H(P)
Attwood and Sutton, 1965	U.K.	H(P)
Canning, 1965	U.K.	H(P)
Petri, 1965	Denmark	H(P)
Canning, 1967	U.K.	H(P)
Chalupský, Vávra and Bedrník, 1979	Czechoslovakia	IFAT
Gannon, 1980	U.K.	IPT; IFAT
Kellett and Bywater, 1980	Switzerland	CIA; IFAT
Singh, Kane, Mackinlay, Quaki, Yap, Ho, Ho and Lim, 1982 (low titres: doubtful positives)	Singapore	IFAT
Majeed and Zubaidy, 1982	U.K.	H(P)
Rats (wild)		
Tazaki, 1956	Japan	H(P)
Iino, 1959	Japan	SP
Gannon (quoted by Wilson, 1979)	U.K.	Not stated
Multimammate rats (*Mastomys natalensis*) (laboratory)		
Lainson, Garnham, Killick-Kendrick and Bird, 1964	U.K.	H(P)
Frenkel (quoted by Shadduck and Pakes, 1971	U.S.A.	Unknown
Muskrats (*Ondatra zibethica*) (captive)		
Wobeser and Schuh, 1979	Canada	H(P)

(Continued)

TABLE 4.1   *Continued*

Authors	Locality	Method of diagnosis[a]
Guinea pigs (*Cavia porcellus*) (laboratory)		
Cowdry and Nicholson, 1924a	U.S.A.	H(L)
Perrin, 1943	U.S.A.	H(P)
Ruge, 1950	Germany	H(P)
Šebek, 1969	Czechoslovakia	H(P)
Moffat and Schiefer, 1973	U.S.A.	H(P)
Chalupský, Vávra and Bedrník, 1979	Czechoslovakia	IFAT
Hamsters (*Mesocricetus auratus*) (laboratory)		
Lainson, Garnham, Killick-Kendrick and Bird, 1964	U.K.	H(P)
Kinzel and Meizer, 1968	Germany	H(P); TC
Meizer, Kinzel and Jírovec, 1971	Germany	H(P)
Chalupský, Vávra and Bedrník, 1979	Czechoslovakia	IFAT
Gannon, 1980	U.K.	IPT
Ground shrews (*Cryptotis parva*) (laboratory)		
Nelson, Mock and Flatt, 1969	U.S.A.	H(P)
Goat		
Khanna and Iyer, 1971	India	H(P)
Waller, Uggla and Bergquist, 1983	Sweden	CIA; IFAT
Sheep		
Singh, Kane, Mackinlay, Quaki, Yap, Ho, Ho and Lim, 1982	Singapore	IFAT
Waller, Uggla and Bergquist, 1983	Sweden	CIA; IFAT
Swine		
Waller, Uggla and Bergquist, 1983	Sweden	CIA; IFAT
Horse		
Waller, Uggla and Bergquist, 1983	Sweden	CIA; IFAT
Carnivora (Canidae)		
Dog (domestic)		
Plowright, 1952	U.K.	H(P)
Plowright and Yeoman, 1952	Tanganyika	H(P)
Basson, McCully and Warnes, 1966	South Africa	H(P)

TABLE 4.1 *Continued*

Authors	Locality	Method of diagnosis[a]
McCully, van Dellen, Basson and Lawrence, 1978	South Africa	H(P)
Shadduck, Bendele and Robinson, 1978	U.S.A.	H(P)
van Dellen, Botha, Boomker and Warnes, 1978	South Africa	H(P)
Botha, van Dellen and Stewart, 1979	South Africa	H(P); TC
Stewart, van Dellen and Botha, 1979	South Africa	TC
Stewart, Botha and van Dellen, 1979	South Africa	IFAT
Cole, Sangster, Sulzer, Pursell and Ellinghausen, 1982	U.S.A.	H(P); TC

N.B.: Reports of parasites resembling microsporidia in dogs (Kantorowicz and Lewy, 1923; Levaditi, Nicolau and Schoen, 1924a,d, 1926; Manouelian and Viala, 1924, 1927; Schuster, 1925; Perdrau and Pugh, 1930; Peters and Yamagiva, 1936) may have been *Encephalitozoon* but this cannot be confirmed.

Red fox (Vulpes vulpes) (wild)

Wilson, 1979b	U.K.	CIA

Arctic or blue fox (Alopex lagopus) (captive)

Vávra, Blažek, Lávička, Koczková, Kalafa and Stehlík, 1971	Prague Zoo	H(P)
Nordstoga, 1972	Norway	H(P); IFAT; CIA
Kangas, 1973	Finland	H(P)
Kull (quoted by Nordstoga, 1972)	Sweden	Unknown
Nordstoga and Westbye, 1976	Norway	H(P)
Mohn and Ødegard, 1977	Norway	IFAT
Arneson and Nordstoga, 1977	Norway	H(P)
Mohn, 1982	Norway	CIA; IFAT

N.B.: The parasites seen by Peters and Yamagiva (1936) in silver foxes, *Vulpes vulpes*, suffering encephalitis may have been *Encephalitozoon* but this cannot be confirmed.

Carnivora (Mustelidae)

Polecat, (Mustela eversmanni satunini) (captive)

Novilla, Carpenter and Kwapien, 1980	Siberia (imported into U.S.A.)	H(P)

N.B.: The infection reported by Levine, Dunlap and Graham (1938) in a ferret is now thought to be due to *Histoplasma* (N. D. Levine, personal communication, quoted by Vávra, Blažek, Lávička, Koczková, Kalafa and Stehlík, 1971.

(Continued)

TABLE 4.1 *Continued*

Authors	Locality	Method of diagnosis[a]
Carnivora (Viverridae)		
Suricate (Suricata suricatta) (captive)		
Vávra, Blažek, Lávička, Koczková and Stehlík, 1971	Prague Zoo	H(P)
Carnivora (Felidae)		
Cats (domestic)		
van Rensburg and du Plessis, 1971	South Africa	H(P)
Waller, Uggla and Bergquist, 1983	Sweden	CIA; IFAT

N.B.: The report of *Encephalitozoon* in the brains of apparently normal cats (Schuster, 1925) cannot be confirmed.

Clouded leopards (Neofelis nebulosa)		
Vávra, Blažek, Lávička, Koczková, Kalafa and Stehlík, 1971	Prague Zoo	H(P)
Puma, mountain lion (Neofelis concolor)		
Blažek, Koczková, Lavička, Vávra and Stehlík, 1972	Berlin Zoo	H(P)
Primates		
Squirrel monkey (Saimiri sciureus)		
Anver, King and Hunt, 1972	U.S.A. (Primate Research Centre)	H(P)
Brown, Hinkle, Trevethan, Kupper and McKee, 1973	U.S.A. (Primate Research Centre)	H(P)
Rhesus monkey		
Lucke, 1925 (Hindsight, based on the appearance of lesions)[b]	U.S.A.	H(L)
Baboon		
Lucke, 1925 (Hindsight, based on the appearance of lesions)[b]	U.S.A.	TC
Shadduck, Kelsoe and Helmke, 1979 (A microsporidian contaminant of a primary cell culture derived from baboon placental cells were ultrastructurally similar to *E. cuniculi*)[c]	U.S.A.	TC

TABLE 4.1 *Continued*

Authors	Locality	Method of diagnosis[a]
Man (Homo sapiens)		
Matsubayashi, Koike, Mikata, Takei and Hagiwara, 1959[d]	Japan	P
Stewart, van Dellen and Botha, 1981	South Africa	IFAT
Singh, Kane, Mackinlay, Quaki, Yap, Ho, Ho and Lim, 1982[e]	Europe, Ghana, Nigeria, Malaysia	IFAT
WHO Weekly Epidemiological Record, 1983[f]	Sweden	Serology
Bergquist, Stintzing, Smedman, Waller and Andersson, 1984[g]	Colombian (living in Sweden)	CIA; IFAT; P
Bergquist, Morfeldt-Mänon, Pehrson, Petrini and Wasserman, 1984[h]	Sweden	IFAT

N.B.: The parasite in the eye of a boy from Ceylon (Ashton and Wirasinha, 1973) has spores larger than *E. cuniculi* and cannot positively be attributed to *Encephalitozoon*. It is considered in Sect. III,F. Also the parasite in the cornea of a woman in Botswana, Africa (Pinnolis, Egbert, Font and Winter, 1981) has spores larger than those of *E. cuniculi* and with a greater number of coils of the polar tube. It is considered in Sect. III,H.

[a]H(P), histology (parasites seen); H(L), histology (lesions only); TC, tissue culture; ST, skin test; IFAT, indirect fluorescent antibody test; CIA, carbon immuno-assay (India ink test); IPT, immuno-peroxidase test; MAT, micro-agglutination test; SP, serial passage through host; P, parasites recovered from urine.
[b]See also Sect. IIIA,7b (Histopathology, p. 223).
[c]See also Sect. 6d (Tissue Culture, p. 213).
[d]See also Sect. 7a (Clinical Signs, p. 222); Sect. IV (Medical Importance, p. 239).
[e]See also Sect. IV (Medical Importance, p. 240).
[f]See also Sect. IV (Medical Importance, p. 239).
[g]See also Sect. 7a (Clinical Signs, p. 222); Sect. IV (Medical Importance, p. 239).
[h]See also Sect. IV (Medical Importance, p. 240).

reinstated the genus *Encephalitozoon*. *Encephalitozoon cuniculi* Levaditi, Nicolau and Schoen, 1923, is the correct name of this organism.

1970–present: Increasingly frequent reports of infection from a wide range of host species, including carnivores and primates; numerous studies on development, ultrastructure, pathogenicity, relationships with malignant tumours, transmission, tissue culture, serology and immunology; several review articles (Petri, 1969; Shadduck and Pakes, 1971; Wilson, 1979a).
2. BINOMIAL SYNONYMS: *Encephalitozoon rabiei* Manouelian and Viala, 1924; *Glugea lyssae* Levaditi, Nicolau and Schoen, 1924 (1924e); *Glugea rabiei* Levaditi, Nicolau and Schoen, 1926; *Encephalitozoon negrii* Manouelian and Viala, 1927; *Encephalitozoon muris* Garnham and Roe, 1954;

Nosema cuniculi (Levaditi, Nicolau and Schoen, 1923) Weiser 1964; *Nosema cuniculi* (Levaditi and Nicolau, 1923) Lainson, Garnham, Killick-Kendrick and Bird, 1964; *Nosema cuniculi* (Levaditi, Nicolau and Schoen, 1923) Petri 1969; *Nosema muris* Weiser, 1965; *Encephalitozoon matsubayashii* Sprague, 1977. Possibly also *Encephalitozoon chagasi* Torres, 1927 (1927b).

3. HOSTS AND LOCALITIES: The principal isolates, the geographical localities and method(s) of diagnosis are given in Table 4.1.

4. MORPHOLOGY AND DEVELOPMENT: The first stages of development, after ingestion of spores by the host have not been observed. Lainson, Garnham, Killick-Kendrick and Bird (1964) suggested that there might be a cycle of development in gut epithelial cells before invasion of other tissues. Cox, Hamilton and Attwood (1979) found no evidence of division stages in the gut wall of rabbits and suggested that viable spores might be taken up from the lumen of the gut by pinocytotic activity of the lymphoid tissue or by phagocytosis by white cells in the gut lumen and, thereby, enter directly into the cells in which they develop.

When the parasite is introduced intraperitoneally into mice, there is an acute phase of development with multiplication in macrophages, especially those of the peritoneal fluid, often with accumulation of ascitic fluid (Nelson, 1962; Lainson, Garnham, Killick-Kendrick and Bird, 1964). However, Gannon (1980b) found that when mice are infected by the oral route, the liver is the first organ to show infection and that peritoneal macrophages are not infected, at least until a late stage. He concluded that the invasive stages cross the gut wall and are carried to the liver in the hepatic portal vein. Cox, Hamilton and Attwood (1979) also found that the liver is infected together with the lungs and kidney, at an early stage in rabbits. A particular feature of the infection in carnivores is localisation of parasites in vascular endothelial cells, especially when the animals are infected by the transplacental route (Nordstoga and Westbye, 1976; Shadduck, Bendele and Robinson, 1978). In the chronic stages of infection, parasites tend to localise in large aggregates in the brain and kidney. In the kidney the parasites develop in vascular endothelial cells, Bowman's capsule and the tubule cells.

Development occurs within the host cells in vacuoles bounded by a membrane of presumed host origin (Weidner, 1975) (Figs. 4.8, 4.9). Morphologically the parasites appear the same, regardless of host origin. The life cycle has been studied in smears of kidney tubule cells (Ray and Raghavachari, 1941), mouse peritoneal macrophages (Lainson, Garnham, Killick-Kendrick and Bird, 1964) rat sarcoma cells (Petri, 1969) and in monolayers of cells in tissue culture (Vávra, Bedrník and Činátl, 1972). Ultrastructural studies on developmental stages and/or spores have been carried out using rat sarcoma cells (Petri and Schiødt, 1966) mouse peritoneal macrophages (Lainson, Garnham, Killick-Kendrick and Bird, 1964; Arison, Cassaro and Pruss, 1966; Akao, 1969; Sprague and Vernick, 1971; Barker, 1975; Weidner, 1975), tissues of rabbits (Cali, 1971; Hunt, King and Foster, 1972), dog (McCully, van Dellen, Basson and Lawrence, 1978; van Dellen, Botha, Boomker and Warnes, 1978), squirrel monkeys (Anver, King and Hunt, 1972; Brown, Hinkle, Trevethan, Kupper and McKee, 1973) and tissue cultures infected with isolates of *E. cuniculi* from rabbits, mice and hamsters (Pakes, Shadduck and Cali, 1975) rabbits (Hamilton, Cox and Pye, 1977) blue foxes (Mohn, Landsverk and Nordstoga, 1981), dog (Shadduck, Bendele and Robinson, 1978) and baboon (Shadduck, Kelsoe and Helmke, 1979).

Merogony: Typically, meronts, with one or two nuclei, are rounded oval or slightly irregular bodies, measuring 2–6 × 1–3 μm. These divide repeatedly by binary fission. Highly elongate meronts and occasional tetranucleate meronts have been seen (Fig. 4.10) but the multinucleate meronts recorded by Ray and Raghavachari (1941) are probably a misinterpretation.

At the ultrastructural level meronts can be seen lying isolated in small vacuoles or at the edge of larger vacuoles in close contact with the vacuolar membrane (Fig. 4.9). They have a simple plasma membrane, isolated nuclei and sparse cytoplasmic organelles.

Sporogony: Typically, binucleate sporonts divide into two uninucleate sporoblasts (Fig. 4.11). An increase in endoplasmic reticulum and division of the nucleus precedes cytoplasmic fission. Delayed cytokinesis and chain formation of sporoblasts have been seen (Vávra, Bedrník and Činátl, 1972).

A dense wall is laid down at the surface, first in bands then as a complete covering. The process begins on the free surface, away from the vacuolar membrane and is completed when the sporont lies free in the centre of the vacuole (Fig. 4.9). Sporoblasts are usually seen as crenated cells in which elements of the spore organelles are laid down (Fig. 4.12).

Spores: When fresh are ellipsoid, without obvious polar vacuole (Fig. 4.14) and measure about 2.5 × 1.5 μm (Lainson, Garnham, Killick-Kendrick and Bird, 1964; Petri, 1969). Measurements given by different authors vary from 1 × 2 μm (Gordon, 1940) to 1.5 × 3.5 μm (Ruiz, 1964). In the first description, Wright and Craighead (1922) said that the length and thickness were "probably never more than 4 and 1.5 microns respectively". The length of the extruded polar tube has been variously given as 20–25 μm (Nelson, 1962), 24 μm (Lainson *et al.*, 1964) and 30–35 μm (Petri, 1969). In stained spores a single nucleus (Fig. 4.13) and PAS positive polar cap are visible.

Ultrastructural characteristics are a corrugated exospore, thick endospore, 4.5 or 5 coils of the polar tube and often a posterior vacuole (Figs. 4.15, 4.16). McCully, van Dellen, Basson and Lawrence (1978) and Shadduck, Bendele and Robinson (1978) found 6 coils and 5–7 coils (Fig. 4.17) respectively in their isolates from dogs.

5. TRANSMISSION

a. *Oral and intranasal*. The kidney is an organ universally infected in the different hosts of *E. cuniculi*. The parasites invade the tubule cells, are discharged via the tubule lumina and can readily be detected in urine, directly (Goodman and Garner, 1972) or by immunofluorescence (Cox, Walden and Nairn, 1972; Cox and Pye, 1975; Cox, Hamilton and Attwood, 1979). Excretion of spores is sporadic (Cox and Pye, 1975) but they are nevertheless passed out from the kidneys to contaminate the environment and food, and spores must therefore constitute a common source of infection for horizontal transmission between animals in nature.

The oral route of infection has been established experimentally: Levaditi *et al.,* (1924d) used spores, concentrated from urine, to infect rabbits by stomach tube, as did Cox, Hamilton and Attwood (1979). Petri (1969) fed ascites sarcoma cells, infected with *E. cuniculi,* to rats and Nelson (1967) allowed weanling and nursling mice to suckle fluid containing spores from a pipette.

Infection may also be possible by the intranasal route with parasites entering directly into the lungs. Perrin (1943) transmitted the parasite to mice by instillation of spores into the nasal cavities and Nelson (1967) found that pulmonary lesions developed in 80% of mice after he had placed drops of infected ascitic fluid or lung suspensions on to the external nares. In both these cases spores could have been swallowed rather than inhaled. Cox, Hamilton and Attwood (1979) improved on the technique by injecting the spores intratracheally into rabbits: although they found groups of parasites in the lungs at 31 days post infection, even heavier pulmonary infections were obtained after the same interval, when the spores had been given orally. They concluded that inhalation of spores could lead to infections but that the pulmonary route was less likely than the oral route.

Infection by simple contact between animals has not been easy to achieve. Levaditi, Nicolau and Schoen (1924d) got positive results with rabbits, only when contact was prolonged (117–147 days). Perrin (1943) obtained only one infection, in an adult mouse, by contact but more readily transmitted the infection to nursling mice born of an infected mother. Nelson (1967) also experienced difficulty in obtaining contact transmission between mice. He failed to infect weanlings and obtained variable results with nurslings, which had been kept in contact with adult mice, which had been infected orally or by intraperitoneal inoculation. In contrast, and Nordstoga (1982) readily infected blue fox cubs by contact with infected foster mothers or other cubs.

Carnivores have access to infective stages by predation, the large aggregates of spores in the brain of the prey providing a more concentrated supply of infective stages for them, than does the spore-contaminated environment for the rodents. Mohn, Nordstoga and Møller (1982), Mohn and Nordstoga (1982) and Mohn, Nordstoga and Dishington (1982) have infected blue fox vixens by feeding them with mice infected with *E. cuniculi*.

b. *Transplacental.* The possibility that *E. cuniculi* can be transmitted from mother to offspring, via the placenta, has been suspected ever since the discovery of the parasite but it has been difficult to prove. When the litters born of infected mothers were examined, results were generally inconclusive. Sometimes, some litters became infected, others not, as found by Levaditi, Nicolau and Schoen (1924d). Sometimes there was no transmission to litters as found by Nelson (1967) using mice and by Petri (1969) using rats. Owen and Gannon (1980) traced the serological and histological history of rabbit dams and their litters, following natural infection in the does before mating and following experimental infection in specific pathogen free (SPF) does after mating. They concluded that none of the young had been infected transplacentally. Antibody titres in the neonates, which dropped steeply at 48 days, were attributed to maternal antibodies passed on to the young in the milk, while rising antibody titres in two nurslings at 8–10 weeks were attributed to infection by contamination. The experiments provided confirmatory evidence that contact infections are difficult to acquire and slow to develop.

The most convincing evidence for vertical transmission comes from experiments with gnotobiotic animals. Hunt, King and Foster (1972) found *E. cuniculi* in three rabbits that had been delivered by Caesarian section and reared in germ-free isolators and Innes, Zeman, Frenkel and Borner (1962) reported 50% prevalence in a colony of mice, established from four pairs, delivered by Caesarian section and fostered to germ-free rats: ultimately it was discovered that all four pairs of mice had cerebral lesions typical of *E. cuniculi*.

Experiments using foster mothers have generally given inconclusive results. Perrin (1943) transferred litters from infected mice to uninfected foster mothers and vice versa but, in one experiment, where the litter fostered by the uninfected mother became infected, he had not swapped the mice until two days after birth and thereby allowed time for postnatal transmission to have occurred. In another experiment the corresponding litter did not become infected and, in any case, the supposedly uninfected foster mother was subsequently shown to have a natural infection.

The occurrence of virulent infections in newborn dogs (Plowright, 1952; Plowright and Yeoman, 1952; Basson, McCully and Warnes, 1966) and blue foxes (Nordstoga, 1972) has added weight to the probability of transplacental transmission of *E. cuniculi,* as has the death from encephalitozoonosis, within 24 hours of birth, of a squirrel monkey baby (Anver, King and Hunt, 1972). Epidemiologically, the disease is interesting in blue foxes because it has only been manifested in cubs, the mothers of which appeared clinically healthy (Sect. 7,a). This suggests that when infections are acquired perorally by adult foxes, the disease is mild and undetected but, after transmission transplacentally from the vixen to the offspring, there is a severe pathogenesis resulting in high morbidity and mortality. Mohn, Nordstoga and Møller (1982) pursued the technique of cross fostering of cubs to investigate transmission across the placenta. In their experiments the cubs from three vixens were delivered by Caesarian section and transferred to foster mothers. Spores were not demonstrated histologically in any cubs examined immediately after the births. However, the mouse ascites test (Sect. 6,c) revealed infection in two cubs, both from the same vixen, which were culled immediately after the Caesarian birth and in six out of eleven cubs from the same vixen's litter, which had been killed by the foster mother two days after the transfer. This foster mother's own cubs were all negative. Only three cubs survived the fostering longterm: two of these developed severe disease with lesions typical of encephalitozoonosis, including demonstrable parasites; the other remained healthy. Parasites were demonstrated directly, or by inoculation into mice, in all thirteen placentae from the only vixen examined. These results provided strong evidence for transplacental transmission in the foxes, yet were somewhat vitiated by the manifestation of infection in one of the foster mothers and in four out of eight of her own cubs. The authors concluded that this foster mother had acquired the infection naturally on the farm and that infection in her cubs was further evidence of transplacental transmission.

It is likely that transplacental transmission can occur in all host species under the correct conditions, which may depend on the time that the infection was acquired in relation to pregnancy and the degree of infection. However, it appears to be common amongst the carnivores and rare in rodents and rabbits.

6. DIAGNOSIS

a. *Histological.* Direct histological examination of smears or sections of tissues, particularly kidney and brain may reveal parasites but, more commonly reveals only the inflammatory lesions which remain after destruction of parasites. The lesions are characteristic but not specific. Stains which are particularly useful for demonstrating the spores are Giemsa, which stain them pale blue and Gram, Goodpasture and Ziehl Neelsen, which stain them in characteristic shades of violet to reddish-purple. A comprehensive list of stains and reactions which distinguish *E. cuniculi* from *Toxoplasma gondii* and *Pneumocystis carinii* has been given by Wilson (1979). Disadvantages of the histological methods are that they can only be performed on biopsy or post-mortem material, are time consuming and are unreliable when infections are light.

b. *Urine sampling.* Urine sampling provides a method for diagnosis in the living animal. Spores passed in the urine can be detected by conventional stains as recommended by Goodman and Garner (1972) or by immunofluorescence (Cox, Walden and Nairn, 1972) but, again, this technique is unreliable unless successive urine samples are tested because release of spores from the kidneys is sporadic.

c. *Mouse ascites test.* The mouse ascites test, first demonstrated by Nelson (1967) is a useful diagnostic tool, when there is difficulty in finding low numbers of parasites by direct examination of tissues. Homogenates of tissues or urine sediments containing spores, when inoculated into mice, stimulate the production of ascitic fluid containing infected macrophages. Thus, abdominal distention, even without direct observation of parasitised cells is indicative of infection in the donor animal and, as urine can be used as a source of spores, the donor can be kept alive. Care has to be taken to ensure that the mouse recipient does not carry a latent infection. Although the results are not available for 2–3 weeks after inoculation of mice, the test is less labour intensive than direct histological examination.

Immunosuppressive drugs will activate latent infections. With the resultant renewal of infection in peritoneal macrophages, the drugs can be used to enhance the efficacy of mouse passage and reveal latent infections in apparently normal animals. The technique has been used successfully by Innes, Zeman, Frenkel and Borner (1962) using 2.5 mg cortisone acetate, Bismanis (1970) using hydrocortisone acetate and Kaneda (1969) using cyclophosphamide.

d. *Tissue culture.* Tissue culture has, on occasion, been a means of demonstrating latent infections of *E. cuniculi.* Parasites have appeared spontaneously in cells set up from laboratory animals. Thus, in cultures of rabbit choroid plexus cells (RCP), set up from an apparently normal rabbit, parasites were noticed at the seventh passage and thereafter growth of the organisms was prolific (Shadduck, 1969). Armstrong, Ke, Breinig and Ople (1973) found abundant parasites, even in primary cultures set up from rabbit kidney cells for assay of interferon: as many as four out of six rabbits from one colony gave contaminated cultures. An interesting isolate was obtained in baboon placental cells, when an organism, closely resembling *E. cuniculi* in its ultrastructure, was observed in the cells, after they had been through eight subcultures over a period of twelve weeks. Abundant developmental stages and spores were seen by electron microscopy but presumably had not been seen while the cultures were alive (Shadduck, Kelsoe and Helmke, 1979).

Cultured *E. cuniculi* have also been useful in studies on the development of the parasite (Vávra, Bedrník and Činátl, 1972) but by far the most important application of the culture system is as a prolific source of antigen for serological tests. Numerous tests have been devised which have simplified the diagnosis of the disease and have enabled whole stocks of laboratory animals to be screened, on a scale which could not previously have been contemplated.

Successful tissue culture systems are summarised in Table 4.2. The cells were either infected spontaneously or experimentally by adding mouse ascitic fluid, tissue homogenates or spores purified from urine or from the cells and medium of other tissue cultures. Some cell types support the growth of *E. cuniculi* better than others. In a study of isolates of *E. cuniculi* from rabbit, mouse and hamster, Montrey, Shadduck and Pakes (1973) compared the replication indices (total number of organisms recovered from the cultures in a given time, divided by the number introduced) in a variety of primary and established cell cultures. In several cell types the indices were less than 1 but in rabbit choroid plexus cells (RCP) there was 100-fold replication and in canine embryo cells it was 50-fold. Prolific growth of *E. cuniculi* in RCP cells, compared with other cell types, was also obtained by Vávra, Bedrník and Činátl (1972) and it led to destruction of the cultures in 1–3 weeks. Excellent multiplication has also been obtained in a human diploid foetal tongue cell line (CSL 300) and in a human foetal lung cell line (W138) by Cox and Pye (1975). Although perhaps not the most prolific of spore producers, the canine kidney cell line MDCK is an ideal system because, once the level of infection in the monolayer has stabilised at about 10 weeks from inoculation, a weekly output of 10^7 or even nearly 10^8 spores can be obtained from a monolayer of 75 cm^2 surface area. An equilibrium is established in the MDCK cells, whereby destroyed cells are continuously replaced by new cells, and the cultures can be maintained indefinitely without addition of new cells (Waller, 1975). Thus, from the point of view of convenience the MDCK cells are superior to RCP cells. Bismanis (1970) also found that an equilibrium was established between *E. cuniculi* and a mouse embryonic cell line but, at only a 2% level of cells infected, the cultures were poor producers.

The number of parasites needed to infect cultures is quite small, of the order of 200–500 organisms (Pye and Cox, 1977) but routine inoculations of about 1×10^7 organisms per culture flask are used, to reduce the time before maximum output. Optimum temperature is 37°C but Bedrník and

TABLE 4.2   Cell types supporting growth of *Encephalitozoon cuniculi*

Cell types	Origin of parasites	Reference
1. Mouse Lymphosarcoma MB III	Mouse ascites	Morris, McCown and Blount, 1956
2. Mouse fibroblast (L-cells)	Mouse ascites	Kaneda, 1969; Vávra, Bedrník and Činátl, 1972; Montrey, Shadduck and Pakes, 1973
3. Mouse embryonic fibroblasts	Mouse ascites	Chumakov, Viting, Konosh and Ashmarina, 1970
4. Mouse embryonic cells	Mouse ascites	Bismanis, 1970
5. Hamster, plasmacytoma cells	Hamster plasmacytoma	Kinsel and Meizer, 1968; Meizer, Kinzel and Jírovec, 1971
6. Baby hamster kidney (BHK)	Rabbit brain	Montrey, Shadduck and Pakes, 1973
7. Rabbit choroid plexus cells (RCP)	Spontaneous and rabbit brain	Shadduck, 1969; Vávra, Bedrník and Činátl, 1972; Montrey, Shadduck and Pakes, 1973; Shadduck, Watson, Pakes and Cali, 1979
8. Weanling albino rabbit cells	Contaminant from rabbit	Armstrong, Ke, Breinig and Ople, 1973
9. Primary glial cells (rabbits, mice and hamsters)	Spontaneous infection in RCP cells, rabbit brain, mouse ascites and hamster plasmacytoma	Montrey, Shadduck and Pakes, 1973
10. Primary kidney cells (rabbits, mice and hamster)		
11. Rabbit corneal cells (SIRC)		
12. Rabbit kidney cell line (RK13)		
13. Rabbit kidney cell line (RK13)	Mouse ascites from rabbit isolate	Waller, 1975
14. Rabbit fibroblasts	Rabbit brain	Niederkorn, Shadduck and Weidner, 1980
15. Pig kidney cells (PK)	Mouse ascites	Vávra, Bedrník and Činátl, 1972
16. Ovine choroid plexus cells	Blue fox tissues	Mohn and Ødegaard, 1977
17. Bovine kidney cells (BK)	As 13	Waller, 1975
18. Canine embryo cells	As 9-12	Montrey, Shadduck and Pakes, 1973
19. Primary canine kidney	Canine brain, kidney and spleen	Shadduck, Bendele and Robinson, 1978
20. Canine embryo fibroblasts	As 19	As 19

(Continued)

TABLE 4.2 *Continued*

Cell types	Origin of parasites	Reference
21. Primary canine embryo fibroblasts	Urine sediments from rabbits	Cox and Pye, 1975
22. Canine embryo fibroblast cell line	As 21	As 21
23. Canine kidney cell line (MDCK)	As 13 Mouse ascites Blue fox brain and kidney	Waller, 1975 Gannon, 1978 Mohn, Landsverk and Nordstoga, 1981
24. Feline lung cell line (FL)	As 13	Waller, 1975
25. Human uterine cancer cells (HeLa)	Mouse ascites	Tsuneo, 1960; Kaneda, 1969; Vávra, Bedrník and Činátl, 1972
26. Ehrlich cancer cells	Mouse ascites	Tsuneo, 1960
27. Human embryonic fibroblasts	Mouse ascites	Chumakov, Viting, Konosh and Ashmarina, 1970
28. Human foetal lung cell line (W138)	Culture medium from canine embryo cultures (rabbit isolate)	Cox and Pye, 1975
29. Human diploid foetal cell line (CSL300)	As 28	Cox and Pye, 1975
30. Primary baboon placental cells	Contaminant from baboon placenta	Shadduck, Kelsoe and Helmke, 1979
31. Chick embryonic fibroblasts	Mouse ascites	Kalyakin and Akinshina, 1970
32. Chick embryonic kidney cells	Mouse ascites	Vávra, Bedrník and Činátl, 1972
33. Fathead minnow (fish) cells, *Pimephales promeles*	Not stated	Bedrník and Vávra, 1972

Vávra (1972) found that in RCP cells growth continued at a slower rate at 25°C and, in fathead minnow cells, even at 18°C the parasite could complete its life cycle and produce spores. Other factors affecting the replication rate may be the number of passages of the cell line and the age of the cells after passage, as Shadduck and Polley (1978) found that the susceptibility of RCP cells increased with successive passages and that cells were less susceptible in the first week after passage than later. Antibiotics do not inhibit the growth of *E. cuniculi*.

e. *Skin hypersensitivity test* (SHT). A skin test for use with rabbits was designed by Pakes, Shadduck and Olsen (1972). Organisms harvested from RCP cells, purified and disrupted, are used as the antigen (encephalito-zoonin) and inoculated into the skin of the abdomen. Induration and erythema of at least 3mm diameter and 1mm thickness, lasting 72 h, are criteria for a positive reaction.

The test can used reliably to demonstrate infection. In the hands of Pakes, Shadduck and Olsen (1972) infections were confirmed in 24 experimentally infected rabbits and naturally occurring encephalitozoonosis was revealed in 16 out of 60 apparently healthy rabbits from diverse rabbit farms. The results correlated well with the histological picture and there seemed to be no problems with false positives or cross reactions with coccidia or *Toxoplasma*. The disadvantages of the test are the large quantity of organisms required for preparation of the inoculum, equivalent to 5×10^6 spores per rabbit, the necessity of exposing the animals to antigen and the delay in reading the test.

f. *Serological tests*. Several established serological tests have been modified for use with *E. cuniculi,* each with its particular advantages and disadvantages. A microprecipitation test in agar gel (Hübner, Uhliková, Bedrník and Vávra, 1973) has not been explored since the preliminary communication. All of the following serological tests require less antigen than the SHT and are reasonably simple to perform.

Indirect Fluorescent Antibody Test (IFAT)
Carbon Immuno-Assay (CIA)
Complement Fixation Test (CFT)
Immuno-Peroxidase Test (IPT)
Enzyme Linked Immuno-Sorbent Assay (ELISA)
Indirect Microagglutination Test (IMT)

The IFAT was first introduced into studies of *E. cuniculi* by Chalupský, Bedrník and Vávra (1971) and Chalupský, Vávra and Bedrník (1973) who used spores from tissue culture as antigen to detect infection in rabbits. As the problem of antigen availability had been solved by concurrent developments in tissue culture of the organism, numerous authors confirmed the reliability of the IFAT by using experimentally infected rabbits (Cox and Pye, 1975; Cox and Gallichio, 1977), blue foxes (Mohn and Ødegaard, 1977) and dogs (Stewart, Botha and van Dellen, 1979). Wosu, Shadduck, Pakes, Frenkel, Todd and Conroy (1977) found that the IFAT and SHT were equally effective in detecting infections. Cox, Walden and Nairn (1972) had used the IFAT as a sensitive staining method for revealing spores in urine and in tissue smears and sections of rabbits. These authors also used organisms, concentrated from urine, to detect serum antibody.

The simplicity of the test has made it possible to survey large numbers of animals for infection. Thus Cox and Ross (1980) found no infections in 175 wild rabbits in England and Scotland and Singh, Kane, Mackinlay, Quaki, Yap, Ho, Ho, and Lim (1982) obtained high titres to *E. cuniculi* in rabbits and low or moderate titres in sheep, rats and in human sera from Ghanaian, Nigerian and Malaysian patients (Sect. IV). In dogs in South Africa the prevalence rate was found to be 18% in serum samples submitted for various clinical pathological examinations and 70% from kennels where *E. cuniculi* infection had been confirmed (Stewart, van Dellen and Botha, 1981).

An advantage of this test is that serum antibodies can be demonstrated early in infection, for example at least two weeks before the organisms can be demonstrated histologically in rabbits and at least four weeks before there are pathological lesions or spores passed in the urine (Cox and Gallichio, 1978). The fact that rabbits become seropositive before they can transmit the infection helped Cox, Gallichio, Pye and Walden (1977) to establish an *Encephalitozoon*-free rabbit colony.

Disadvantages of the test are the subjective nature of the end point determination and the requirement of expensive equipment, namely the fluorescence microscope, for reading the results.

The CIA, modified by Waller (1977) for *E. cuniculi,* from a simple India ink reaction previously used for enteric bacteria, is cheap and easy to perform and uses formalin-fixed organisms. Waller (loc. cit.) found complete concordance of the CIA, SHT and IFAT on the same rabbit sera, and when applied to rabbits, the CIA revealed prevalences of 9.1% to 81.9% in different colonies. Kellett and Bywater (1978) and Bywater and Kellett (1978a,b) applied the test to rabbits and, by culling animals immediately they became seropositive, quickly eliminated the parasite from a specific pathogen free colony.

The test, which relies on adherence of microscopic carbon particles from India ink to rabbit IgG, was used for the simultaneous diagnosis of encephalitozoonosis and toxoplasmosis in rabbits (Waller and Bergquist, 1982). Kellett and Bywater (1980) found that the simple test was not suitable for use with rat and mouse serum and developed an indirect carbon immunoassay for use with these animals. This indirect test has been extended to the simultaneous diagnosis of toxoplasmosis and encephalitozoonosis in sera from cats, dogs, goats, horses, rabbits, sheep and swine (Waller, Uggla and Bergquist, 1983) and, with IFAT provided the first serological evidence for the involvement of *E. cuniculi* in clinical illness in a human patient (Weekly Epidemiological Record, 1983). The CIA has also been used to demonstrate that antibodies to *E. cuniculi* are common in commercially available rabbit antisera and serum products (Bywater, Kellett and Waller, 1980). A further refinement of the CIA was in the detection of antibodies in blood collected on filter paper (Waller, Lyngset and Morein, 1979) and a technique for photographing the reaction was published by Kellett and Bywater (1979).

The immunological basis of this reaction is not fully understood, although it is known that the carbon particles are adsorbed non-specifically to the heavy chains of rabbit IgG. The early humoral response to *E. cuniculi* in rabbits is largely by IgM production (Waller, Morein and Fabiansson, 1978) and thus the CIA is unlikely to pick up positives in the early stages of infection. This was shown by Mohn (1982), who found that antibodies could be detected in infected blue foxes at 20 days by IFAT but not until 35 days by CIA. Another disadvantage of the CIA is that not all brands of India ink are suitable.

The CFT developed by Wosu, Olsen, Shadduck, Koestner and Pakes (1977) was found to be a sensitive test, capable of detecting infections in rabbits as early as 15 days. The antibody response is easy to quantify by this method but there are disadvantages in the lengthy processing of the antigen and in that the different immunoglobulin classes cannot be distinguished.

The IPT (Gannon, 1978) has advantages, similar to those of the CIA, in that the preparation can be kept indefinitely and requires no special apparatus to read but care is required in interpreting the results, as there is a continuous reduction in the brown staining of the spores with increasing dilution rather than a definite cut off point. The ELISA test used by Cox, Horsburgh and Pye (1981) and Hollister and Canning (1985) is a major advance, in that it enables the procedure to be standardised completely and quantified spectrophotometrically, thereby eliminating operator subjectivity. The major disadvantage is the requirement for expensive equipment, especially an ELISA reader, for processing large numbers of plates.

The IMT used by Shadduck and Geroulo (1979) using as substrate hydrophilic beads to which rabbit IgG was coupled gave serum titres almost invariably within one dilution of those obtained by IFAT on the same sera. Although the test is cheap and simple to perform, the preparation is as time-consuming as the IFAT and vigorous shaking throughout the incubation of sera with the beads is essential. Owing to some non-specific adherence of *E. cuniculi* spores to the beads, counting of large numbers of beads bearing agglutinated parasites is necessary, until experience is gained: thereafter simple inspection for a background virtually free of spores and clumps of spores adherent to the beads is sufficiently reliable. The IMT as well as the IFAT, IPT and ELISA can be used to measure the levels of the different immunoglobulin classes separately.

7. PATHOGENESIS

a. *Clinical signs*. The disease encephalitozoonosis varies markedly according to host species. In rodents and rabbits *E. cuniculi* generally causes a mild, chronic disease without clinical signs. Often, characteristic lesions with or without parasites, can be demonstrated histologically in apparently healthy animals. Occasionally, overt disease is expressed. In the first paper describing the organism in rabbits, Wright and Craighead (1922) reported a progression from drowsiness, through tremor, to slight or marked, local or general paralysis. Cameron and Maitland (1924) described a similar disease in rabbits, manifested by early fever, followed by dullness, stupor and some paralysis. In an outbreak among broiler rabbits (Pattison, Clegg and Duncan, 1971) there was morbidity in about 15% of young rabbits from the age of 5 weeks, especially ataxia and paralysis, but low mortality: survivors continued to exhibit a variety of neurological signs. A mortality rate of 50%, in the 0–16 weeks age group, was recorded in a laboratory colony (Cox, Walden and Nairn, 1972) where some rabbits also developed opisthotonus, torticollis, hyperaesthesia and paralysis. Torticollis is well illustrated in Julini (1983). Several authors have described stunting, emaciation and roughened pelts (Smith and Florence, 1925; Robinson, 1954; Reddy, 1963) and a mild suppurative endometritis in females has been described by Yost (1958). However, by and large, rabbits remain outwardly healthy, even in colonies where transmission must be facilitated and prevalences can be very high. The average weight of the meat carcass of positive animals is, however, 11% lower than that of negative ones (Vávra, Chalupský, Oktábec and Bedrník, 1980). Symptoms probably appear only when the animals are under stress.

Mice, when inoculated intraperitoneally but not usually when other routes are used, exhibit marked abdominal distention at 10–14 days, due to accumulation of ascitic fluid (Nelson, 1967). At this time there may be a low mortality rate but the fluid is resorbed in the survivors and the mice return to normal over a period of weeks. Anomalous behaviour has not been reported in mice and rats.

The picture is very different in carnivores. The disease is most fully described in blue foxes and dogs, in which many more cases of encephalitozoonosis have been described than in other carnivores. Outbreaks in blue foxes in Scandinavian countries, where the fur industry is an important economic development, have been documented by Nordstoga (1972); Kangas (1973); Nordstoga, Mohn and Loftsgard (1974); Mohn, Nordstoga, Krogsrud and Helgebostad (1974); Nordstoga and Westbye (1976); Nordstoga, Mohn, Aamdal and Helgebostad (1978); Mohn, Nordstoga and Møller (1982) and Mohn, Nordstoga and Dishington (1982). There have been reports from dogs in England (Plowright, 1952), Tanganyika (Tanzania) (Plowright and Yeoman, 1952), South Africa (Basson, McCully and Warnes, 1966; McCully, van Dellen, Basson and Lawrence, 1978; van Dellen, Botha, Boomker and Warnes, 1978; Botha, van Dellen and Stewart, 1979; Stewart, van Dellen and Botha, 1979), and from the U.S.A.

(Shadduck, Bendele and Robinson, 1978; Cole, Sangster, Sulzer, Pursell and Ellinghausen, 1982).

In blue foxes severe clinical signs only appear in cubs, which acquire the infections *in utero* from infected vixens. In these cubs the mortality rate can be very high, as for example in an outbreak in Norway, affecting 20 farms and causing the death of 1500 cubs representing 33% of the litters. Within a few weeks of birth the cubs show signs of reduced appetite and growth, ataxia and posterior weakness, then blindness due to cataract and severe neurological disorder, in the form of convulsions and circling behaviour. Cubs infected neonatally sometimes show subclinical disease with roughened pelts and stunted growth, from which they recover to reach normal weights, with good fur quality at pelting time.

There is almost certainly a similar dependence on transplacental transmission for induction of clinical disease in cats and dogs. The disease, described as an "encephalitis-nephritis syndrome" (a misnomer because other organs are consistently parasitised) has been reported only in littermate puppies, of which the parents were clinically healthy. The puppies exhibit a range of symptoms including nephritis, anorexia, stunting and emaciation, weakness and depression, ataxia, tremors, blindness, convulsions and aggressive behaviour. Sometimes almost all puppies in the litters die. The disease in two littermate puppies in Tanganyika was likened to rabies (Plowright and Yeoman, 1952) and the differential diagnosis from distemper, "fading-puppy" syndrome and other canine diseases may be difficult (Botha, van Dellen and Stewart, 1979).

Severe encephalitozoonosis which occurred in a litter of Siberian polecats, *Mustela eversmanni satunini,* born of an apparently healthy dam, which was one of 46 polecats imported into the U.S.A. from the U.S.S.R., provides yet further evidence that disease is linked with transplacental transmission (Novilla, Carpenter and Kwapien, 1980).

Of only two case reports of encephalitozoonosis in cats, one deals only with histopathology (Schuster, 1925) and the other describes littermate Siamese kittens, which became ill simultaneously and exhibited spasms and twitching of muscles (van Rensburg and du Plessis, 1971). The CIA test was positive to *E. cuniculi* in 2 out of 22 cats in Sweden, some of which had exhibited neurological disorders (Waller, Uggla and Bergquist, 1983). A microsporidium, which damaged the cornea of an otherwise healthy adult cat (Buyukmihci, Bellhorn, Hunziker and Clinton, 1977), differs from *E. cuniculi* in morphology and is considered to be a distinct species (Sect. G, p.235).

In contrast to the above strong evidence that clinical disease is limited to transplacentally infected animals, there was an outbreak in wild (presumably adult) carnivores at the Prague Zoo (Vávra, Blažek, Lávička, Koczková, Kalafa and Stehlík, 1971), all of which had exhibited neurological signs before death, which were linked with pathologic changes in the brain and the presence of agglomerations of *E. cuniculi*, 60–120 μm in diameter (Figs. 4.18, 4.19). Infection was thought to have been derived from rodents fed to the carnivores.

There are only four records of diseases in primates, which can be ascribed to *E. cuniculi* with any certainty. Organisms resembling *E. cuniculi* were isolated on the 5th day of illness from the cerebro-spinal fluid and on the 11th-13th days from the urine of a 9-year-old Japanese boy (Matsubayashi, Koike, Mikata, Takei and Hagiwara (1959). The boy had been admitted to hospital unconscious and with fever, after suffering spasmic convulsions. Vomiting and headache were recurrent after recovery of consciousness. The patient was discharged from hospital 24 days after the onset of illness and apparently made a full recovery. Doubts have been cast on the aetiology because passage into mice had been used as an aid to diagnosis. However, the demonstration of the organisms themselves is sufficiently conclusive.

Another child, a Colombian boy about 3 years old, adopted by a Swedish family, suffered a similar convulsive illness (Bergquist, Stintzing, Smedman, Waller and Andersson (1984). He suffered a series of convulsions, responded to anticonvulsive therapy but later developed stiffness and painful joints which were successfully treated with antiinflammatory drugs. His health has remained excellent since 1983. Spores were isolated from his urine and all sera showed high IgG titres (1/2560) with one sample, taken several months before the onset of the illness and examined retrospectively, also showing IgM antibodies (1/160).

Two squirrel monkeys, *Saimiri sciureus,* born in captivity died at 25 hours and 2 months respectively (Anver, King and Hunt, 1972; Brown, Hinkle, Trevethan, Kupper and McKee, 1973). The first was found dead but the second had suffered frequent *petit mal* seizures from about one month old until death and had failed to make normal weight gains. Encephalitozoonosis was confirmed on post mortem examination, by demonstration of characteristic lesions and presence of parasites in the brains of both animals and in the liver and kidney of the older monkey. In size and ultrastructure the spores were typical of *E. cuniculi*.

b. *Histopathology*. The parasites develop in aggregates within parasitophorous vacuoles in a variety of cells. The largest, most conspicuous aggregates, containing thousands of organisms, are in the brain (Figs. 4.18, 4.20), but they are also very numerous in the kidney tubule cells which become hypertrophic and release spores into the tubule lumina. The parasites infect endothelial cells and macrophages and thus, can be found in most tissues, including liver, spleen, suprarenal glands, pancreas, lungs, myocardium and placenta, but less abundantly than in the brain and kidney. While the host cell remains intact there is no host reaction but breakdown and release of organisms is followed by an intense cellular reaction (Figs. 4.20–4.23) leading to lesions which persist after the disappearance of the organisms themselves. Hence, lesions on their own can be strongly indicative of infections and it has been found that seropositivity correlates well with the histological demonstration of lesions. Many early accounts of spontaneous encephalitis in animals, which include several before the organism itself was described (see History, Sect. IIIA,1), are typical of encephalitozoonosis. Particularly interesting are the necropsy reports on three monkeys (two baboons and one Rhesus monkey) by Lucke (1925), which are strongly suggestive of lesions due to *E. cuniculi*. If so, it may be that infections are commoner in primates than is suspected at present. However, Stewart, van Dellen and Botha (1981) found no evidence of natural infection in 127 vervet monkeys from the Kruger National Park in South Africa.

Neural lesions can be present at all levels of the brain and spinal cord, in grey or white matter. The leptomeninges may be focally or diffusely infiltrated by large and small mononuclear cells, including plasma cells, especially around the blood vessels. The encephalitis is manifested by microgranuloma formation after rupture of the host cell and release of individual spores (Fig. 4.20). The granulomata are focal rather than diffuse and are characterised by infiltrations of mononuclear cells including lymphocytes, plasma cells and macrophages, often around a necrotic centre. In an ultrastructural study of the histopathology in dogs, van Dellen, Botha, Boomker and Warnes, (1978) concluded that the granuloma is a vascular lesion, even though the lumen of the vessel may sometimes no longer be apparent. It is always vascular endothelial cells which are parasitised, while reactive perithelial cells transform into epithelioid cells or macrophages in concentric layers and are the major contributors to the perivascular inflammation. Viable organisms never appear other than in endothelial cells, those in perithelial cells and glial macrophages becoming degenerate, after phagocytosis and degradation. Lymphoid cells infiltrate between the perithelial cells, and glial cells are reactive around the vasculitis. These authors concluded that vasculitis, in the form of perivascular cuffing, is the underlying lesion of canine encephalitozoonosis.

Kidney damage is often indicated in rabbits and carnivores at the surface by small indented grey areas. In all hosts tubule cells in both cortex and medulla, can be infected, as well as cells of the glomerular capillaries and Bowman's capsule (Canning, 1967; McCully, van Dellen, Basson and Lawrence, 1978). Destruction of these cells usually leads to an acute interstitial nephritis (Fig. 4.22) with lymphocytes, some plasma cells and numerous macrophages, although this is not typical of mice and rats. Nephritis progresses to fibrosis (Fig. 4.21), which results in the cortical scarring sometimes seen at the surface. Granuloma formation is also common in liver (Fig. 4.23). Similar inflammatory reactions can be found according to host in most, if not all, other organs, including the placenta, where endometritis was observed in rabbits (Yost, 1958) and in blue fox vixens, which had transitted *E. cuniculi* to their offspring (Mohn, Nordstoga and Dishington, 1982).

Cardiovascular changes are of almost universal occurrence. Perivascular cuffings, involving infiltration of inflammatory cells around the blood vessels, are widespread and account for lesions in the liver, myocardium and lungs as well as for those in the brain. Severe vasculitis is more common in carnivores but arteritis, with extensive degenerative changes in the media, thickening of the arterial walls and focal myocardial necrosis has been recorded in rabbits (Robinson, 1954).

Polyarteritis nodosa is a particular feature of encephalitozoonosis in fox cubs, which have acquired the infection *in utero* (Nordstoga, 1972; Nordstoga and Westbye, 1976; Mohn and Nordstoga, 1982). These thickened and nodular vessels, altered by cellular infiltrations and/or fibrotic changes in all layers (Fig. 4.24), are consistently found in the small or medium sized arteries of the viscera but the most prominent lesions are in the superficial coronary arteries where they can be seen macroscopically. Sometimes the lumen is totally occluded. Similar nodular lesions of the blood vessels have been observed in dogs (Shadduck, Bendele and Robinson, 1978). The polyarteritis nodosa is invariably accompanied by plasma cell proliferation and hypergammaglobulinaemia (Mohn and Nordstoga, 1975; Nordstoga and Westbye, 1976). These authors have suggested that the pathological condition may be the result of a hypersensitivity reaction, but van Dellen, Botha, Boomker and Warnes (1978) proposed that simple occlusion of the vessel may be responsible. In contrast to the polyarteritis, thickening of the walls of the myocardial arteries, by fibrocellular changes in the intima only, is a feature of orally infected, rather than transplacentally infected foxes (Mohn and Nordstoga, 1982).

The commonly occurring neurological signs of encephalitozoonosis are clearly a reflection of the brain lesions, and abnormal thirst reflects renal damage. Death in the early stage of the disease in fox cubs was attributed to cerebral haemorrhages, while renal dysfunction was proposed as the direct cause of death when the disease followed a protracted course with severe interstitial nephritis (Mohn, 1983).

c. *Ocular encephalitozoonosis*. In the case reports of ocular involvement in encephalitozoonosis, perivascular lesions of the interocular arteries and cataractal changes of the lenses have been described. In a rat, lesions and parasites were restricted to the retina (Perrin, 1943). In a rabbit, parasites were found only in the lens (Fig. 4.25) but inflammatory reactions were seen both in the lenses and their capsules, in the adjacent ciliary bodies and in the surface layers of the retina, especially in association with blood vessels (Ashton, Cook and Clegg, 1976). In a blue fox, parasites and lesions of the polyarteritis nodosa type were found in the walls of the posterior ciliary arteries (Fig. 4.26) and their intraocular branches and huge numbers of spores were present in the cataractous lenses (Fig. 4.27) (Arnesen and Nordstoga, 1977).

The ocular lesions reported by Ashton and Wirasinha (1973), Buyuk-mihci, Bellhorn, Hunzinger and Clinton (1977) and Pinnolis, Egbert, Font and Winter (1981) were caused by microsporidia, which in the present work are considered distinct from *E. cuniculi* (Sects. F-H, pp. 234, 235).

8. MECHANISMS OF RESISTANCE IN *E. cuniculi* INFECTIONS

In common with many other parasites *E. cuniculi* is able to persist in its hosts in the face of an immune response. In rabbits and rodents the infection is chronic and clinical signs are rarely expressed, which indicates that a balance has been struck between parasite multiplication and the immune response. Uncontrolled multiplication would result in the death of the host but an overreactive immune response by the host might itself cause tissue damage, by the formation of immune complexes. This probably happens in blue fox encephalitozoonosis: hypergamma-globulinaemia always accompanies clinical disease in cubs when they are infected trans-placentally (Mohn and Nordstoga, 1975). Polyarteritis nodosa, the severe disease exhibited by these cubs is a condition which is usually associated with an immunological disturbance, possibly by a type III hypersensitivity reaction (Nordstoga and Westbye, 1976; Mohn, 1983).

The survival and multiplication of *E. cuniculi* in its host cells, which include macrophages, is due to a failure of lysosomes to fuse with par-asitophorous vacuoles (Weidner, 1975).

a. *Humoral response*. A humoral response to *E. cuniculi* is mounted within a few weeks and persists for many months if not for the duration of the infection (Bywater and Kellett, 1979). In rabbits IgM antibodies appear early in infection, at about 20 days post inoculation and the IgG response comes a little later, at about 35 days (Waller, Morein and Fa-biansson, 1978). A similar pattern of response is assumed to occur in blue foxes because immunofluorescence, which detects both immunoglobulin classes, reveals antibodies earlier than the india ink reaction which detects only IgG (Mohn, 1982) However Gannon (1980b) using mice, found very little difference, if any, in the time when the two classes of antibodies appeared.

Lyngset (1980) investigating a rabbit colony, in which there was a 73% prevalence of encephalitozoonosis, showed that maternal antibodies are passively transmitted to their offspring and persist for up to 4 weeks. Further, antibodies have been found to persist in rabbits for at least two days after death: they are always present in pleural fluid and in extracts of kidney and lung taken *post mortem* and are sometimes present in extracts from other organs (Waller, Lyngset, Elvander and Morein, 1980).

The humoral response to *E. cuniculi* in euthymic mice is specific and anamnestic (Schmidt and Shadduck, 1983). The response is expressed by (a) lower parasite burdens when previously-infected mice are challenged, compared with burdens in previously-uninfected mice given the same dose; (b) specific antibodies; (c) specific delayed type hypersensitivity; and (d) a secondary anamnestic humoral response with higher antibody titres and shorter latent period than in a primary response.

The role of the antibodies is not entirely clear. If *E. cuniculi* spores are treated with immune serum and complement and are then added to cell cultures, there is a reduction in the number of cells which become infected. Spores have a thick chitinous wall, almost certainly impervious to complement mediated lysis, so it is unlikely that antibodies and complement kill the spores. Rather these serum components may act by blocking the ligand–receptor interactions which enable the parasites to enter host cells (Schmidt and Shadduck, 1984). Some antibodies in mice have been shown to have an opsonising effect *in vitro*, which enables macrophages to kill the parasites on entry (Schmidt and Shadduck, 1984).These authors suggested that complement fixation possibly attracts macrophages to the site of the opsonised parasites, which are then ingested and degraded. Opsonising antibodies have also been demonstrated in rabbits by Niederkorn and Shadduck (1980): the opsonised parasites, when taken up by the macrophages, are no longer able to prevent lysosomal fusion with the parasitophorous vacuoles and are thus digested in the vacuoles.

In contrast to euthymic mice, no antibody response is mounted by athymic (nude) mice (Gannon, 1980; Schmidt and Shadduck, 1983). The infections progress rapidly and the mice die at about 25 days post inoculation. Immune serum from sensitised euthymic mice transferred to athymic mice does not halt the progression of the *E. cuniculi* infection nor extend the survival time of the mice. The fine balance between parasite multiplication and host survival in euthymic animals is therefore not governed by antibodies alone.

b. *Cell mediated immunity*. The role of T-cells in controlling infections was investigated by Schmidt and Shadduck (1984). Resistance to lethal disease was found to be T-cell dependent in experiments with T-cell transfers, i.e. that T-cell enriched spleen cells, but not T-cell depleted spleen cells, from sensitised donor mice will protect athymic mice infected with *E. cuniculi.* Also spleen cells, from sensitised donor mice, which are B-cell or macrophage depleted still protect athymic mice. T-cells and macrophages may co-operate to prevent lethal disease, since fewer T-cells are required to protect athymic mice in the presence of macrophages. The mechanism of the co-operation appears to be activation of the macrophages by T-cell lymphokines. Culture supernatants, containing *E. cuniculi*-induced lymphokines, when added to cultures of macrophages infected with *E. cuniculi,* reduce the number of viable organisms recoverable. Schmidt and Shadduck (1984) felt that the ability of corticosteroid treatment to reveal latent *E. cuniculi* infections (Innes, Zeman, Frenkel and Borner, 1962; Bismanis, 1970) could be interpreted in terms of the lymphocyte factors which activate macrophages, i.e. that lymphokine activation is blocked by hydrocortisone, allowing the *E. cuniculi* to survive within non-activated macrophages.

No evidence was obtained by Schmidt and Shadduck (1984) that there was any action by cytotoxic T-cells against either *E. cuniculi* or macrophages infected with *E. cuniculi.*

9. INTERACTIONS OF *E. cuniculi* WITH CONCURRENT INFECTIONS

There are several examples of non-specific resistance induced by *E. cuniculi.* Lepine and Sauter (1949) reported that mice naturally infected with *E. cuniculi* are resistant to infection with strains of *lymphogranulomata venereum.* Similarly Armstrong, Ke, Breinig and Ople (1973) found that cultures of rabbit kidney cells, contaminated with *E. cuniculi,* are resistant to vesicular stomatitis virus and they suggested that this is due to the elaboration of an interferon-like molecule by the kidney cells when infected with *Encephalitozoon.*

In contrast, *E. cuniculi* depresses the humoral antibody response of rabbits to *Brucella abortus* (Cox, 1977b): there is a depression of IgG response from the fifth week and an elevation of the IgM response from the eighth week. The *E. cuniculi* spores constantly released from degenerate cells, might be in competition with *B. abortus* for "uncommitted" macrophages and lead to a shortage of macrophages to process the *B. abortus* antigen. Also the development of *E. cuniculi* within macrophages would cause them to become defective and render them unable to process or present antigens to lymphocytes. Both factors might inhibit a T-cell mediated switch from IgM to IgG synthesis and account for the altered immune response.

Huldt and Waller (1974) found that double infections of *Toxoplasma gondii* and Maloney virus inoculated into newborn mice precipitated heavy infections of *E. cuniculi.* The mice were immunologically immature when they received the inoculum and were further immunosuppressed by the double infection. This allowed the *E. cuniculi* to overcome the natural balance.

These reponses emphasise that latent *E. cuniculi* infections in laboratory animals can severely interfere with the interpretation of experiments using these animals and, therefore, that all laboratory colonies should be *E. cuniculi* free.

10. RELATIONSHIP BETWEEN *E. cuniculi* AND TUMOURS

a. *Effect on tumour growth.* The interactions between *E. cuniculi* and mammalian tumour cells have been observed in rats and mice by Petri (1965, 1966, 1967, 1968a,b, 1969), in hamsters by Kinzel and Meiser (1968) and Meiser, Kinzel and Jírovec (1971) and in mice by Arison, Cassaro and Pruss (1966). These studies have been summarised by Petri (1976).

Animals infected by *E. cuniculi* generally show increased survival times after subjection to Yoshida, Ehrlich, Sarcoma 180 and Krebs-2 ascites transplantable tumours, apparently by retardation of the growth of the tumours. There is a similar effect if mice are given *E. cuniculi,* within four months of chemical induction of a fibrosarcoma but not at five months from induction. The mechanism of retardation is obscure. Although the Yoshida ascites tumour cells became heavily infected (Petri, 1966, 1969), *E. cuniculi* organisms remained rare in the Ehrlich ascites tumour in mice (Petri, 1969) and in the plasmacytoma tumour cells in hamsters (Kinzel and Meiser 1968) and were completely absent from the neoplastic tissue in the experimental tumours studied by Arison, Cassaro and Pruss (1966). The slower growth rates of the tumours cannot, therefore, be attributed directly to destruction of the cells by *E. cuniculi.*

The *E. cuniculi*-infected Yoshida ascites tumour maintains its malignancy and transplantability and, usually, the tumour growth, although retarded, causes the death of the rats. Less usually the tumour grows, then regresses leaving a fibrous agglutination in the peritoneal cavity. Very unusually, the ascites, with viable sarcoma cells and *E. cuniculi,* persists and the rats survive up to 6 months. In the survivors with regressed tumours, there is a high incidence, after about 18 months, of rhabdomyosarcomas in the scar tissue around the organs in the peritoneal cavity. Another effect of *E. cuniculi* is that the Yoshida tumour, when transplanted subcutaneously, does not adopt its usual solid form (Petri, 1966, 1968a).

Hamsters infected with *E. cuniculi* do not survive the plasmacytoma tumour. When transplanted subcutaneously solid tumors are formed in which no *E. cuniculi* are found. Death ensues within 10–12 days and there is thus no time for generalised spread of the microsporidium to the solid tumour cells (Meiser, Kinzel and Jírovec, 1971).

b. *Natural killer cell activity enhanced by E. cuniculi.* Natural killer (NK) cells represent a sub-population of lymphocytes derived from bone marrow, which are able to lyse spontaneously a variety of tumour cells. These cells possess the asialo GM-1 determinant, they are non adherent and non-phagocytic and their activity can be augmented by agents which induce interferon production.

In a series of experiments using euthymic and athymic mice, T-cell and B-cell depletions and antiserum to the asialo GM-1 determinant plus complement to render the NK cells non-functional, Niederkorn, Brieland and Mayhew (1983) investigated NK cell activity. Enhanced splenic NK cell activity against YAK-1 tumour cells was demonstrated when spleen cells from *E. cuniculi*-infected mice were added to the tumour cells *in vitro*. The augmented cytolysis of tumour cells was found to be transient, dose dependent and dependent on the strain of mouse. The NK cell response was not induced by inoculation of dead spores, only by a true infection.

These results are in accord with the resistance of *E. cuniculi* infected kidney cells to vesicular stomatitis virus, in that the protozoon may have induced the cells to produce interferon (Sect. 9). The discovery of NK cell activity associated with *E. cuniculi* throws light on the resistance of *E. cuniculi*-infected rodents to transplantable tumours (Sect. 10,a).

11. CONTROL AND THERAPY OF *E. cuniculi*

Chemicals and drugs are needed to effect the destruction of free spores in the environment, to help maintain *Encephalitozoon*-free colonies of animals which are housed under intensive conditions, and for chemotherapy of infections in valuable animals and man. The spores remain viable for at least 4 weeks at ambient temperatures and a temperature of 56°C has to be maintained for more than 1 hour to ensure their total destruction (Shadduck and Polley, 1978). Animal houses, pens, kennels and cages are best treated with disinfectants, of which there are several which are 100% effective at low concentrations (Waller, 1979). However, because of transplacental transmission, it is necessary to diagnose and cull all infected animals concurrently with environmental treatment.

Only Waller (1979) has attempted to evaluate drugs for the treatment of encephalitozoonosis. Of the 8 commonly used drugs that he tested *in vitro* only chloroquine and oxytetracycline reduced the spore harvest by more than 50%, the figures being 69% reduction for cloroquine and 58% for oxytetracycline. There is clearly a need for further research in this field.

B. *Thelohania apodemi* Doby, Jeannes and Rault, 1963 (Fig. 4.28)

HOST AND LOCALITY: *Apodemus sylvaticus* collected from several localities in France. Prevalence was about 10% (15 out of 153) (Doby, Jeannes and Rault, 1963, 1965).

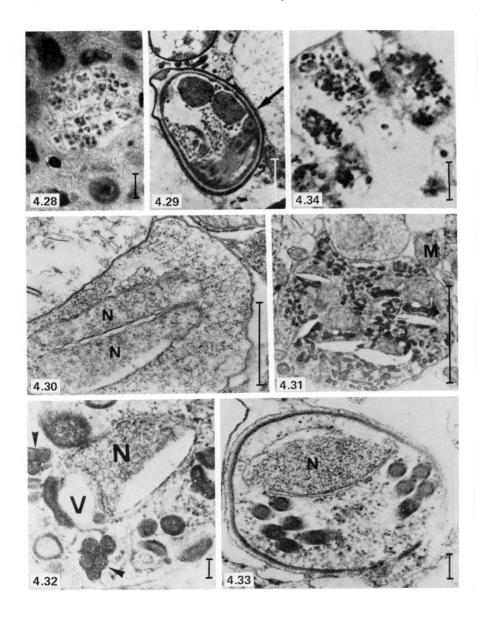

LESIONS: In the brain, as spherical or ovoid colonies, 30–100 μm in diameter, with no tissue reaction round them (Fig. 4.28), or as free or phagocytised spores in small granulomata. In one field mouse, without brain cysts, there was a single colony of parasites, measuring 20 x 200 μm, in the muscle (Doby and Jeannes, 1968).

MORPHOLOGY AND DEVELOPMENT

Merogony: Stages referable to a merogonic cycle were not described.

Sporogony: Early sporonts are small cells, measuring 3–4 μm, with a single voluminous nucleus measuring 2.5–3 μm. After growth and nuclear division, octonucleate sporonts are formed, which retract within SPOVs of wall thickness 0.5 μm, and divide into eight sporoblasts. The SPOVs appear polyhedral when compressed against one another in the colony and the number making up the colony ranges from 12 to nearly 1000.

Spores: Ovoid or pyriform, measuring 4–5 × 2–2.5 μm (probably in the fixed and stained state), have distinct anterior and posterior vacuoles when stained. One nucleus (sometimes two were said to occur) in the

---

Fig. 4.28 *Thelohania apodemi:* colony of organisms in the brain of field mouse, *Apodemus sylvaticus;* the spores are segregated into groups of 8 in compartments which may be SPOVs. Giemsa/Colophonium stain. Scale bar = 10 μm (approximately). (Photograph from material provided by Professor J. M. Doby.)

Fig. 4.29 *Nosema connori* from human infant: electron micrograph of binucleate spore (arrow) lying in direct contact with host cell cytoplasm. Scale bar = 1 μm. (Reproduced from Shadduck *et al.,* 1979, with permission of the authors and publisher.)

Figs. 4.30–4.33 *Enterocytozoon bieneusi:* electron micrographs of stages from small intestinal epithelium of an immunodeficient human patient suffering with AIDS. (Figs. 4.30–4.33 reproduced from Desportes *et al.,* 1985, with permission of the authors and publisher.)

Fig. 4.30 Binucleate meront with diplokaryon (N) and simple surface membrane. Scale bar = 1 μm.

Fig. 4.31 Sporogonial plasmodium showing isolated nuclei, each associated with a vesicle and developing coils of the polar tube; M = host cell mitochondrion. Scale bar = 1 μm.

Fig. 4.32 Part of a sporogonial plasmodium showing one nucleus (N) with associated vesicle (V) and coils of the polar tube (arrowheads). Scale bar = 0.1 μm.

Fig. 4.33 Sporoblast nearing maturity showing thin spore wall lacking the electron lucent endospore of most microsporidia, single nucleus (N) and low number of polar tube coils. Scale bar = 0.1 μm.

Fig. 4.34 *Microsporidium simiae* spores in desquamated jejunal epithelium of monkey, *Callicebus moloch.* Giemsa stain. Scale bar = 10 μm. (Reproduced from Seibold and Fussell, 1973, with permission of the authors and publisher.)

cytoplasm and traces of the polar tube and a polar granule are visible in the vacuole at the wider end. An intense yellow-green autofluorescence is emitted by fresh spores subjected to ultraviolet radiation (Doby, Rault and Barker, 1975). Although the authors attributed this to the presence of chitin in the spore wall, fluorescence is not typical of microsporidia.

C. *Nosema connori* Sprague, 1974 (Fig. 4.29)

HOST AND LOCALITY: An immunodefective, 4-month-old male human infant in the U.S.A.

LESIONS: An overwhelming, disseminated infection correlated with a markedly defective lymphoid system. Abundant spores were present in smooth and cardiac muscles and were, thus, found in the *muscularis* of the stomach and intestine, the media of the arterial walls of many organs, in the myocardium and diaphragm. Organisms were also found in the kidney tubules, adrenal cortex, liver and lungs. The spleen was not infected and examination of the central nervous system had not been permitted. The only sign of an interstitial inflammatory reaction of mononuclear cells was in the diaphragm.

After an illness of some 4-months duration, characterised by diarrhoea and maladsorption, the patient died. Respiratory distress was linked with a concurrent infection of *Pneumocystis carinii*. The case history was described by Margileth, Strano, Chandra, Neafie, Blum and McCully (1973).

MORPHOLOGY: The only stages of the parasite which were recognised were sporoblasts and immature and mature spores, some of which lay in a cyst bounded by a membrane. Spores oval, 4×2 μm, with a PAS positive polar granule (Margileth, Strano, Chandra, Neafie, Blum and McCully, 1973), $4-4.5 \times 2-2.5$ μm (Sprague, 1974). Two nuclei, seen by Sprague using Heidenhain's stain, in sporoblasts and immature spores were confirmed, by electron microscopy, by Shadduck, Kelsoe and Helmke (1979). The polar tube, with about eleven coils, was shown in electron micrographs by Margileth, Strano, Chandra, Neafie, Blum and McCully (1973) and Shadduck, Kelsoe and McCully (1979) (Fig. 4.29).

NOTES: In the absence of information about developmental stages, the parasite was named and assigned tentatively to the genus *Nosema,* on the basis of the two nuclei in the sporoblasts and spores.

Margileth, Strano, Chandra, Neafie, Blum and McCully (1973) presumed that invasion had taken place through the intestinal epithelium and had been disseminated from that site in the bloodstream.

D. *Enterocytozoon bieneusi* Desportes, Le Charpentier, Galian, Bernard, Cochand-Priollet, Lavergne, Ravisse and Modigliani, 1985
(Figs. 4.30–4.33)

HOST AND LOCALITY: A 29-year-old Haitian man who had lived in France for 4 years and in whom the acquired immune deficiency syndrome (AIDS) had been diagnosed.

LESIONS: About half of the enterocytes of the villi from duodeno-jejunal and ileal biopsies had developmental stages in the supranuclear cytoplasm. The patient had a 5-month history of diarrhoea, weight loss, fever and epigastric pain and there was a concurrent infection of *Giardia lamblia* with the microsporidium. There was an accumulation of neutral fat in the intercellular spaces of the epithelium and in the perivascular area of the *lamina propria*. The case history was described by Modigliani, Bories, Le Charpentier, Salmeron, Messing, Galian, Rambaud and Desportes (1984).

MORPHOLOGY AND DEVELOPMENT: All stages are in direct contact with the host cell cytoplasm; early stages have nuclei in diplokaryotic arrangement and later stages have isolated nuclei.

Merogony: Meronts were not positively recognised but proliferation of diplokaryotic stages may occur before these enter sporogony. Small diplokaryotic cells, 3–4 μm in diameter with a simple plasma membrane, which were referred by the authors to the sporogonic cycle, were probably meronts (Fig. 4.30).

Sporogony: Multinucleate, probably octonucleate, sporogonial plasmodia (Fig. 4.31). The putative exospore coat is laid down on the plasma membrane and the nuclei become isolated. In association with each nucleus are differentiated the primordia of the anchoring disc of the polar tube and the polaroplast and, at a slightly later stage, 4–5 coils of the polar tube (Fig. 4.32). Only when these elements of the extrusion apparatus are fully differentiated does the sporogonial plasmodium divide into 8 sporoblasts. Fission may be effected by enlargement and fusion of flattened vesicles in the cytoplasm of the plasmodium. The uninucleate sporoblasts (Fig. 4.33) thus already bear a complete set of spore organelles by the time they are separated from the sporogonial plasmodium and maturation into spores is simply a process of condensation of cytoplasm and deposition of a thin, electron-lucent endospore layer beneath the exospore. A lucent vacuole, also recognisable at the plasmodial stage, lies alongside the nucleus.

Spores: 1.5 × 0.5 μm in sections. They differ from other microsporidia of the order Microsporida in having a very thin endospore. They are uninucleate and have about 5 coils of the polar tube.

NOTES: The origin of this opportunistic parasite in the patient with a profound immune deficiency is unknown. Desportes, Le Charpentier, Galian, Bernard, Cochand-Priollet, Lavergne, Ravisse and Modigliani (1985) called attention to an infection similar to *E. bieneusi* in the enterocytes of a *Callicebus* monkey (Sect. E) and speculated that the two might be related, implying presumably that the infection was of vertebrate origin.

E. *Microsporidium simiae* n. sp. (Fig. 4.34)

A parasite described by Seibold and Fussell (1973) as a "microsporidian" from the jejunal epithelium of a *Callicebus* monkey was listed by Sprague (1977) as *Microsporidium* sp. Seibold and Fussell, 1973. Although the genus cannot be determined, specific characters are available and it is here named as a new taxon *Microsporidium simiae* n.sp.

HOST AND LOCALITY: Adult female *Callicebus moloch cupreus,* obtained from South America and held at a primate centre in the U.S.A.

LESIONS: Many epithelial cells (enterocytes) from a region of jejunum taken at necropsy contained clusters of microsporidian spores (Fig. 4.34). The epithelium of this region was totally desquamated, in contrast to other regions of the intestine. The monkey had been in captivity for about 6 months, then lost weight rapidly and died after a further month.

MORPHOLOGY: Spores are round to oval or elliptical in sections and measure 2–4 μm. Probably the smaller round sections were viewed transversely. The description of 2 clear areas separated by a biconcave wall of basophilic material suggests anterior and posterior vacuoles. There are 7 coils of the polar tube.

NOTES: Seibold and Fussell (1973) speculated that ingested arthropods, may have been the source of infection, when declining health predisposed the monkey to attack by unusual pathogens. This is unlikely, as microsporidia of invertebrate origin do not survive at vertebrate blood temperature. Desportes, Le Charpentier, Galian, Bernard, Cochand-Priollet, Lavergne, Ravisse and Modigliani (1985) have speculated that this microsporidium may be related to *Enterocytozoon bieneusi* (Sect. D).

F. *Microsporidium ceylonensis* n. sp. (Figs. 4.35, 4.36)

Ashton and Wirasinha (1973) presented a case report on an 11-year-old Tamil boy with a microsporidian infection of the cornea. Beyond reporting that *Nosema helminthorum* and *Encephalitozoon cuniculi* had been suggested to them, the authors did not attempt an identification. Sprague (1977) listed the parasite as *Nosema* sp. Ashton and Wirasinha, 1973. It is here transferred to the collective group *Microsporidium* and given a specific name.

HOST AND LOCALITY: 11-year-old Tamil boy in Ceylon (Sri Lanka).
LESIONS: Cornea of the right eye was scarred and vascularised and granulation tissue was present over the conjunctival surface of the upper lid. In sections of the affected cornea there were numerous spores, free and in macrophages, in the deep corneal tissue above Descemet's membrane (Fig. 4.35). There was an area of inflammatory tissue underlying Descemet's membrane but the parasites had not penetrated into this area. The boy suffered defective vision.
MORPHOLOGY: Oval spores measuring 3.5 × 1.5 μm in sections. There is a prominent polar vacuole in fixed spores (Fig. 4.36).
NOTES: The spores are longer than those of *Encephalitozoon cuniculi*. The boy had been gored by a goat 6 years previously but the diagnosis of *Nosema helminthorum,* a parasite of cestodes which may parasitise goats, is unlikely because the spores of *N. helminthorum* are considerably longer (5.8–6.8 × 3.5 μm fresh, given by Dissanaike, 1965).

G. *Microsporidium buyukmihcii* n. sp. (Figs. 4.37, 4.38)

Buyukmihci, Bellhorn, Hunziker and Clinton, 1977 reported on a microsporidium in the cornea of a cat. Only the spores were seen and the large number of coils of the polar tube clearly differentiate the species from *E. cuniculi*. As the genus is indeterminate, the parasite is here placed in the collective group *Microsporidium* and given a specific name.
HOST AND LOCALITY: 3.5-year-old male short-haired cat in the U.S.A.
LESIONS: Opacities of the right cornea with moderate inflammation of the cornea, conjunctiva and anterior uvea. The corneal epithelium was patchily degenerate and there was a diffuse inflammatory reaction in the corneal stroma with polymorph and mononuclear cells.
MORPHOLOGY: Spores, measuring 1.5 × 4.0 μm were present in the stroma of the cornea and between the stromal fibres (Fig. 4.37). There were 15 or 16 coils of the polar tube shown in the electron micrograph of the spore (Fig. 4.38).
NOTES: No characters were given from which a generic diagnosis could be made. The large number of coils of the polar tube is close to the number in *M. africanum* (Sect. H) but the spore measurements differ. The spore structure differentiates this parasite from *E. cuniculi*.

H. *Microsporidium africanum* n. sp. (Fig. 4.39, 4.40)

Pinnolis, Egbert, Font and Winter (1981) presented a case report of a 26-year-old woman from Botswana, who underwent enucleation of her left eye, made blind and painful by a perforated corneal ulcer. The lesion

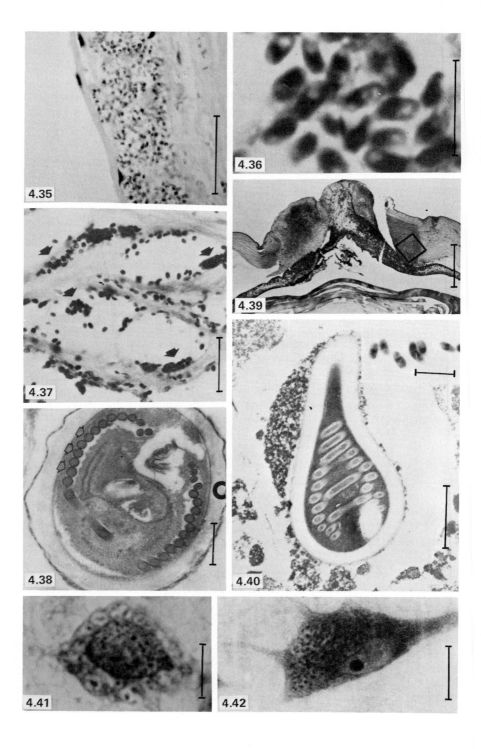

was caused by a microsporidium with spores quite distinct from those of *E. cuniculi* and probably different from those of *M. ceylonensis* (Sect. F). The author's identification of the parasite, as a *Nosema* without specific assignation, was not made on well founded taxonomic principles and it is here transferred to the collective group *Microsporidium* and named as a new species.

HOST AND LOCALITY: 26-year-old woman in Botswana.

LESIONS: Perforated corneal ulcer in left eye (Fig. 4.39) with incarceration of the iris. A marked inflammatory reaction in the corneal stroma with mononuclear, neutrophil and epithelioid cells and necrosis adjacent to the perforation. Spores were present mainly in the cytoplasm of histiocytes in the cornea, located just anterior to Descemet's membrane at one edge of the perforation.

Fig. 4.35 *Microsporidium ceylonensis:* ocular lesion in human male. Deep cornea showing abundant microsporidia on the corneal side of Descemet's membrane. Giemsa stain. Scale bar = 10 μm.

Fig. 4.36 *Microsporidium ceylonensis* spores in the deep cornea of human eye. Methylene Blue. Scale bar = 10 μm. (Reproduced from Ashton and Wirasinha, 1973, with permission of the authors and publisher.)

Fig. 4.37 *Microsporidium buyukmihcii:* ocular lesion in cornea of cat. Diffuse infiltrations and aggregates of spores (arrows) in the corneal stroma. Scale bar = 100 μm. (Reproduced from Buyukmihci *et al.,* 1977, with permission of the authors and publisher.)

Fig. 4.38 *Microsporidium buyukmihcii:* electron micrograph of spore showing thick wall and 15 coils of the polar tube (arrows). Scale bar = 1 μm. (Reproduced from Buyukmihci *et al.,* 1977, with permission of the authors and publisher.)

Fig. 4.39 *Microsporidium africanum:* ocular lesion in human female. Perforated central corneal ulcer. Spores are concentrated within necrotic deep corneal lamellae (within square). H. & E. Scale bar = 1 mm. (Reproduced from Pinnolis *et al.,* 1981, with permission of the authors and publisher.)

Fig. 4.40 *Microsporidium africanum:* electron micrograph of posterior half of mature spore showing 11 coils of the polar tube; the spore lies isolated. Inset: light microscope view of uninucleate spores, some slightly curved. Acid fast stain. Scale bar = 1 μm; inset = 10 μm. (Reproduced from Pinnolis *et al.,* 1981, with permission of the authors and publisher.)

Fig. 4.41 Neurone from brain of man who had died of acute toxoplasmosis; bodies similar to those of Viting's patient (Fig. 4.42) are present in the cytoplasm. Scale bar = 10 μm. (Photograph by Dr. R. Killick-Kendrick.)

Fig. 4.42 Neurone from brain stem of human patient with multiple sclerosis: cytoplasm packed with bodies surrounded by a light border superficially resembling *E. cuniculi*. Ziehl Nielsen stain. Scale bar = 10 μm. (Reproduced from Viting, 1965, with permission of the publisher.)

MORPHOLOGY: Oval, sometimes crescentic, spores measuring 4.5–5 × 2.5–3 μm. A polar granule and a single nucleus were observed. The spore wall measured 158–184 nm thickness and the polar tube showed 11–13 coils (Fig. 4.40).

NOTES: The diplokaryotic nuclei and disporoblastic sporogony, characteristic of *Nosema* were not seen. One spore seen in an electron micrograph appears to be in direct contact with the cell cytoplasm, a feature which distinguishes it from *Encephalitozoon,* which develops in groups in a parasitophorous vacuole: many spores were lying individually and those that were in groups may have been phagocytosed, as the host cells were histocytes. The genus is indeterminate.

IV. THE MEDICAL IMPORTANCE OF MICROSPORIDIA

The early papers naming new species of *Encephalitozoon* from man can mostly be dismissed as misidentifications (see Sect. IIIA,1: *E. chagasi, E. brumpti* and *E. hominis*). The convulsive illness in the baby with *E. chagasi* is consistent with encephalitozoonosis but it could equally have been of other aetiology and conclusions can no longer be drawn. Its status has been discussed by Weiser (1976).

An association between microsporidian infection and multiple sclerosis (MS) was proposed by Viting (1965, 1969). She found small and larger bodies, measuring 1–1.5 μm and 1.5–2 μm respectively, which she considered were similar to *E. cuniculi,* in the cytoplasm of enlarged and altered neurones of MS patients (Fig. 4.42). Chalupský, Lenský, Bedrník and Vávra (1972) found that sera from 121 patients suffering from MS were either completely negative or insignificantly positive for *E. cuniculi* by immunofluorescence. The serological results cannot be taken as conclusive, since we now know that microsporidia other than *E. cuniculi* can infect man. However, E. U. Canning and R. Killick-Kendrick (unpublished observations) found bodies similar to those illustrated by Viting (1965) in the neurones of a patient who had died of acute toxoplasmosis (Fig. 4.41) and of another who had died of cerebral malaria. The nature of these formations was not determined but they were not microsporidia. Vávra (personal communication) found variations in size, from cell to cell, of the inclusions in Viting's material. These findings, coupled with the absence of *Encephalitozoon* from all other histological investigations of MS patients, indicate that Viting's interpretations were incorrect.

Marcus, van der Walt and Burger (1973) described the occurrence of microsporidia in the cytoplasm of tumour cells of a 58-year-old South

African man, who had died of a pancreatic adenocarcinoma with metastases in many organs. I have examined necropsy material kindly sent to me by Dr. Marcus and am satisfied that the organism involved is not a microsporidium. This is in agreement with Petri (1976) and Sprague (1977).

Bywater (1979), after summarising the evidence then available for infection of man, doubted whether *E. cuniculi* could be regarded as a human pathogen, even as an opportunistic pathogen. However, in spite of the exclusions discussed above, there is no longer any doubt that microsporidia, including *E. cuniculi*, can infect man. On two occasions *E. cuniculi* has been isolated from patients—both children, one Japanese (Matsubayashi, Koike, Mikata, Takei and Hagiwara, 1959) and the other Colombian (Bergquist, Stintzing, Smedman, Waller and Andersson, 1984)—and in the latter case there was also serological evidence of high titres of antibodies to *E. cuniculi*. Both these children exhibited clinical signs of severe neurological illness but each made a complete recovery. Sprague (1977) considered the parasite isolated by Matsubayashi, Koike, Mikata, Takei and Hagiwara (1959) to be a separate species which he named *E. matsubayashii*, on the basis of its larger spores: he quoted 2.7–3 × 1.2–1.8 μm for *E. matsubayashii* and 2.0–2.5 × 1.0 μm for *E. cuniculi*. In fact, the measurements given by different authors for *E. cuniculi* vary considerably (Sect. IIIA,4) and the variation is likely to be indicative of the difficulties of obtaining accurate measurements of these very small organisms. The measurements of the spores from the urine of the Colombian boy (Bergquist, Stintzing, Smedman, Waller and Andersson, 1984) were 1.5–3.0 × 0.7–1.0 μm and this child was seropositive for *E. cuniculi*. We have chosen to consider *E. matsubayashii* Sprague, 1977 as a synonym of *E. cuniculi* Levaditi, Nicolau and Schoen (1923) because we are not convinced of the significance of the differences in spore measurements and because there is some (though not incontrovertible) evidence that spores from the Japanese boy were infective to mice.

Shadduck, Watson, Pakes and Cali (1979) were unable to isolate organisms or find lesions attributable to *E. cuniculi* in 6 Rhesus monkeys inoculated with a rabbit isolate but Lucke (1925) reported lesions in one Rhesus monkey, which are very like those formed as a reaction to *E. cuniculi*. Perhaps infection in primates depends on the immunological status.

The positive identification of *E. cuniculi* in the Colombian boy (Bergquist, Stintzing, Smedman, Waller and Andersson, 1984) stimulated a human population survey in 1982 in Sweden for antibodies to *E. cuniculi* (W.H.O. Weekly Epidemiological Record, World Health Organisation—anonymous report). The results were surprising, especially that there was a strong correlation between tropical disease, even tropical experience, and *E. cuniculi,* i.e. 38% of malaria patients, 14.8% of Chagas disease patients, 12.2% of healthy travellers returning from the tropics and 8.7% of filariasis patients were positive for *E. cuniculi* antibodies, whereas sera from people who had never been outside.Europe were consistently negative. The meaning of these results is not entirely clear—whether they represent true positives or cross reactions with other microsporidia, for instance those of mosquito origin, which may be inoculated with the saliva when the mosquito takes a blood meal. Niederkorn, Shadduck and Weidner (1980) found that there were antigenic similarities between *E. cuniculi* and three other microsporidia, including the mosquito microsporidium *Nosema algerae.*

However, it is possible that nonspecific immunodepression, arising from the polyclonal lymphocyte stimulation associated with malaria and other infectious diseases, allows *E. cuniculi* to establish itself, whereas immunologically normal individuals are resistant. This explanation becomes more plausible than that of cross reactions since Bergquist, Morfeldt-Manson, Pehrson, Petrini and Wasserman (1984) found IgM antibodies to *E. cuniculi* at low titre, in 33% (10 out of 30) of homosexual men belonging to the group at risk for AIDS. All had travelled to the tropics and their lowered immunological status might have modified their susceptibility to *Encephalitozoon.* Whether *E. cuniculi* is more common in the tropics remains to be determined.

In another survey, carried out in Singapore (Singh, Kane, Mackinlay, Quaki, Yap, Ho, Ho and Lim, 1982) on sera of Asian, African and European origin, low titres to *E. cuniculi* of between 1/16 and 1/256 were found. Of particular interest were the high percentages of positives among Ghanaian malaria, Nigerian tuberculosis and Malaysian (Orang Asli) filariasis patients (35.9%, 42.7% and 18.6% respectively). Also, only one out of 25 healthy animal handlers showed positivity, indicating that *E. cuniculi* is not readily transferable from animals to man by contact. This was also found by Cox and Pye (1975), who attributed low positivity in 3 out of 4 rabbit handlers in the immunofluorescence reaction, to inhalation of antigenic material.

The fulminating infections, due to microsporidia other than *E. cuniculi,* in the immunocompromised infant (Margileth, Strano, Chandra, Neafie, Blum and McCully, 1973) and the AIDS patient (Desportes, Le Charpentier, Galian, Bernard, Cochand-Priollet, Lavergne, Ravisse and Modigliani, 1985) are clear warnings that microsporidia can be opportunistic parasites. The occurrence of microsporidia of unknown origin in the cornea of two patients (Ashton and Warasinha, 1973; Pinnolis, Egbert, Font and Winter, 1981) is also interesting. A lot might be learned of their origins and likely mode of infection by attempting more accurate identification when these cases are presented.

Attention should be drawn to microsporidia as possible complicating factors in patients on immunodepressive therapy (including transplant patients) as these patients undoubtedly constitute a group at risk. Efforts should also be directed towards finding satisfactory drugs for the treatment of microsporidiosis, which at present are virtually lacking.

5. Techniques

Vávra and Maddox (1976) have set out in some detail many of the methods used for studying microsporidia. Their account deals mainly with microsporidia of invertebrate origin but most techniques can easily be adapted to species from vertebrates. No details are given here of common histological methods. The chapter is intended as a guide to the techniques which have been found to give good results with microsporidia and references are given, so that the reader may obtain further information on the specialist techniques (eg. immuno-biochemical techniques), which have been adapted for the study of microsporidia.

I. COLLECTION OF MATERIAL AND DIAGNOSIS

Very often infected hosts show no external signs or behavioural abnormalities. The prevalence of infection in a host population may be low and detection depends on careful examination of tissues, fresh or in stained smears and sections. Many of the species in fish are localised in cysts, some superficial, some in internal organs and these can readily be detected by eye.

A. Poikilothermic Vertebrates

No serological methods have yet been developed. Common signs of infection are: whitish cysts or nodules protruding from the surface; swellings at the surface due to cysts in the deeper tissues; hypertrophy and opacity of internal tissues, including thickening of the gastro-intestinal wall and overt cyst formation; whitish, opaque streaks in the musculature. Particularly large cysts may cause gross deformities of the body. Small cysts may be almost undetectable by eye. A technique which may be widely applicable to the detection of microsporidian lesions in fish musculature is the candling method used by Fischthal (1944) for the commercial detection of *Pleistophora macrozoarcidis* in eelpout, *Macrozoarces americanus*. Lesions larger than 1 cm long can be detected easily, and smaller ones with difficulty, by pulling the fillets rapidly across the surface of a frosted glass plate illuminated from beneath.

Care has to be taken not to confound other parasitic, or even non-parasitic formations, with abnormalities due to microsporidia. Some commonly occurring structures which can cause confusion are lipoid tumours, cysts caused by trematodes, myxosporidia and even yeasts.

B. Homoeothermic Vertebrates

Routine examination of tissues by standard histological methods must be used for most species. *Encephalitozoon cuniculi,* which infects the kidney of its mammalian hosts, may be detected in living animals by examination of urine sediments for spores by conventional staining (Goodman and Garner, 1972) or by immunofluorescence (Cox, Walden and Nairn, 1972) but the methods are unreliable as spores are only passed sporadically (Chap. 4, Sect. IIIA,6b). The mouse ascites test (Nelson, 1967) (Chap. 4, Sect. IIIA,6c), which depends on the development of ascites in mice a few weeks after inoculation of infective material from a suspect host, has been largely supplanted by serological tests. The development of culture techniques for *E. cuniculi* for large-scale antigen production has facilitated diagnosis of the infection in large populations of hosts, by means of a skin test and a variety of serological tests. Details of the methods can be obtained from the original papers as follows:

a. Skin Hypersensitivity Test (Chap. 4, Sect. IIIA,6e, p. 216): Pakes, Shadduck and Olsen, 1972
b. Indirect Fluorescent Antibody Test (Sect. 6f, p. 217): Chalupský, Vávra and Bedrník (1973); Cox, Walden and Nairn (1972); Cox and Pye (1975); Mohn and Ødegaard (1977); Wosu, Shadduck, Pakes, Frenkel, Todd and Conroy (1972); Stewart, Botha and van Dellen (1979).
c. Carbon Immuno-Assay (India Ink Test) (Sect. 6f, p. 218): direct test, Waller (1977); indirect test, Kellett and Bywater (1980).
d. Complement Fixation Test (Sect. 6f, p. 219): Wosu, Olsen, Shadduck, Koestner and Pakes (1977).
e. Immuno-Peroxidase Test (Sect 6f, p. 219): Gannon (1978).
f. Enzyme-linked Immuno-Sorbent Assay (Sect 6f, p. 219): Cox, Horsburgh and Pye (1981).
g. Indirect Micro-Agglutination Test (Sect. 6f, p. 219): Shadduck and Geroulo (1979).

The advantages and disadvantages of the various tests have been discussed in Sect. 6f, p. 219.

II. SELECTIVE STAINING METHODS

A. Periodic Acid-Schiff Reaction (PAS) for Detection of Polysaccharides

A distinct purple-red granule, the polar cap, is revealed at the anterior end of the spore.

B. Gram's Stain

Microsporidian spores are Gram-positive, in contrast to all other protozoal cysts or spores. Even single spores become visible in smears or sections against an unstained background.

C. Giemsa Stain for Smears and Giemsa Colophonium for Sections

Methanol or osmium tetroxide vapour can be used as fixatives for dry smears.

Meronts and sporonts are stained particularly well in smears, with pale blue cytoplasm and red nuclei. Bray and Garnham's (1962) modification of the Giemsa Colophonium technique, with Carnoy's fluid as fixative, is useful for staining spores in sections but pre-spore stages are not easily recognised. With both Giemsa techniques, the spores stain a characteristic pale blue but the nuclei are often masked. Hydrolysis in 1 N HCl at 60°C before staining improves the visibility of the nuclei.

D. Goodpasture's Aniline-carbol-fuchsine

Goodpasture's aniline-carbol-fuchsine also stains spores well in sections fixed with Zenker fixative or 10% formalin, as long as the latter is rinsed for 2 days in 2.5% $K_2CR_2O_7$. Paraffin embedding is satisfactory. Sections fixed in Zenker must be treated with iodine, then sodium thiosulphate before staining. Stain in Goodpasture-Perrin's stain (30% ethanol-100 ml; basic fuchsine-0.39 g; aniline-1 ml; crystalline phenol-1 g) for 5 min at 70°C or 10 min at room temperature. Rinse in distilled water. Differentiate by dropping 40% formaldehyde on to the slide until it ceases to release the stain. Rinse as before. Counterstain in saturated aqueous picric acid for 1 min. Dehydrate in 2 changes each of 95% ethanol and 100% ethanol. Clear in xylene and mount in synthetic resin.

Spores stain blue-black; tissue nuclei are pale red and cytoplasm is yellow.

III. RECORDING MICROSPORIDIAN SPORES

A. Photography and Measurement

Spores should be photographed and measured when fresh. This is best achieved by examining spores immobilised in monolayers.

1. At a Paraffin-water Interface

A small drop of a concentrated spore suspension on a coverslip is inverted on to a drop of liquid paraffin on a slide. The spores are trapped in places between the liquids.

2. In an Agar Layer

A drop of melted 1.5% agar is spread over a warm slide to produce a thin even layer. A drop of concentrated spore suspension is spread in the central area of a coverslip, which is then carefully inverted on to the agar. The spores are held firmly with a minimum amount of water.

Slides can be stored, preferably cooled, for several weeks by trimming the agar from the edge of the coverslip and sealing with paraffin wax or nail varnish.

For photomicrography high contrast film is recommended. Measurements may be calculated from the prints or taken directly, using an eyepiece micrometer or the image-splitting eyepiece (Vickers Instruments).

B. Mucous Envelopes

The presence or absence of a mucous envelope (‘‘mucocalyx’’) around the spores can be demonstrated by mixing a drop of India ink with the spore suspension. The ink particles are excluded from the mucous envelopes and they appear bright against a dark background. So far these envelopes have not been found in microsporidia of vertebrates.

C. Polar Tube Extrusion

Spores need activation (priming) before they will extrude their polar tubes. Spores of several species from invertebrates will hatch in neutral conditions, after exposure to high pH conditions in the presence of alkali metal ions (Undeen, 1978). The techniques can be adapted, by varying the ionic concentrations or pH values, to suit most species. More artifical methods, such as mechanical pressure or 5–10% H_2O_2 can be used in difficult cases but may result in incomplete extrusion of the tube. The length of the polar tube is a useful character for species diagnosis but must be used with caution if complete extrusion cannot be guaranteed.

D. Purification

Spores can be harvested from their hosts by maceration of the infected tissues or cysts in distilled water and filtration through several layers of muslin to remove debris. By repeated centrifugation of the filtrate, re-suspension and filtration, a good degree of purity can be obtained. Further purification can be obtained by gradient density centrifugation using, e.g. Ludox (Undeen and Alger 1971; Kelly and Knell, 1979) or Percoll (Jouvenaz, 1981). Spores purified for infection of tissue cultures must be processed with sterile precautions and treated with antibiotics.

IV. STORAGE OF SPORES

Spores harvested from host tissues or from cultures can be stored for at least several months in distilled water at 4°C with antibiotics (usually 100 i.u. penicillin and 100 μg streptomycin).

If very small quantities are to be stored the use of a sealed glass capillary tube is recommended. The suspension of spores is allowed to half fill a capillary tube of about 1 mm inner diameter by capillary attraction; the tube is tilted to position the suspension in the centre and both ends are sealed.

Cryopreservation in liquid nitrogen is often successful and has been used for a variety of microsporidia from vertebrates and invertebrates. Dimethyl sulfoxide (DMSO) or glycerol are suitable cryoprotectants. Several microsporidia of invertebrate origin have been lyophilised successfully but the technique has not been explored fully for those in vertebrates.

V. EXPERIMENTAL INFECTION

Many species of microsporidia in vertebrates can be transmitted when mature spores are fed to the hosts, forcibly or in the food. Examples are *Glugea* spp. and *Microsporidium takedai* in fish, *Pleistophora myotrophica* in toads and *Encephalitozoon cuniculi* in mammals. Spores of *Glugea anomala* should be "seasoned" in pure pond water for one month before use. The spores can be given to fish mixed with pelletted food or fed via crustaceans in a food chain. *Pleistophora hyphessobryconis* has produced infections in neon tetras when inoculated intramuscularly.

Encephalitozoon cuniculi can be obtained from chronically infected mice by induction of patent infections with cortisone given weekly at a rate of 2.5–5 mg per mouse (Bismanis, 1970). The mice develop ascites containing macrophages with large aggregates of spores. The ascitic fluid can be inoculated into recipient mice and the strain maintained by serial passage.

VI. *IN VITRO* CULTIVATION

Encephalitozoon cuniculi is the only microsporidium of vertebrate origin which has been grown in tissue culture. The range of cells which have been infected is given in Table 4.2. When rabbit choroid plexus cells (RCP) are used, the parasite may grow spontaneously if the cell donor is infected. Cell lines can be infected by addition of sterile infected ascitic fluid from mice which have been infected by serial passage or by activation of latent infections (Chap. 4, Sect. IIIA,6d). Sterile spores harvested from cultures can be used to infect new cultures. The most successful cell systems are RCP cells which become very heavily infected and canine kidney MDCK cells which maintain a constant level of infection after reaching a maximum so that spores can be harvested regularly from long-term cultures.

VII. BIOCHEMICAL AND IMMUNOLOGICAL TECHNIQUES

Biochemical and immunological techniques are being explored not only to investigate fundamental properties of microsporidia but also to differentiate between species.

Polyacrylamide disc gel electrophoresis used by Fowler and Reeves (1974a,b) for characterisation of microsporidia on hydrophobic and hydrophilic protein spectra have been superceded by sodium dodecyl sulphate–polyacrylamide gel electrophoresis (SDS-PAGE). Unique electrophoretic profiles of spore polypeptides obtained with this technique can be used to identify species of microsporidia, when disrupted spores are run on the gels (Street and Briggs, 1982). Disruption of spores can be achieved with a cell homogeniser and small glass beads (c̄ 0.5 mm diameter) at 4°C. A natural sequel of this technique would be immunoblot analysis of SDS-PAGE polypeptides with polyclonal or monoclonal antisera when these become available.

Immunological differentiation of species has been achieved by double immunodiffusion (Knell and Zam, 1978) and may also be possible using an enzyme-linked immunosorbent assay (ELISA) on spore homogenates as developed by Greenstone (1983) or even on whole spores. However, cross reactivity between genera was found by immunofluorescence (Niederkorn, Shadduck and Weidner, 1980) and this technique may not be useful in identification of microsporidian isolates.

References

Agapova, A. L. (1966). "Parasites of fish of Kazakhstan reservoirs". Nauka, Alma-Ata, U.S.S.R. pp. 342 (in Russian).

Akao, S. (1969). Studies on the ultrastructure of *Nosema cuniculi*, a microsporidian parasite of rodents. *Jap. J. Parasitol.* **18**, 8–20.

Akhmerov, A. K. (1946). Microsporidiosis of pond smelt *Hypomesus olidus* Pallas (Teleostei). *Tr. Leningradsk. obshch estestvoznan.* **69**, 3–6 (in Russian).

Akhmerov, A. K. (1951). Some data on the parasites of Alaskan pollock. *Izv. Tikhookeans nauchno-issled. Inst. rybnogo khoz. Okeanograf.* **30**, 99–104 (In Russian).

Alvarez, P. H. J. (1971). *Nosema cuniculi,* parasito en el sistema nervioso central del raton. *Patologia,* **9**, 35–41.

Andódi, L. and Frank, S. (1969). "Aquarium Fishes". Obzor, Bratislava, pp. 210 (In Slovak).

Anenkova-Khlopina, N. P. (1920). Contribution to the study of parasitic diseases of *Osmerus eperlanus. Izv. otd. rybovad. nauch-prom. issled.* **1,2.** (In Russian).

Anigstein, L. (1925). Cycle évolutif de l'*Encephalitozoon,* (Levaditi) des souris. *C.R. Soc. Biol. (Paris).* **92**, 993–995.

Anver, M. R., King, N. W. and Hunt, R. D., (1972). Congenital encephalitozoonosis in a squirrel monkey (*Saimiri sciureus*). *Vet. Pathol.* **9**, 475–480.

Arison, R. N., Cassaro, J. A. and Pruss, M. P. (1966). Studies on a murine ascites-producing agent and its effect on tumor development. *Cancer Res.* **26**, 1915–1920.

Armstrong, J. A., Ke, Y. -H., Breinig, M. C. and Ople, L. (1973). Virus resistance in rabbit kidney cell cultures contaminated by a protozoan resembling *Encephalitozoon cuniculi. Proc. Soc. Exp. Biol. Med.* **142**, 1205–1208.

Arnesen, K. and Nordstoga, K. (1977). Ocular encephalitozoonosis (nosematosis) in blue foxes. Acta Ophthalmol. **55**, 641–651.

Ashton, N., Cook, C. and Clegg, F. (1976). Encephalitozoonosis (Nosematosis) causing bilateral cateract in a rabbit. *Br. J. Ophthalmol.* **60**, 618–631.

Ashton, N. and Wirasinha, P. A. (1973). Encephalitozoonosis of the cornea. *Br. J. Ophthalmol.* **57**, 669–674.

Attwood, H. D. and Sutton, R. D. (1965). *Encephalitozoon* granulomata in rats. *J. Pathol. Bacteriol.* **89**, 735–738.

Auerbach, M. (1910). Zwei neue Cnidosporidien aus cyprinoiden Fischen. *Zool. Anz.* **36**, 440–441.

Awakura, T. (1974). Studies on the microsporidian infection in salmonid fishes. *Sci. Rep. Hokkaido Fish Hatchery,* **29**, 1–96.

Awakura, T. (1978). A new epizootic of a microsporidiosis of salmonids in Hokkaido, Japan. *Fish. Pathol.* **13**, 17–18 (In Japanese).

Awakura, T. and Kurahashi, S. (1967). Studies on the *Plistophora* disease of salmonid fish. III. On prevention and control of the disease. *Sci. Rep. Hokkaido Fish Hatchery*, **22**, 51–68.

Awakura, T., Kurahashi, S. and Matsumoto, H. (1966). Studies on the *Plistophora* disease of salmonid fish. II. Occurrence of the microsporidian disease in a new district. *Sci. Rep. Hokkaido Fish Hatchery*, **21**, 1–12.

Awakura, T., Tanaka, M. and Yoshimiz, M. (1982). Studies on parasites of masu salmon, *Oncorhynchus masou* - IV. *Loma* sp. (Protozoa: Microsporea) found in the gills. *Sci. Rep. Hokkaido Fish Hatchery*, **37**, 49–55.

Bangham, R. V. (1941). Parasites from fish of Buckeye Lake, Ohio. *Ohio J. Sci.* **41**, 441–448.

Barker, R. J. (1974). The nature of *Encephalitozoon brumpti* Coulon, 1924. *J. Parasitol.* **60**, 542–544.

Barker, R. J. (1975). Ultrastructural observations on *Encephalitozoon cuniculi* from mouse peritoneal macrophages. *Folia Parasitol.* (Praha), **22**, 1–9.

Basson, P. A., McCully, R. M. and Warnes, W. E. J. (1966). Nosematosis: report of a canine case in the Republic of South Africa. *J. Sth. Afr. Vet. Med. Assoc.* **37**, 3–9.

Bauer, O. N. (1948). Parasites of fish of the river Yeniseĭ. *Izv. Gosniorch*, **27**, 97–156 (In Russian).

Bazikalova, A. (1932). Additions to parasitology of Murmansk fishes. *Sbor nauchno-prom. rab. Murman*. CNIRC. pp.136–153 (In Russian).

Bedrník, P. and Vávra, J. (1972). Further observations on the maintenance of *Encephalitozoon cuniculi* in tissue culture. *J. Protozool.* **19** (suppl.), 75 (abstract).

Behkti, M. (1981). Étude comparative des espèces des microsporidies parasites des poissons téléostéens du littoral languedocien. Rapport D. E. A., Univ. Sci. et Tech. Languedoc, Montpellier. 32 pp.

Behkti, M. (1984). Contribution à l'étude des microsporidioses des poissons des côtes méditérranéenes: les genres *Loma* et *Glugea*, biologie et relations hôte-parasite. *Thèse, Univ. Sci. Tech. Languedoc, Montpellier*, pp. 208.

Behkti, M. and Bouix, G. (1985). *Loma salmonae* (Putz, Hoffman and Dunbar, 1965) et *Loma diplodae* n.sp., microsporidies parasites des branchies des poissons téléostéens : implantation et données ultrastructurales **21**, 47–59.

Bell, E. T. and Hartzell, T. B. (1919). Spontaneous nephritis in rabbits and its relation to chronic nephritis in man. *J. Infect. Dis.*, **24**, 628–635.

Bender, L. (1925). Spontaneous central nervous system lesions in the laboratory rabbit. *Amer. J. Pathol.* **1**, 653–656.

Bergquist, R., Morfeldt-Månon, L., Pehrson, P. O., Petrini, B. and Wasserman, J. (1985). Antibody against *Encephalitozoon cuniculi* in Swedish homosexual men. *Scand. J. Infect. Dis.* (in press).

Bergquist, N. R., Stintzing, G., Smedman, L., Waller, T. and Andersson, T. (1984). Diagnosis of encephalitozoonosis in man by serological tests. *Br. Med. J.* **288**, 902.

Berkin, S. and Kahraman, M. M. (1983). *Encephalitozoon cuniculi* infection of rabbits in Turkey. *Ankara Univ. Vet. Fak. Derg.* **30**, 397–406 (In Turkish).

Berrebi, P. (1978). Contribution à l'étude biologique des zones saumatres du littoral méditérranéen francais. Biologie d'une microsporidie: *Glugea atherinae* n. sp., parasite de l'athérine : *Atherina boyeri* Risso, 1810 (Poisson - Téléostéen) des étangs cotiers. *Thèse, Univ. Sci. Tech. Languedoc, Montpellier*, pp.196.

Berrebi, P. (1979). Étude ultrastructurale de *Glugea atherinae* n. sp., microsporidie de l'athérine, *Atherine boyeri* Risso, 1810 (Poisson, Téléostéen) dans les lagunes du Languedoc et de Provence. *Z. Parasitenk.* **60**, 105–122.

Berrebi, P. and Bouix, G. (1978). Premières observations sur une microsporidiose de l'athérine des étangs languedociens, *Atherina boyeri* Risso, 1810 (Poissons Téléostéens). *Ann. Parasitol. Hum. Comp.* **53**, 1–20.

Bismanis, J. E. (1970). Detection of latent murine nosematosis and growth of *Nosema cuniculi* in cell cultures. *Can. J. Microbiol.* **16**, 237–242.

Blasiola, G. C. Jr. (1979). *Glugea heraldi* n.sp. (Microsporida, Glugeidae) from seahorse *Hippocampus erectus* Perry. *J. Fish. Dis.* **2**, 493–500.

Blažek, K., Koczková, I., Lávička, M., Vávra, J. and Stehlík, M. (1972). Nosematosis (Encephalitozoonosis) bei Karnivoren. Verhandlungsberichte XIV. Internat. Symp. Enkrankungen der Zootiere, Acad. Verlag DDR, 369–371.

Bogdanova, E. A. (1957). The microsporidian *Glugea hertwigi* Weissenberg in the stint (*Osmerus eperlanus* m. *spirinchus*) from Lake Ylyna-yarvi. *In* "Parasites and diseases of fish". ed. G. K. Petrushevskii. *Izv. Vsesoyuz. nauch-issled. Inst. Ozer. Rech. Ryb. Khoz.* Leningrad, p.328, (In Russian).

Bogdanova, E. A. (1961). The parasite fauna of some commercial fish of the Volga river before the Volgograd dam lake construction. *Tr. Sov. Ikhtyol. Kom.* AN SSSR, **11**, 169–177.

Bond, F. F. (1937). A microsporidian infection of *Fundulus heteroclitus* (Linn). *J. Parasitol.* **23**, 229–230.

Bond, F. F. (1938). Cnidosporidia from *Fundulus heteroclitus* (Linn). *Trans. Am. Microsc. Soc.* **57**, 107–122.

Bosanquet, W. C. (1910). Brief notes on two myxosporidian organisms (*Pleistophora hippoglossoideos* n.sp. and *Myxidium mackiei* n.sp.). *Zool. Anz.* **35**, 434–438.

Botha, W. S. van Dellen, A. F. and Stewart, C. G. (1979). Canine encephalitozoonosis in South Africa. *J. Sth. Afr. Vet. Med. Ass.*, **50**, 135–144.

Branstetter, D. G. and Knipe, S. M. (1982). Microsporidian infection in the lovebird *Agapornis roseicollis. Micron,* **13**, 61–62.

Bray, R. S. and Garnham, P. C. C. (1962). The Giemsa Colophonium technique for staining tissue sections. *Ind. J. Malariology* **16**, 153–155.

Brown, R. J., Hinkle, D. K., Trevethan, W. R., Kupper, J. L. and McKee, A. E. (1973). Nosematosis in a squirrel monkey (*Saimiri sciureus*). *J. Med. Prim.* **2**, 114–123.

Bückmann, A. (1952). Infektionen mit *Glugea stephani* und mit *Vibrio anguillarum* bei Schollen (*Pleuronectes platessa* L.) *Kurze Mitteilung. Fischbiol. Abt. Max Planck Inst. Meeresbiol.* **1**, 1–7.

Bull, C. G. B. (1917). The pathologic effects of streptococci from cases of poliomyelitis and other sources. *J. Exp. Med.* **25**, 557–580.

Buyukmihci, N., Bellhorn, R. W., Hunziker, J. and Clinton, J. (1977). *Encephalitozoon (Nosema)* infection of the cornea in a cat. *J. Am. Vet. Med. Ass.* **171**, 355–357.

Bywater, J. E. C. (1979). Is encephalitozoonosis a zoonosis? *Lab. Anim.* **13**, 149–151.

Bywater, J. E. C. and Kellett, B. S. (1978a). *Encephalitozoon cuniculi* antibodies in a specific pathogen free rabbit unit. *Inf. Immun.* **21**, 360–364.

Bywater, J. E. C. and Kellett, B. S. (1978b). The eradication of *Encephalitozoon cuniculi* from a specific pathogen-free rabbit colony. *Lab. Anim. Sci.* **28**, 402–404.

Bywater, J. E. C. and Kellett, B. S. (1979). Humoral immune response to natural infection with *Encephalitozoon cuniculi* in rabbits. *Lab. Anim.* **13**, 293–297.

Bywater, J. E. C., Kellett, B. S. and Waller, T. (1980). *Encephalitozoon cuniculi* antibodies in commercially-available rabbit antisera and serum reagents. *Lab. Anim.* **14**, 87–89.

Cali, A. (1971). Morphogenesis in the genus *Nosema*. *Proc. IVth Int. Coll. Insect. Pathol.* Maryland, 1970. 431–438.

Cameron, G. C. and Maitland, H. B. (1924). A description of parasites in spontaneous encephalitis of rabbits. *J. Path. Bact.* **27**, 329–333.

Canning, E. U. (1965). An unusually heavy natural infection of *Nosema cuniculi* (Levaditi *et al.*) in a laboratory rat. *Trans. Roy. Soc. Trop. Med. Hyg.* **59**, 371 (abstract).

Canning, E. U. (1967). Vertebrates as hosts to microsporidia, with special reference to rats infected with *Nosema cuniculi*. *Protozoology (Suppl. J. Helminthol.)* **2**, 197–205.

Canning, E. U. (1976a). Microsporidia in vertebrates; host parasite relations at the organismal level. *In* Bulla, L. A. and Cheng, T. C., eds. *Comparative Pathobiology*, Vol. 1. Biology of the Microsporidia. Plenum Press, New York and London, pp. 137–161.

Canning, E. U. (1976b). The microsporidian parasites of Platyhelminthes : their morphology, development, transmission and pathogenicity. *Commonwlth. Inst. Helm. Misc. Publ.* **2**, pp.32.

Canning, E. U. (1981). *Encephalitozoon lacertae* n.sp., a microsporidian parasite of the lizard *Podarcis muralis*. *In* "Parasitological Topics" Society of Protozoologists Spec. Publ. 1, 57–64.

Canning, E. U. (1986). Phylum Microspora. In L. Margulis, D. Chapman and J. D. Corliss eds. Handbook of Protoctists, Science Books International (in press).

Canning, E. U. and Elkan, E. (1963). Microsporidiosis in *Bufo bufo*. *Parasitology*. **53**, 11P–12P. (abstract).

Canning, E. U., Elkan, E. and Trigg, P. I. (1964). *Plistophora myotrophica* spec. nov. causing high mortality in the common toad *Bufo bufo* L., with notes on the maintenance of *Bufo* and *Xenopus* in the laboratory. *J. Protozool.* **11**, 157–166.

Canning, E. U. and Hazard, E. I. (1982). Genus *Pleistophora* Gurley, 1893 : an assemblage of at least three genera. *J. Protozool.* **29**, 39–49.

Canning, E. U., Hazard, E. I. and Nicholas. J. P. (1979). Light and electron microscopy of *Pleistophora* sp. from skeletal muscle of *Blennius pholis*. *Protistologica*. **15**, 317–332.

Canning, E. U. and Landau, I. (1971). A microsporidian infection of *Lacerta muralis*. *Trans. Roy. Soc. Trop. Med. Hyg.* **65**, 431 (abstract).

Canning, E. U., Lom, J. and Nicholas, J. P. (1982). Genus *Glugea* Thélohan, 1891 (Phylum Microspora) : redescription of the type species *Glugea anomala* (Moniez, 1887) and recognition of its sporogonic development within sporophorous vesicles (pansporoblastic membranes). *Protistologica* **18**, 193–210.

Canning, E. U. and Nicholas, J. P. (1980). Genus *Pleistophora* (Phylum Microspora) : redescription of the type species, *Pleistophora typicalis* Gurley, 1893 and ultrastructural characterisation of the genus. *J. Fish Dis.* **3**, 317–338.

Casaubon, H. M. T. (1975). Nosematosis in a rabbit, a case report. *Veterinaria Mex.* **6**, 78–79.

Caullery, M. (1953). Appendice aux sporozoaires class des Haplosporidies. *In Traité de Zoologie* 1(2) ed. P. P. Grassé pp. 922–934. Masson et Cie, Paris.

Caullery, M. and Mesnil, P. (1905). Sur des haplosporidies parasites de poissons marins. *C. R. Soc. Biol. Paris.* **58**, 640–642.

Cépède, C. (1906). Sur une microsporidie nouvelle, *Pleistophora macrospora,* parasite des loches frances du Dauphiné. *C. R. Acad. Sci. Paris.* **142**, 56–58.

Cépède, C. (1924). *Mrazekia piscicola* n.sp. microsporidie parasite du Merlan (*Gadus merlangus*). *Bull. Soc. Zool. France* **49**, 109–113.

Chalupský, J., Bedrník, P. and Vávra, J. (1971). The indirect fluorescent antibody test for *Nosema cuniculi*. *J. Protozool.* **18**, 47 (abstract).

Chalupský, J., Lenský, P., Bedrník, P. and Vávra, J. (1972). An attempt to demonstrate the connection between multiple sclerosis and *Encephalitozoon cuniculi* infection. *J. Protozool.* **19** (suppl), 76–77.

Chalupský, J., Vávra, J. and Bedrník, P. (1973). Detection of antibodies to *Encephalitozoon cuniculi* in rabbits by the indirect immunofluorescent antibody test. *Folia Parasitol.* (Praha). **20**, 281–284.

Chalupský, J., Vávra, J. and Bedrník, P. (1979a). The use of the India ink immunoreaction in the diagnosis of rabbit encephalitozoonosis. *J. Protozool.* **26**(3), 70A–71A.

Chalpuský, J., Vávra, J. and Bedrník, P. (1979b). Encephalitozoonosis in labratory animals - a serological survey. *Folia Parasitol.* (Praha). **26**, 1–8.

Chatton, E. (1920). Un complexe xéno-parasitaire morphologique et physiologique *Neresheimeria paradoxa* chez *Fritillaria pellucida*. *C.R. Acad. Sci. Paris*. **171**, 55–57.

Chatton, E. and Courrier, R. (1923). Formation d'un complexe xénoparasitaire géant avec bordure en brosse, sous l'influence d'une microsporidie, dans le testicule de *Cottus bubalis*. *C. R. Soc. Biol. Paris*. **89**, 579–583.

Chen, C. L. (1956). The protozoan parasites from four species of Chinese pond fishes: *Ctenopharyngodon idellus, Mylopharyngodon piceus, Aristhichthys nobilis* and *Hypophthalmichthys molithrix II*. The protozoan parasites of *Mylopharyngodon piceus*. *Acta Hydrobiol. Sinica*. **1**, 19–42.

Chen, C. L. and Hsieh, S. (1960). Studies on sporozoa from the freshwater fishes *Ophiocephalus maculatus* and *O. argus* of China. *Acta Hydrobiol. Sinica*. **2**, 171–196.

Chen, M. and Power, G. (1972). Infection of American smelt in Lake Ontario and Lake Erie with the microsporidian parasite *Glugea hertwigi* (Weissenberg). *Can. J. Zool.* **50**, 1183–1188.

Chumakov, M. P., Viting, A. I., Konosh, O. V. and Ashmarina, E. E. (1970). The cultivation of the microsporidian *Encephalitozoon (Nosema) cuniculi* in human and mouse embryo fibroblast cultures. *Med. Parazitol. Parazitarn Bolezni*. **39**, 643–647 (in Russian).

Clemens, W. A. (1920). Histories of new food fishes. *IV* The Honfish. *Biol. Board Can. Bull.* **4**, 1–12.

Corbel, M. J. (1975). The immune response in fish : a review. *J. Fish Biol.* **7**, 539–563.

Cole, J. R. Jr., Sangster, L. T., Sulzer, C. R., Pursell, A. R. and Ellinghausen, H. C. (1982). Infections with *Encephalitozoon cunculi* and *Leptospira interrogans*, serovars *grippotyphosa* and *ballum*, in a kennel of foxhounds. *J. Am. Vet. Med. Ass.* **180**, 435–437.

Coulon, G. (1929). Présence d'un nouvel *Encephalitozoon* (*Encephalitozoon brumpti* n.sp.) dans le liquide céphalo-rachidien d'un sujet atteint de méningite suraigue. *Ann. Parasit. Hum. Comp.* **7**, 449–452.

Cowdry, E. V. and Nicholson, F. M. (1924a). Meningo-encephalitic lesions and protozoan-like parasites in brains of apparently normal laboratory animals commonly employed for experimentation. *J. Amer. Med. Ass.* **82**, 545.

Cowdry, E. V. and Nicholson, F. M. (1924b). The co-existence of protozoan-like parasites and meningoencephalitis in mice. *J. Exp. Med.* **40**, 51–62.

Cox, J. C. (1977a). Isolation of *Encephalitozoon cuniculi* from urine samples. *Lab. Anim.* **11**, 233–234.

Cox, J. C. (1977b). Altered immune responsiveness associated with *Encephalitozoon cuniculi* infection in rabbits. *Inf. Immun.* **15**, 392–395.

Cox, J. C. and Gallichio, H. A. (1977). An evaluation of indirect immunofluorescence in the serological diagnosis of *Nosema cuniculi* infection. *Res. Vet. Sci.* **22**, 50–52.

Cox, J. C. and Gallichio, H. A. (1978). Serological and histological studies on adult rabbits with recent, naturally acquired encephalitozoonosis. *Res. Vet. Sci.* **24**, 260–261.

Cox, J. C., Gallichio, H. A., Pye, D. and Walden, N. B. (1977). Application of immuno-fluorescence to the establishment of an *Encephalitozoon cuniculi*-free rabbit colony. *Lab. Anim. Sci.* **27**, 204–209.

Cox, J. C., Hamilton, R. C. and Attwood, H. D. (1979). An investigation of the route and progression of *Encephalitozoon cuniculi* infection in adult rabbits. *J. Protozool.* **26**, 260–265.

Cox, J. C., Horsburgh, R. and Pye, D. (1981). Simple diagnostic test for antibodies to *Encephalitozoon cuniculi* based on enzyme immunoassay. *Lab. Anim.* **15**, 41–43.

Cox, J. C. and Pye, D. (1975). Serodiagnosis of nosematosis by immunofluorescence using cell culture grown organisms. *Lab. Anim.* **9**, 297–304.

Cox, J. C. and Ross, J. (1980). A serological survey of *Encephalitozoon cuniculi* infection in the wild rabbit in England and Scotland. *Res. Vet. Sci.* **28**, 396.

Cox, J. C., Walden, N. B. and Nairn, R. C. (1972). Presumptive diagnosis of *Nosema cuniculi* in rabbits by immunofluorescence. *Res. Vet. Sci.*, **13**, 595–597.

Crandall, T. and Bowser, P. R. (1981). A microsporidian infection in the mosquitofish, *Gambusia affinis*. *J. Fish. Dis.* **4**, 317–324.

Da Fano, C. (1924). Spontaneous and experimental encephalitis in rabbits. *Med. Sci.* **10**, 355–379.

Danilewsky, B. (1891). Über die Myoparasiten der Amphibien and Reptilien. *Zentralb. Bakt. Parasitenk.* **9**, 9–10.

Debaisieux, P. (1919a). Hypertrophie des cellules animales parasitées par des Cnidosporidies. *C. R. Soc. Biol.* Paris, **82**, 867–869.

Debaisieux, P. (1919b). Etudes sur les Microsporidies II. *Glugea danilewskyi* L. Pfr. III *Glugea mulleri* L. Pfr. *Cellule,* **30**, 153–183.

Debaisieux, P. (1920). Etudes sur les microsporidies IV. *Glugea anomala* Monz. *Cellule.* **30**, 215–245.

Dechtiar, A. O. (1965). Preliminary observations on *Glugea hertwigi*, Weissenberg, 1911 (Microsporidia; Glugeidae) in American smelt, *Osmerus mordax* (Mitchill) from Lake Erie. *Can. Fish. Cult.* **34**, 35–38.

Delisle, C. (1965). A study in the mass mortality of the stunted smelt population, *Osmerus eperlanus mordax* at Heney Lake, Gatineau Co. Quebec. *Proc. 10th Tech. Sess. Ontario Res. Found,* Kingston Ontario, Canada, 25–27.

Delisle, C. (1969). Bimonthly progress of a non-lethal infection by *Glugea hertwigi* in young-of-the-year smelt, *Osmerus eperlanus mordax*. *Can. J. Zool.* **47**, 871–876.

Delisle, C. E. (1972). Variations mensuelles de *Glugea hertwigi* (Sporozoa : Microsporidia) chez différents tissues et organes de l'éperlan adulte dulcicole et consequences de cette infection sur une mortalité massive annuelle de ce poisson. *Can. J. Zool.* **50**, 1589–1600.

Delisle, C. and Veilleux, C. (1969). Répartition géographique de l'éperlan arc-en-ciel *Osmerus eperlanus mordax* et de *Glugea hertwigi* (Sporozoa : Microsporidia) en eau douce, au Québec. *Nat. Can.* (Québec) **96**, 337–358.

Delphy, M. J. (1916). Scoliose abdominale chez le *Mugil auratus* Risso et présence d'une myxosporidie parasite de ce poisson. *C. R. Acad. Sci. Paris* **163**, 71–73.

Desportes, I., Le Charpentier, Y., Galian, A., Bernard, F., Cochand-Priollet, B., Lavergne, A., Ravisse, P. and Modigliani, R. (1985). Occurrence of a new microsporidian : *Enterocytozoon bieneusi* n.g., n.sp. in the enterocytes of a human patient with AIDS. *J. Protozool* **32**, 250–254.

Dissanaike, A. S. (1957). The morphology and life cycle of *Nosema helminthorum*. Moniez, 1887. *Parasitology,* **47**, 335–346.

Doby, J. -M., Jeannes, A. and Rault, B. (1963). *Thelohania apodemi* n.sp., première microsporidie du genre *Thelohania* observée chez un Mammifère. *C. R. Acad. Sci. Paris,* **257**, 248–251.

Doby, J. -M. and Jeannes, A. (1968). Présence de *Thelohania apodemi* (Doby, Jeannes et Rault, 1963) (Microsporidie) dans le tissu musculaire du mulot. *Ann. Parasitol. Hum. Comp.* **43**, 619–622.

Doby, J. -M., Jeannes, A. and Rault, B. (1965). Systematical research of toxoplasmosis in the brain of small mammals by a histological method. *Ceskoslov. Parazitol.* **12**, 133–143.

Doby, J. -M., Rault, B. and Barker, R. (1975). Phénomènes de fluorescence chez *Thelohania apodemi* Doby, Jeannes et Rault, 1963. microsporidie parasite de l'encéphale du mulot *Apodemus sylvaticus. C. R. Soc. Biol. Paris.* **169**, 1053–1056.

Doerr, R. and Zdansky, E. (1923a). Zur Aetiologie der Encephalitis epidemica. *Schweizer Med. Wochenschr.* **53**, 349–351.

Doerr, R. and Zdansky, E. (1923b). Weitere parasitologische Befunde in Gehirn von Kaninchen. *Schweizer Med. Wochenschr.* **53**, 1189–1190.

Doflein, F. (1898) Studien zur Naturgeschichte der Protozoen III. Über Myxosporidien. *Zool. Jahr. Abt. Anat.* **11**, 281–350.

Doflein, F. and Reichenow, E. (1927–1929). Lehrbuch der Protozoenkunde. G. Fischer. Jena. pp. 1262.

Dogiel, V. A. (1936). Parasites of cod from the relic lake Mogilny. *Tr. Leningrad, Univ. Biol.* **7**, 123–133. (In Russian).

Dogiel, V. A. and Bychovskiĭ, B. E. (1939). Parasites of fishes of Caspian Sea. *Trudy.po Kompletnom. Izuchenii. Kaspiiskogo. Morya.* 7. Publ. Ac. Sci. U.S.S.R., Moscow - Leningrad, pp30. (in Russian).

Drew, G. H. (1909). Some notes on parasitic and other diseases of fish. *Parasitology,* **2**, 193–201.

Drew, G. H. (1910). Some notes on parasitic and other diseases of fish, 2nd series. *Parasitology,* **3**, 54–62.

Dyková, I. and Lom, J. (1978). Tissue reaction of the three-spined stickleback *Gasterosteus aculeatus* L. to infection with *Glugea anomala* (Moniez, 1887). *J. Fish. Dis.* **1**, 83–90.

Dyková, I. and Lom, J. (1980a). Tissue reaction to *Glugea plecoglossi* infection in its natural host, *Plecoglossus altivelis. Fol. Parasitol.* (Praha), **27**, 213–216.

Dyková, I. and Lom, J. (1980b). Tissue reaction to microsporidian infection in fish hosts. *J. Fish Dis.* **3**, 265–283.

Egusa, E. S. (1982). A microsporidian species from yellowtail juveniles, *Seriola quinqueradiata,* with "Beko" disease. *Fish. Pathol,* **16**, 187–192 (In Japanese).

Elkan, E. (1963). A microsporidium affecting the common toad (*Bufo bufo* L.). *Br. J. Herpetol.* **3**, 89.

Fantham, H. B., Porter, A. and Richardson, L. R. (1941). Some microsporidia found in certain fishes and insects in eastern Canada. *Parasitology,* **33**, 186–208.

Finn, J. P. and Nielson, N. O. (1971). The effect of temperature variation on the inflammatory response of rainbow trout. *J. Pathol.* **105**, 257–268.

Fischthal, J. H. (1944). Observations on a sporozoan parasite of the eelpout, *Zoarces anguillaris,* with an evaluation of candling methods for its detection. *J. Parasitol.* **30**, 35–36.

Flatt, R. E. and Jackson, S. J. (1970). Renal nosematosis in young rabbits. *Path. Vet.,* **7**, 492–497.

Fowler, J. L. and Reeves, E. L. (1974a). Detection of relationships among microsporidian isolates by electrophoretic analysis : hydrophobic extracts. *J. Invert. Pathol.* **23**, 3–12.

Fowler, J. L. and Reeves, E. L. (1974b). Detection of relationships among microsporidian isolates by electrophoretic analysis : hydrophilic extracts. *J. Invert. Pathol.* **23**, 63–69.

Gaïevskaya, A. V. and Kovaleva, A. A. (1975). Diseases of commercial fishes in the Atlantic ocean. *Kaliningradskoe Knizhnoe Idatelstvo*, pp. 124. (In Russian).

Gannon, J. (1978). The immunoperoxidase test for diagnosis of *Encephalitozoon cuniculi* in rabbits. *Lab. Anim.* **12**, 125–127.

Gannon, J. (1980a). A survey of *Encephalitozoon cuniculi* in laboratory animal colonies in the United Kingdom. *Lab. Anim.* **14**, 91–94.

Gannon, J. (1980b). The course of infection of *Encephalitozoon cuniculi* in immunodeficient and immunocompetent mice. *Lab. Anim.* **14**, 189–192.

Garnham, P. C. C. and Roe, F. J. C. (1954). *Encephalitozoon muris* in liver and spleen of subinoculated mice. *Trans. Roy. Soc. Trop. Med. Hyg.* **48**, 1.

Gasimagomedov, A. A. and Issi, I. V. (1970). Microsporidian parasites of fishes of the Caspian Sea. *Zool.Zhur.* **49**, 1117–1125 (In Russian).

Ghittino, P. (1974). Present knowledge of the principal diseases of cultured marine fish. *Riv. Ital. Piscic. Ittio.* **9**, 51–56.

Goodman, D. G. and Garner, F. M. (1972). A comparison of methods for detecting *Nosema cuniculi* in rabbit urine. *Lab. Anim. Sci.* **22**, 568–572.

Goodpasture, E. W. (1924). Spontaneous encephalitis in rabbits. *J. Infect. Dis.* **34**, 428–432.

Gordon, F. P. (1940). Parasite resembling *Encephalitozoon* found in a white rat. *Arch. Pathol.* **30**, 824–825.

Goreglyad, Ch. S. (1962). Microsporidiosis of pond carps. *Dok. Akad. Nauk. Belorus.* S.S.R., **6**, 270–271 (In Russian).

Grabda, J. (1961). Zdrawotnośc ryb v potoku Trzebiocha. *Rocz. Nauk. Rolnicz.* **93**, 445–459.

Greenstone, M. H. (1983). An enzyme-linked immunosorbent assay for the Amblyospora sp. of *Culex salinarius* (Microspora:Amblyosporidae). *J. Invert. Pathol.* **41**, 250–255.

Grupcheva, G. and Lom, J. (1980). Protozoan parasites of fishes from Bulgaria. I. *Glugea luciopercae* and the description of three new *Trichodina* species. *Folia Parasit.* (Praha). **27**, 289–294.

Guli, E. and Parisio, B. (1958). Criteri diagnostici differenciali nella Toxoplasmose animale. *Fol. Hered. Pathol.* **8**, 25–36.

Gurley, R. R. (1893). Classification of the Myxosporidia, a group of protozoan parasites infecting fishes. *Bull. U.S. Fish. Comm. for 1891,* **11**, 407–420.

Gurley, R. R. (1894). The Myxosporidia, or psorosperms of fishes, and the epidemics produced by them. *Rep. U.S. Fish. Comm. for 1892,* **26**, 65–203.

Guyénot, E. and Naville, A. (1922a). Recherches sur le parasitisme et l'évolution d'une microsporidie *Glugea danilewskyi* Pfr (?) parasite de la couleuvre. *Rev. Suisse Zool.,* **30**, 1–61.

Guyénot, E. and Naville, A. (1922b). Sur une myxosporidie (*Myxobolus ranae* n.sp.) et une microsporidie. *Rev. Suisse Zool.* **29**, 413–424.

Guyénot, E. and Naville, A. (1924). *Glugea encyclometrae* n.sp. et *G. ghigii* n.sp. parasites de Platodes et leur developpement dans l'hôte vertébré (*Tropidonotus natrix* L.) *Rev. Suisse Zool.* **31**, 75–114.

Guyénot, E., Naville, A. and Ponse, K. (1922). Une larve de Cestode parasitée par une microsporidie. *Compt. Rend. Soc. Biol.*, **87**, 635–636.

Guyénot, E., Naville, A. and Ponse, K. (1925). Deux microsporidies parasites de trématodes. *Rev. Suisse Zool.*, **31**, 399–419.

Guyénot, E. and Ponse, K. (1926). Une microsporidie *Plistophora bufonis,* parasite d'organe de Bidder du Crapaud. *Rev. Suisse. Zool.*, **33**, 213–250.

Hagenmüller, M. (1899). Sur une nouvelle Myxosporidie, *Nosema stephani,* parasite du *Flesus passer* Moreau. *C. R. Acad. Sci.*, Paris, **129**, 836–839.

Haley, A. J. (1952). Preliminary observations on a severe epidemic of microsporidiosis in the smelt, *Osmerus mordax* (Mitchill). *J. Parasitol.*, **38**, 183–185.

Haley, A. J. (1953). Observations on a protozoan infection in the freshwater smelt. *Proc. 32nd Ann. Sess. New Hampshire Acad. Sci* p. 7.

Haley, A. J. (1954). Microsporidian parasite, *Glugea hertwigi,* in American smelt from the Great Bay region, New Hampshire. *Trans. Am. Fish. Soc.* **83**, 84–90.

Hamilton, R. C., Cox, J. C. and Pye, D. (1977). Wall structure of the sporonts of *Encephalitozoon cuniculi* grown in human fibroblasts. *J. Gen. Microbiol.* **98**, 305–307.

Hashimoto, K., Sasaki, Y. and Takinami, K. (1976). Conditions for extrusion of the polar filament of the spore of *Plistophora anguillarum,* a microsporidian parasite of *Anguilla japonica. Bull. Jap. Soc. Sci. Fish.*, **42**, 837–845 (In Japanese).

Hashimoto, K., Sasaki, Y. and Takinami, K. (1979). Inhibitory effect of amines on polar filament extrusion by *Plistophora anquillarum* spores. *Current Microbiol.* **3**, 137–140.

Hashimoto, K. and Takinami, K. (1976). Electron microscopic observations of the spores of *Plistophora anguillarum,* a microsporidian parasite of the eel. *Bull. Jap. Soc. Sci. Fish.* **42**, 411–419. (In Japanese).

Hazard, E. I. and Broodbank, J. W. (1984). Karyogamy and meiosis in an *Amblyospora* sp. (Microspora) in the mosquito *Culex salinarius. J. Invertebr. Pathol.*, **44**, 3–11.

He, Xiaojie (1982). A study on the pleistophorasis of *Priacanthus tayenus,* with description of a new species. *J. Fish, China.* **6**, 97–105. (In Chinese).

Herald, E. S. and Rakowicz, M. (1951). Stable requirements for raising sea horses. *Aquarium J.* **22**, 234–242.

Herman, R. L. and Putz, R. E. (1970). A microsporidian (Protozoa : Cnidospora) in channel catfish (*Ictalurus punctatus*). *J. Wildlife Dis.* **6**, 173.

Hoffman, G. L. (1980). Advances in freshwater fish parasitology. *Proc. Fish Health Sect. Amer. Fish. Soc.,* Seattle, 1–11.

Hollister, W. A. K. and Canning, E. U. (1985). *Encephalitozoon cuniculi:* serological survey for antibodies in human sera. *Trans. Roy. Soc. Trop. Med. Hyg.* **79**, 277–278.

Hong, Songfong, Yu Dengzi, Si Fongwen and Xu Zhaihai (1983). Encephalitozoonosis in rabbits. *Acta Vet. Zootech. Sinica,* **14**, 59–63.

Hoshina, T. (1951). On a new microsporidian, *Plistophora anguillarum* n.sp. from the muscle of the eel, *Anguilla japonica. J. Tokyo Univ. Fish.*, **38**, 35–46.

Howell, J. Mc. C. and Edington, N. (1968). The production of rabbits free from lesions associated with *Encephalitozoon cuniculi. Lab. Anim.* **2**, 143–146.

Hua, Ding-ke and Dong, Hun-ji (1983). A new species of *Pleistophora* from big-eyes, *Priacanthus tayenus* Richardson and *P. macrocanthus* Cuvier and Valenciennes from the South China Sea. *J. Fish. Dis.* **6**, 293–301.

Hübner, J., Uhlíková, M., Bedrník, P. and Vávra, J. (1973). Serological diagnosis of encephalitozoonosis (nosematosis). *J. Protozool* (suppl). **20**, 536.

Huldt, G. and Waller, T. (1974). Accidental nosematosis in mice with impaired immunological competence. *Acta. Path. Microbiol. Scand. B.*, **82**, 451–452.

Hunt, R. D., King, N. W. and Foster, H. L. (1972). Encephalitozoonosis: evidence for vertical transmission. *J. Infect. Dis.* **126**, 212–214.

Iino, H. (1959). Studies on *Encephalitozoon* II. Detection of *Encephalitozoon* from organs of inoculated mice. *Keio Igaku,* **36**, 502–506 (In Japanese).

Innes, J. R. M., Zeman, W., Frenkel, J. K. and Borner, G. (1962). Occult endemic encephalitozoonosis of central nervous system in mice (Swiss Bagg - O'Grady strain). *J. Neuropathol. Exp. Neurol.* **21**, 519–533.

Ishihara, R. (1967). Stimuli causing extrusion of polar filaments of *Glugea fumiferanae* spores. *Can. J. Microbiol.* **13**, 1321–1332.

Ishihara, R. (1968). Some observations on the fine structure of sporoplasm discharged from spores of a microsporidian, *Nosema bombycis. J. Invert. Pathol.* **12**, 245–258.

Iskov, M. P. (1966). Microsporidiosis of the intestine of carps. *Veterinariya,* **6**, 58 (In Russian).

Izyumova, N. A. (1959). On the fluctuation of parasite fauna of fish of Rybinsky reservoir. *Tr. Inst. Biol. Vodokhran.* **2**, 174–190 (In Russian).

Jackson, S. J., Solorzano, R. F. and Middleton, C. C. (1973). An indirect fluorescent antibody test for antibodies to *Nosema cuniculi (Encephalitozoon)* in rabbits. *Proc. 77th Ann. Meeting U.S. Hlth. Ass.,* 480–490.

Jakowska, S. (1966). Infection with neurotropic microsporidians in South American *Lophius. Trans. Am. Microsc. Soc.* **85**, 161–162.

Jensen, L. A., Moser, M. and Heckmann, R. A. (1979). The parasites of the Californian lizardfish, *Synodus lucioceps. Proc. Helm. Soc. Wash.,* **46**, 281–284.

Jensen, H. M. and Wellings, S. R. (1972). Development of the polar filament - polaroplast complex in a microsporidian parasite. *J. Protozool.* **19**, 297–305.

Jírovec, O. (1930). Über eine neue Mikrosporidienart (*Glugea acerinae* n. sp.) aus *Acerina cernua. Arch. Protistenk.* **72**, 198–213.

Jírovec, O. (1932). Ergebnisse der Nuclealfärbung an den Sporen der Mikrosporidien nebst einigen Bemerkungen über Lymphocystis. *Arch. Protistenk.* **77**, 379–390.

Jírovec, O. (1934). *Octosporea machaři* n. sp., aus der Leber von *Dentex vulgaris. Zool. Anz.* **106**, 61–64.

Johnston, T. R. and Bancroft, N. J. (1919). Some sporozoan parasites of Queensland freshwater fish. *J. Proc. Roy. Soc. New South Wales,* **52**, 520–528.

Johnstone, J. (1901). Note on a sporozoan parasite of the plaice *(Pleuronectes platessa). Proc. Liverpool Biol. Soc.,* **15**, 184–187.

Jones, J. B. (1979). Trouble with microsporidia. Proc. Ann. Meeting New Zealand Soc. Parasitol. Wellington, 1979. *New Zealand J. Zool.* **6**, 648 (abstract).

Jordan, K. and Mirick, G. (1965a). An infectious hepatitis of undetected origin in mice. I. Description of the disease. *J. Exp. Med.* **102**, 601–603.

Jordan, K. and Mirick, G. (1965b). An infectious hepatitis of undetermined origin in mice. II Characteristics of the infective agent. *J. Exp. Med.* **102**, 617–630.

Jouvenaz, D. P. (1981). Percoll: an effective medium for cleaning microsporidian spores. *J. Invert. Pathol.* **37**, 319.

Julini, M. (1983). L'encefalitozoonosi dei conigli. *Riv. Coniglicoltura* **20**, 27–30.

Julini, M. and Pelegrino, N. (1981–82). Incidenza della encefalitozoonosi nei conigli macellati. *Ann. Fac. Med. Vet. Torino* **28**, 3–11.

Jungherr, E. (1955). *Encephalitozoon* encephalomyelitis in a rabbit. *J. Am. Vet. Med. Ass.* **127**, 518.

Kabata, Z. (1959). On two little-known microsporidia of marine fishes. *Parasitology* **49**, 309–315.

Kalavati, C. and Lakschminarayana, D. (1982). A new microsporidian *Nosema valamugili* from the gut epithelium of an estuarine fish *Valamugil* sp., *Acta Protozool.* **21**, 251–256.

Kalyakin, V. N. and Akinshina, G. T. (1970). The ability of *Encephalitozoon* to multiply in chick fibroblast cultures. *Med. Parazitol. Parazitar. Bolezni* **39**, 647–649 (In Russian).

Kaneda, Y. (1969). Studies on the effect of endoxan, an anti-tumor substance to promote the growth of *Nosema cuniculi in vivo* and *in vitro. Japan J. Parasitol.* **18**, 294–303.

Kangas, J. (1973). Nosematos hos rav (Nosematosis in the fox). *Stat. Vet. Anst.*, Helsinki.

Kano, T. and Fukui, H. (1982). Studies on *Pleistophora* infection in eel, *Anguilla japonica*. I. Experimental induction of microsporidiosis and fumigillin efficacy. *Fish. Pathol.* **16**, 193–200 (In Japanese).

Kano, T., Okauchi, T. and Fukui, H. (1982). Studies on *Pleistophora* infection in eel *Anguilla japonica*. II. Preliminary test for application of fumigillin. *Fish Pathol.* **17**, 107–114 (In Japanese).

Kantorowicz, R. and Levy, F. H. (1923). Neue parasitologische und pathologish-anatomische Befunde bei der nervösen Staupe der Hunde. *Arch. Tierheilkde.* **49**, 137–157.

Kashkovsky, V. V., Razmashkin, D. A. and Skripchenko, E. G. (1974). Diseases and parasites of fishes in Siberian and Ural fish farms. *Srednie Uralskoe Knizhnoe Izdatelstvo.* pp.160 (In Russian)

Kazieva, N. Sh. and Voronin, V. N. (1981). *Glugea rodei* sp. n. (Protozoa, Microsporidia) from the bitterling (*Rhodeus sericeus amarus*) connective tissue. *Zool. Zh.* **60**, 1254–1256 (In Russian).

Kellett, B. S. and Bywater, J. E. C. (1978). A modified india-ink immunoreaction for the detection of encephalitozoonosis. *Lab. Anim.* **12**, 59–60.

Kellett, B. S. and Bywater, J. E. C. (1979). India ink immunoreaction for encephalitozoonosis; phototechnique. *Lab. Anim.* **13**, 197–198.

Kellett, B. S. and Bywater, J. E. C. (1980). The indirect india-ink immunoreaction for detection of antibodies to *Encephalitozoon cuniculi* in rat and mouse serum. *Lab. Anim.* **14**, 83–86.

Kelly, J. F. and Knell, J. D. (1979). A simple method of cleaning microsporidian spores. *J. Invert. Pathol.* **33**, 252.

Kemp, R. L. and Kluge, J. P. (1975). *Encephalitozoon* sp. in the blue-masked lovebird, *Agapornis personata* (Reichenow): first confirmed report of microsporidian infection in birds. *J. Protozool.* **22**, 489–491.

Khanna, R. S. and Iyer, P. K. R. (1971). A case of *Nosema cuniculi* infection in a goat. *Ind. J. Med. Res.* **59**, 993–995.

King, H. D. (1907). *Bertramia bufonis,* a new parasite of *Bufo lentiginosus. Proc. Acad. Nat. Sci. Philadephia,* **59**, 273–278.

Kinkelin, P. de (1980). Occurrence of a microsporidian infection in zebra danio, *Brachydanio rerio* (Hamilton-Buchanan). *J. Fish. Dis.* **3**, 71–73.

Kinzel, V. and Meiser, J. (1968). Microsporidien-infektion (*Nosema cuniculi*) an einem transplantablen plasmocytom *Verhandl. Deutsch. Ges. Pathol.* **52**, 453–455 (Demonstration).

Knell, J. D. and Zam, S. G. (1978). A serological comparison of some species of Microsporida. *J. Invert. Pathol.,* **31**, 280–288.

Koike, T., Kyo, Y. and Iino, N. (1960). On the spontaneous infection of *Encephalitozoon* in mice. *Jap. J. Parasitol.* **9**, 129–132 (In Japanese).

Koller, L. D. (1969). Spontaneous *Nosema cuniculi* infection in laboratory rabbits. *J. Am. Vet. Med. Ass.* **155**, 1108–1114.

Kramer, J. P. (1960). Observations on the emergence of the microsporidan sporoplasm. *J. Insect. Pathol.* **2**, 433–439.

Kudo, R. R. (1924). A biologic and taxonomic study of the Microsporidia. *Illinois Biol. Monogr.* **9**, 1–268.

Kudo, R. (1944). The morphology and development of *Nosema notablis* Kudo parasitic in *Sphaerospora polymorpha* Davis, a parasite of *Opsanus tau* and *O. beta. Illinois Biol. Monogr.* **20**, 1–59.

Kyo, Y. (1958). Observations on the infection of *Encephalitozoon* in mice with a special note on its parasitaemia. *Nisshin Igaku* **45**, 500–504 (In Japanese).

Labbé, A. (1899). Sporozoa. *In "Das Tierreich"* V. ed. O. Bütschli. Friedlander, Berlin, pp.180.

Lainson, R. (1954). Natural infection of *Encephalitozoon* in the brains of laboratory rats. *Trans. Roy. Soc. Trop. Med. Hyg.* **48**, 5 (abstract).

Lainson, R., Garnham, P. C. C., Killick-Kendrick, R. and Bird, R. G. (1964). Nosematosis, a microsporidial infection of rodents and other animals, including man. *Br. Med. J.* **22**, 470–472.

Laird, M. (1956). Aspects of fish parasitology. *Proc. 2nd Joint Symp. Sci. Soc. Malaya and Malayan Math. Soc.,* 46–54.

Le Danois, E. (1910). Sur une tumeur à Microsporidies chez un Crénilabre. *Bull. Soc. Méd., Rennes,* **19**, 210–211.

Legault, R. O. and Delisle, C. (1967). Acute infection by *Glugea hertwigi* Weissenberg in young-of-the-year rainbow smelt, *Osmerus eperlanus mordax* (Mitchill). *Can. J. Zool.* **45**, 1291–1293.

Legault, R. O. and Delisle, C. (1968). La fraye d'une population d'éperlans géants, *Osmerus eperlanus mordax* (Mitchill) au lac Héney, Comté de Gatineau, Québec. *J. Fish. Res. Board, Canada,* **25**, 1813–1830.

Léger, L. (1905). Deux nouvelles myxosporidies parasites des poissons d'eau douce. *Bull. Mém. Ass. Francaise Avancem. Sci.,* **9**, 330.

Léger, L. and Hesse, E. (1966). *Mrazekia,* genre nouveau de microsporidie à spores tubuleuses. *C. R. Soc. Biol.* Paris, **79**, 345–348.

Lepine, P. and Sautter, V. (1949). Résistance au virus de la lympho-granulomatose vénérienne engendrée chez la souris par *Encephalitozoon cuniculi. Ann. Inst. Pasteur,* **77**, 770–772.

Levaditi, C and Nicolau, S. (1923). Encéphalites du Lapin. *C. R. Soc. Biol.* Paris, **89**, 775–779.

Levaditi, C., Nicolau, S. and Schoen, R. (1923a). L'étiologie de l'encéphalite. *C. R. Acad. Sci.* Paris, **177**, 985–988.

Levaditi, C., Nicolau, S. and Schoen, R. (1923b). L'agent étiologique de l'encéphalite épizootique du lapin (*Encephalitozoon cuniculi*). *C. R. Soc. Biol.* Paris, **89**, 984–986.

Levaditi, C., Nicolau, S. and Schoen, R. (1923c). Nouvelles données sur l'*Encephalitozoon cuniculi. C. R. Soc. Biol.* Paris, **89**, 1157–1162.

Levaditi, C., Nicolau, S. and Schoen, R. (1924a). La microsporidiose du lapin, ses relations avec la rage. *C. R. Acad. Sci.* Paris, **178**, 256–258.

Levaditi, C., Nicolau, S. and Schoen, R. (1924b). Virulence de L'*Encephalitozoon cuniculi* pour la souris. *C. R. Soc. Biol.* Paris, **40**, 194–196.

Levaditi, C., Nicolau, S. and Schoen, R. (1924c). Nouvelles recherches sur l'*Encephalitozoon cuniculi. C. R. Soc. Biol.* Paris, **40**, 662–666.

Levaditi, C., Nicolau, S. and Schoen, R. (1924d). L'étologie de l'encéphalite épizootique du lapin, dans ses rapports avec l'étude expérimentale de l'encéphalite léthargique *Encephalitozoon cuniculi.* (nov. spec.) *Ann. Inst. Pasteur,* (Paris), **38**, 651–711.

Levaditi, C., Nicolau, S. and Schoen, R. (1924e). La nature microsporidienne du virus rabique. *C. R. Soc. Biol.* Paris. **90**, 398–402.

Levaditi, C., Nicolau, S. and Schoen, R. (1926). Recherches sur la rage. *Ann. Inst. Pasteur,* Paris. **40**, 973–1069.

Levine, N. D., Dunlap, G. L. and Graham, R. (1938). An intracellular parasite encountered in a ferret. *Cornell Vet.* **28**, 249–251.

Linton, E. (1901). Parasites of fishes of Woods Hole Region. *Bull. U.S. Fish. Comm.* **19**, 405–492.

Liu, S-K. and King, F. W. (1971). Microsporidiosis in the Tuatara. *J. Am. Vet. Med. Ass.* **159**, 1578–1582.

Lom, J. (1969). Experimental transmission of a microsporidian, *Plistophora hyphessobryconis,* by intramuscular transplantation. *J. Protozool* **16** (suppl.), 17 (abstract).

Lom, J. (1970). Protozoa causing diseases in marine fishes. *In* "A symposium on diseases of fishes and shellfishes," ed. S. F. Snieszko, *Am. Fish. Soc. Washington,* Special Publ. **5**, 101–123.

Lom, J. (1972). On the structure of the extruded microsporidian polar filament. *Z. Parasitenk.* **38**, 200–213.

Lom, J., Canning, E. U. and Dyková, I, (1979). Cell periphery of *Glugea anomala* - xenomas. *J. Protozool.* **26** suppl. 44 (abstract).

Lom, J. and Corliss, J. O. (1967). Ultrastructural observations on the development of the microsporidian protozoon *Plistophora hyphessobryconis* Schaperclaus. *J. Protozool.* **14**, 141–152.

Lom, J., Gaievskaya, A. V. and Dyková, I. (1980). Two microsporidian parasites found in marine fishes in the Atlantic Ocean. *Folia Parasit.,* Praha, **27**, 197–202.

Lom, J. and Laird, M. (1976). Parasitic protozoa from marine and euryhaline fish of Newfoundland and New Brunswick II. Microsporidia. *Trans. Am. Microscop. Soc.* **95**, 569–580.

Lom, J. and Vávra, J. (1961). Niektóre wyniky badaň nad ultrastruktura spor pasožyta ryb *Plistophora hyphessobryconis* (Microsporidia). *Wiadom. Parazytol.* **7**, 828–832.

Lom, J. and Weiser, J. (1969). Notes on two microsporidian species from *Silurus glanis* and on the systematic status of the genus *Glugea* Thélohan. *Folia Parasit.,* Praha, **16**, 193–200.

Lom, J. and Weiser, J. (1972). Surface pattern of some microsporidian spores as seen in the scanning electron microscope. *Folia Parasit.,* Praha, **19**, 359–363.

Lopukhina, A. M. and Strelkov, Yu. A. (1972). Ecological analysis of the parasite fauna of commercial fish of the Verkhnyeye Vervo Lake *Izvest. Gosniorch* **80**, 5–25 (In Russian).

Loubès, C. (1979). Ultrastructure, sexualité, dimorphisme sporogonique des microsporidies (Protozoaires). *Thèse, Univ. Sci. Tech. Languedoc,* Montpellier.

Loubès, C., Maurand, J., Gasc, C., de Buron, I. and Burral, J. (1985). Étude ultrastructurale de *Loma dimorpha* n.sp., microsporidie parasite de poissons Gobiidae languedociens. *Protistologica* (in press).

Loubès, C., Maurand, J. and Ormières, R. (1979). Étude ultrastructurale de *Spraguea lophii* (Doflein, 1898), microsporidie parasite de la Baudroie : essai d'interpretation du dimorphism sporal. *Protistologica* **15**, 43–54.

Loubès, C., Maurand, J. and Walzer, C. (1981). Developpment d'une microsporidie *Glugea truttae* n.sp. dans le sac vitellin de l'alevin de la truite *Salmo faria trutta* L. : étude ultrastructurale. *Protistologica* **17**, 177–184.

Lowenstein, L. J. and Petrak, M. L. (1980). Microsporidiosis in two peach-faced lovebirds. *In* "The comparative pathology of zoo animals," *Proc. Symp. Nat. Zool. Park, Smith-*

sonian Institution, 1978, ed. Montali, R. J. and Migaki, G. J. Smithsonian Institution Press, Washington, D.C., 365–368.

Lucke, B. (1925). Spontaneous cerebral lesions in monkeys, their significance in experimental pathology. *Arch. Neurol. Psychiatr.* Chicago, **10**, 212–225.

Lucký, Z. and Dyk, V. (1964). Cizopasníci ryb v řekách a rybnících povodí Odry a Dyje. *Sbor. Vys. Skoly. Zemed.,* Brno, **12**, 49–73.

Lutz, A. and Splendore, A. (1903). Über Pebrine und verwandte Mikrosporidien. Ein Beitrag zur Kenntnis der brasilianischen Sporozoen. *Z. Bakt. Parasit. Infect. Hyg.* Abt. I. Orig. **33**, 150–157.

Lyngset, A. (1980). A survey of serum antibodies to *Encephalitozoon cuniculi* in breeding rabbits and their young. *Lab. Anim. Sci.,* **30**, 558–561.

Majeed, S. K. and Zubaidy, A. J. (1982). Histopathological lesions associated with *Encephalitozoon cuniculi* (nosematosis) infection in a colony of Wistar rats. *Lab. Anim.* **16**, 244–247.

Maksimova, A. P. (1962). On the parasite fauna of fish in lake Balkhash. In "Parazity dikikh zhivotnykh Kazakhstana, U.S.S.R." **16**, 145–156.

Malherbe, H. and Munday, V. (1958). *Encephalitozoon cuniculi* infection of laboratory rabbits and mice in South Africa. *J. Sth. Afr. Vet. Med. Ass.* **29**, 241–246.

Manouelian, Y. and Viala, J. (1924). *Encephalitozoon rabiei* parasite de la rage. *Ann. Inst. Pasteur.* Paris, **38**, 258–267.

Manouelian, Y. and Viala, J. (1927). *Encephalitozoon negrii* parasite de 'encéphalitomyelite des jeunes chiens. *C. R. Acad. Sci.* Paris, **184**, 630–632.

Marchant, E. H. and Schiffman, S. (1946). The occurrence of a microsporidian in a new host. *Can. Fish. Cult.* **1**, 18–21.

Marcus, P. B., van der Walt, J. J. and Burger, P. J. (1973). Human tumor microsporidiosis. *Arch. Pathol.* **95**, 341–343.

Margileth, A. M., Strano, A. J., Chandra, R., Neafie, R., Blum, M. And McCully, R. M. (1973). Disseminated nosematosis in an immunologically compromised infant. *Arch. Pathol.* **95**, 145–150.

Markowski, S. (1966). The diet and infection of fishes in Cavendish Dock, Barrow-in-Furness. *J. Zool.* London, **150**, 183–197.

Matsubayashi, H., Koike, T., Mikata, I., Takei, H. and Hagiwara, S. (1959). A case of *Encephalitozoon*-like body infection in man. *Arch. Pathol.* **67**, 181–187.

Matthews, R. A. and Matthews, B. F. (1980). Cell and tissue reactions of turbot *Scophthalmus maximus* (L.) to *Tetramicra brevifilum* gen.n., sp.n. (Microspora). *J. Fish. Dis.* **3**, 495–515.

Mawdesley-Thomas, L. E. and Bucke, D. (1973). Tissue repair in a poikilothermic vertebrate, *Carassius auratus* (L.): a preliminary study. *J. Fish. Biol.* **5**, 115–119.

McArn, G. E., Ashley, L. M. and Wellings, S. R. (1969). Microsporidial infestation of English sole. *Am. J. Pathol.* **55**, 85 (abstract).

McCartney, J. E. (1924). Brain lesions in the domestic rabbit. *J. Exp. Med.* **39**, 51–61.

McCully, R. M., van Dellen, A. F., Basson, P. A. and Lawrence, J. (1978). Observations on the pathology of canine microsporidiosis. *Onderstepoort J. Vet. Res.* **45**, 75–92.

McKenzie, K., McVicar, A. H. and Wadell, I. F. (1976). Some parasites of plaice *Pleuronectes platessa* L. in three different farm environments. *Scottish Fish Res. Rep.* **4**, 1–14.

McVicar, A. H. (1975). Infection of plaice *Pleuronectes platessa* L. with *Glugea (Nosema) stephani* (Hagenmüller, 1899). (Protozoa, Microsporidia) in a fish farm and under experimental conditions. *J. Fish. Biol.* **7**, 611–619.

Meiser, J., Kinzel, V. and Jírovec, O. (1971). Nosematosis as an accompanying infection of plasmacytoma ascites in Syrian golden hamsters. *Path. Microbiol.* **37**, 249–260.

Mercier, L. (1921). *Glugea gigantea* Thélohan. Réaction des tissus de l'hôte à l'infection. *C.R. Soc. Biol.* Paris, **84**, 261–263.

Miki, S. and Awakura, T. (1977). The fine structure of *Glugea takedai* Awakura, 1974 (Microsporidia, Nosematidae). *Sci. Rep. Hokkaido Fish Hatch.* **32**, 1–19 (In Japanese).

Modigliani, R., Bories, C., Le Charpentier, Y., Salmeron, M., Messing, B., Galian, A., Rambaud, J. C. and Desportes, I. (1985). Diarrhoea and malabsorption in acquired immune deficiency syndrome: a study of four cases with special emphasis on opportunistic protozoan infestations. *Gut* (in press).

Modin, J. C. (1981). *Microsporidium rhabdophilia* n.sp. from rodlet cells of Salmonid fishes. *J. Fish. Dis.* **4**, 203–211.

Moffatt, R. E. and Schiefer, B. (1973). Microsporidiosis (encephalitozoonosis) in the guinea pig. *Lab. Anim. Sci.* **23**, 282–284.

Mohn, S. F. (1982). Encephalitozoonosis in the blue fox. Comparison between the india-ink immunoreaction and the indirect fluorescent antibody test in detecting *Encephalitozoon cuniculi* antibodies. *Acta. Vet. Scand.* **23**, 99–106.

Mohn, S. F. (1983). Encephalitozoonosis in the blue fox (*Alopex lagopus*). Transmission, diagnosis and control. *Thesis, National Veterinary Institute,* Olso, General discussion and Summary, page 5.

Mohn, S. F., Landsverk, T. and Nordstoga, K. (1981). Encephalitozoonosis in the blue fox - morphological identification of the parasite. *Acta. Pathol. Microbiol. Scand.* (B) **89**, 117–122.

Mohn, S. F. and Nordstoga, K. (1975). Electrophoretic patterns of serum proteins in blue foxes with special reference to changes associated with nosematosis. *Acta. Vet. Scand.* **16**, 297–306.

Mohn, S. F. and Nordstoga, K. (1982). Experimental encephalitozoonosis in the blue fox. Neonatal exposure to the parasite. *Acta. Vet. Scand.,* **23**, 344–360.

Mohn, S. F., Nordstoga, K. and Dishington, I. W. (1982). Experimental encephalitozoonosis in the blue fox. Clinical, serological and pathological examination of vixens after oral and intrauterine inoculation. *Acta. Vet. Scand.* **23**, 490–502.

Mohn, S. F., Nordstoga, K., Krogshud, J. and Helgebostad, A. (1974). Transplacental transmission of *Nosema cuniculi* in the blue fox (*Alopex lagopus*). *Acta Path. Microbiol. Scand.,* B, **82**, 299–300.

Mohn, S. F., Nordstoga, K. and Møller, O. M. (1982). Experimental encephalitozoonosis in the blue fox. Transplacental transmission of the parasite. *Act. Vet. Scand.* **23**, 211–220.

Mohn, S. F. and Ødegaard, Ø. A. (1977). The indirect fluorescent antibody test for the detection of *Nosema cuniculi* antibodies in the blue fox *Alopex lagopus.* *Acta. Vet. Scand.,* **18**, 290–292.

Möller, H. (1974). Untersuchungen über die Parasiten der Flunder (*Patichthys flesus* L.) in der Kieler Förder. *Ber. Deutsch Wissensch. Komm. Meeresforsch.* **23**, 136–149.

Möller, H. and Anders, K. (1983). Krankheiten und Parasiten der Meeresfische. Verlag Heino Möller, Kiel, 258 pp.

Moniez, R. (1887). Observations pour la revision des microsporidies. *C. R. Acad. Sci.* Paris, **104**, 1312–1314.

Montrey, R. D., Shadduck, J. A. and Pakes, S. P. (1973). In-vitro study of host range of three isolates of *Encephalitozoon (Nosema)*. *J. Infect. Dis.* **127**, 450–454.

Morris, J., McCown, J. M. and Blount, R. E. (1956). Ascites and hepatosplenomegaly in mice associated with protozoon-like cytoplasmic structures. *J. Infect. Dis.* **98**, 306–311.

Morrison, C. M. (1983). The distribution of the microsporidian *Loma morhua* in tissues of the cod *Gadus morhua* L. *Can. J. Zool.* **61**, 2155–2161.

Morrison, C. M. and Sprague, V. (1981a). Microsporidian parasites in the gills of salmonid fishes. *J. Fish. Dis.* **4**, 371–386.

Morrison, C. M. and Sprague, V. (1981b). Electron microscope study of a new genus and new species of microsporidia in the gill of Atlantic cod *Gadus morhua* L. *J. Fish. Dis.* **4**, 15–32.

Morrison, C. M. and Sprague, V. (1981c). Light and electron microscopic study of microsporidia in the gill of haddock, *Melanogrammus aeglefinus* (L.). *J. Fish. Dis.* **4**, 179–184.

Morrison, C. M. and Sprague, V. (1983). *Loma salmonae* (Putz, Hoffman and Dunbar, 1965) in the rainbow trout, *Salmo gairdneri* Richardson, and *L. fontinalis* sp. nov. (Microsporidia) in the brook trout, *Salvelinus fontinalis* (Mitchill). *J. Fish. Dis.* **6**, 345–394.

Morrison, C., Hoffman, G. L., and Sprague, V. (1985). *Glugea pimephales* (Fantham *et at.*, 1941) n. comb. (Microsporidia, Glugeidae) in the fathead minnow *Pimephales promelas. Can. J. Zool.*, **63**, 380–391.

Morrison, C. M., Marryat, V. and Gray, B. (1985). Structure of *Pleistophora hippoglossoideos* Bosanquet in the American plaice *Hippoglossoideos platessoides* (Fabricius). *J. Parasitol.* (in press).

Mrázek, A. (1899). Sporozoenstudien II. *Glugea lophii* Doflein. *Sitzungsber. Böhm. Ges. Wiss. Math. - Naturwiss. Classe,* 1–8.

Muradian, Z. (1972). Contribution à la connaissance de la parasitofaune des clupéides (Clupeidae, Pisces) de Roumanie. *Trav. Mus. Hist. Nat. "G. Antipa,"* **12**, 11–24.

Nagel, M. L. and Summefelt, R. C. (1977a). Nitrofurazone for control of the microsporidian parasite. *Pleistophora ovariae* in golden shiners. *Progr. Fish. Cult.* **39**, 18–23.

Nagel, M. L. and Summerfelt, R. C. (1977b). Apparent immunity of goldfish to *Pleistophora ovariae. Proc. Oklahoma Acad. Sci.* **57**, 61–63.

Naidenova, N. N. (1974). Parasite fauna of fishes of the family Gobiidae of the Black and Azov Seas. Publ. House "Naukova Dunka," Kiev, U.S.S.R. pp.182 (in Russian).

Nakajima, K. and Egusa, S. (1976). An attempt to use the scanning electron microscope for the identification of microsporidian spores. *Fish. Pathol.* **11**, 167–170 (in Japanese).

Narasimhamurti, C. C. and Kalavati, C. (1972). Two new species of microsporidian parasites from a marine fish *Saurida tumbil. Proc. Ind. Acad. Sci.,* (B) **4**, 165–170.

Narasimhamurti, C. C. and Sonabai, R. (1977). A new microsporidian *Pleistophora carangoidi* n.sp. from the body muscles of the marine fish, *Carangoides malabaricus. Proc. 5th Int. Congr. Protozool.,* New York, 192.

Nelson, J. B. (1962). An intracellular parasite resembling a microsporidian, associated with ascites in Swiss mice. *Proc. Soc. Exp. Biol. Med.* **109**, 714–717.

Nelson, J. B. (1967). Experimental transmission of a murine microsporidian in Swiss mice. *J. Bacteriol.* **94**, 1340–1345.

Nelson, L. R., Mock, O. B. and Flatt, R. E. (1969). Nosematosis in the shrew. *20th Ann. Sci. Session, Am. Ass. Lab. Anim. Sciences.*

Nemeczek, A. (1911). Beiträge zur Kenntnis der Myxo-und Microsporidien der Fische. *Arch. Protistenk.* **22**, 144–163.

Nepszy, S. J., Budd, J. and Dechtiar, A. O. (1978). Mortality of young-of-the-year rainbow smelt (*Osmerus mordax*) in Lake Erie associated with the occurrence of *Glugea hertwigi. J. Wildl. Dis.* **14**, 233–239.

Nepszy, S. J. and Dechtiar, A. O. (1972). Occurrence of *Glugea hertwigi* in Lake Erie rainbow smelt (*Osmerus mordax*) and associated mortality in adult smelt. *J. Fish. Res. Board, Canada.* **29**, 1639–1641.

Niederkorn, J. Y., Brieland, J. K. and Mayhew, E. (1983). Enhanced natural killer activity in experimental murine encephalitozoonosis. *Inf. Immun.* **41**, 302–307.

Niederkorn, J. Y. and Shadduck, J. A. (1980). Role of antibody and complement in the control of *Encephalitozoon cuniculi* infection by rabbit macrophages. *Inf. Immun.* **27**, 995–1002.

Niederkorn, J. Y., Shadduck, J. A. and Schmidt, E. C. (1981). Susceptibility of selected inbred strains of mice to *Encephalitozoon cuniculi. J. Inf. Dis.* **144**, 249–253.

Niederkorn, J. Y., Shadduck, J. A. and Weidner, E. (1980). Antigenic cross reactivity among different microsporidian spores as determined by immunofluorescence. *J. Parasitol.* **66**, 675–677.

Nigrelli, R. F. (1946). Parasites and diseases of the ocean pout, *Macrozoarces americanus. Bull. Bingham Oceanogr. Collect. Yale Univ.* **9**, 187–202.

Nigrelli, R. F. (1953). Two diseases of the neon tetra, *Hyphessobrycon innesi. Aquarium J.* San Francisco **24**, 203–208.

Noble, E. R. and Collard, S. B. (1970). The parasites of midwater fishes. *In* "A symposium on diseases of fishes and shellfishes." ed. S. F. Snieszko, *Am. Fish. Soc., Washington* Special Publ. **5**, 57–68.

Nordstoga, K. (1972). Nosematosis in blue foxes. *Nord. Vet. Med.* **24**, 21–24.

Nordstoga, K., Mohn, S. F., Aamdal, J. and Helgebostad, A. (1978). Nosematosis (Encephalitozoonosis) in a litter of blue foxes after intrauterine injection of *Nosema* spores. *Acta. Vet. Scand.* **19**, 150–152.

Nordstoga, K., Mohn, S. F. and Loftsgard. G. (1974). Nosematose hos blårev (Nosematosis in the blue fox). *Proc. 12th Nordic Vet. Congr.* Reykjavik, 183–186.

Nordstoga, K. and Westbye, K. (1976). Polyarteritis nodosa associated with nosematosis in blue foxes. *Acta. Pathol. Microbiol. Scand.* **84**, 291–296.

Novilla, M. N., Carpenter, J. W. and Kwapien, R. (1980). Dual infection of Siberian polecats with *Encephaltizoon cuniculi* and *Hepatozoon mustelis* n.sp. *In* "Comparative Pathology of Zoo Animals" ed. R. J. Montali and G. Migaki. *Proc. Symp. Nat. Zool. Park, Smithsonian Institution, 1978.* 353–363.

Novilla, M. N. and Kwapien, R. P. (1978). Microsporidian infection in the pied peach-faced lovebird (*Agapornis roseicollis*). *Avian Dis.* **22**, 198–204.

Oliver, J. (1924). Spontaneous chronic meningoencephalitis of rabbits. *J. Infect. Dis.* **30**, 91–94.

Olsen, Y. H. and Merriman, D. (1946). Studies on the marine resources of southern New England. IV The biology and economic importance of the ocean pout, *Macrozoarces americanus* (Bloch and Schneider). *Bull. Bingham Oceanogr. Coll. Yale, Univ.* **9**, 132–184.

Olson, R. E. (1975). Transmission of the microsporidian *Glugea stephani* in the laboratory. *J. Protozool.* (suppl.) **22**, 29 (abstract).

Olson, R. E. (1976). Laboratory and field studies on *Glugea stephani* (Hagenmüller), a microsporidian parasite of pleuronectid flatfishes. *J. Protozool.* **23**, 158–164.

Olson, R. E. and Pratt, I. (1973). Parasites as indicators of English sole (*Parophrys vetulus*) nursery grounds. *Trans. Am. Fish. Soc.* **102**, 405–411.

Ophüls, W. (1910–1911). Occurrence of spontaneous lesions in kidneys and livers of rabbits and guinea pigs. *Proc. Soc. Exp. Med. Biol.* **8**, 75–77.

Opitz, H. (1942). Mikrosporidienkrankheit *Plistophora* auch bei *Hemigrammus ocellifer* und *Brachydanio rerio. Wochenschr. Aq. Terrarienkd.* **39**, 83–85.

Oshima, K. (1937). On the function of the polar filament of *Nosema bombycis. Parasitology* **29**, 220–224.

Osmanov, S. O. (1971). Parasites of fishes of Uzbekistan. FAN Publishing House of the Uzbeck S.S.R. pp.532.

Otte, E. (1964). Mikrosporidien bei Donaufischen. *Wien. tierärzt. Monatsschrift.* **51**, 316–319.

Owen, D. G. and Gannon, J. (1980). Investigation into the transplacental transmission of *Encephalitizoon cuniculi* in rabbits. *Lab. Anim.* **14**, 35–38.

Pace, D. (1908). Parasiten und Pseudoparasiten der Nervenzelle. Vorläufige Mitteilungen über vergleichende Parasitologie des Nervensystems. *Z. Hyg. Infekt.* **60**, 62–74.

Pakes, S. P., Shadduck, J. A. and Cali, A. (1975). Fine structure of *Encephalitozoon cuniculi* from rabbits, mice and hamsters. *J. Protozool.,* **22**, 481–488.

Pakes, S. P., Shadduck, J. A. and Olsen, R. G. (1972). A diagnostic skin test for encephalitozoonosis (nosematosis) in rabbits. *Lab. Anim. Sci.* **22**, 870–877.

Pattison, M., Clegg, F. G. and Duncan, A. L. (1971). An outbreak of encephalomyelitis in broiler rabbits caused by *Nosema cuniculi*. *Vet. Rec.* **88**, 404–405.

Perdrau, J. R. and Pugh, L. P. (1930). The pathology of disseminated encephalomyelitis of the dog (the "nervous form" of canine distemper). *J. Pathol. Bacteriol.* **33**, 79–91.

Perrin, T. L. (1943). Spontaneous and experimental *Encephalitozoon* infection in laboratory animals. *Arch. Pathol.* **36**, 559–567.

Peters, G. and Yamagiva, S. (1936). Zur Histopathologie der Staupe-encephalitis der Hunde und der epizootischen Encephalitis der Silberfüsche. *Arch. Tierheilkd.* **20**, 138–152.

Petri, M. (1965). A cytologic parasite in the cells of transplantable, malignant tumours. *Nature* (London) **205**, 302.

Petri, M. (1966). The occurrence of *Nosema cuniculi* (*Encephalitozoon cuniculi*) in the cells of transplantable, malignant ascites tumours and its effect upon tumour and host. *Acta Path. Microbiol. Scand.* **66**, 13–30.

Petri, M. (1967). Persistence and possibly late recurrences of tumour tissue in rats with Nosema-infected Yoshida sarcoma. *Acta. Path. Microbiol. Scand.* **187** (suppl.) 86 (abstract).

Petri, M. (1968a). High incidence of intraperitoneal rhabdomyosarcomas in rats after growth and regression of *Nosema cuniculi*-infected Yoshida ascites sarcoma. *Acta Path. Microbiol. Scand.* **73**, 1–12.

Petri, M. (1968b). Mikrosporidiose. En ny sydgom hos varmblodede dyr og mennesket. *Nord. Med.* **80**, 1686–1690.

Petri, M. (1969). Studies on *Nosema cuniculi* found in transplantable ascites tumours with a survey of microsporidiosis in mammals. *Acta. Path. Microbiol. Scand.* **204** (suppl.) 1–91.

Petri, M. (1976). Microsporidia and mammalian tumours. *In* "Biology of the Microsporidia" *Comparative Pathobiology* Vol. 1. eds. L. A. Bulla and T. C. Cheng, Plenum Press, New York and London, pp. 239–245.

Petri, M. and Schiødt, T. (1966). On the ultrastructure of *Nosema cuniculi* in the cells of the Yoshida rat ascites sarcoma *Acta. Pathol. Microbiol. Scand.* **66**, 437–466.

Petruschevsky, G. K. and Shulman, S. S. (1958). Parasitic diseases in fish in water reservoirs of the USSR. *In* "Parasitology of Fishes" eds. V. A. Dogiel, G. K. Petrushevsky and Yu I. Polyansky. Leningrad Univ. Press. U.S.S.R. pp. 301–320 (in Russian).

Pfeiffer, A. (1891). "Die Protozoen als Krankheitserreger" Fischer, Jena 2nd ed. pp.216.

Pfeiffer, A. (1895a). "Die Protozoen als Krankheitserreger, Nachträge." Fischer, Jena. pp.122.

Pfeiffer, L. (1895b). Protozoären Krankheiten (Suppl) pp 38, 54–60, 72 (quoted by Sprague, 1977).

Pflugfelder, O. (1952). Die sog Restkörper der Cyprinodontidae und Cyprinidae als Abwehrreaktionen gegen Mikrosporidienbefall. *Z. Parasitenk.* **15**, 321–334.

Pinnolis, M., Egbert, P. R., Font, R. L. and Winter, F. C. (1981). Nosematosis of the cornea. Case report, including electron microscopic studies. *Arch. Ophthalmol.* **99**, 1044–1047.

Plehn, M. (1924). Praktikum der Fischkrankheiten. E. Schweizerbartsche Verlagsbuchhandlung, Stuttgart, pp.179.

Plowright, W. (1952). An encephalitis - nephritis syndrome in the dog, probably due to congenital *Encephalitozoon* infection. *J. Comp. Pathol.* **62**, 83–92.

Plowright, W. and Yeoman, G. (1952). Probable encephalitozoonosis infection of the dog. *Vet. Rec.* **64**, 381–383.

Polyansky, Yu, I. (1955). Additions to parasitology of fishes of northern seas of the U.S.S.R. Parasites of Barents Sea fishes. *Tr. Zool. Inst. Akad. Nauk. SSSR.* **19**, 5–170.

Polyansky, Yu, I. and Kulemina, I. V. (1963). On parasite fauna of young cods in Barents Sea. *Uch. Zap. Leningrad Obshch. Univ. Ser. Biol.* **18**, 12–21.

Porter, A. and Vinall, H. F. (1956). A protozoan parasite (*Ichthyosporidium* sp.) of the neon fish *Hyphessobrycon inessi*. *Proc. Zool. Soc.* London, **126**, 397–402.

Price, R. L. (1982). Incidence of *Pleistophora cepedianae* (Microsporidia) in gizzard shad (*Dorosoma cepedianum*) on Carlyle Lake, Illinois. *J. Parasitol.*, **68**, 1167–1168.

Priebe, K. (1971). Zur Verbreitung des Befalls des Seeteufels (*Lophius piscatorius*) mit *Nosema lophii* auf Fischfangplätzen im östlichen Nordatlantik. *Arch. Fischereiwiss.* **22**, 98–102.

Putz, R. E. (1964). Parasites of freshwater fish II. Protozoa. I. Microsporidea of fish. *Fishery Leaflet (U.S.Dept. Inter. Washington)*, **571**, 1–4.

Putz, R. E., Hoffman, G. L. and Dunbar, C. E. (1965). Two new species of *Plistophora* from North American fish with a synopsis of Microsporidea of freshwater and euryhaline species. *J. Protozool.* **12**, 228–236.

Putz, R. E. and McLaughlin, J. J. A. (1970). Biology of Nosematidae (Microsporida) from freshwater and euryhaline fishes. *In* "A Symposium on Diseases of Fishes and Shellfishes" ed. S. F. Snieszko. *Am. Fish. Soc. Washington*, **5**, 124–132.

Pye, D. and Cox, J. C. (1977). Isolation of *Encephalitozoon cuniculi* from urine samples. *Lab. Anim.* **11**, 233–234.

Raabe, H. (1935). Un microsporidium dans les Lymphocystis chez les plies. *Bull. Inst. Océanogr.* **665**, 1–11.

Raabe, H. (1936). Etudes de micro-organismes parasites des poissons de mer. I. *Nosema ovoideum*. Thél. dans le foie des Rougets. *Bull. Inst. Océanogr.* **696**, 1–12.

Radulescu, I. and Vasiliu-Suceveanu, N. (1956). Contributiuni la cunoasterea parazitilor pestilor din complexul lagunar Razelm-Sinoe. *An. Inst. Cercet. Piscic. Romanei*, **1**, 309–333.

Ralphs, J. R. (1984). Hepatic microsporidiosis of juvenile grey mullet *Chelon labrosus* with particular reference to parasite development and transmission. Ph.D. Thesis, University of London.

Ralphs, J. R. and Matthews, R. A. (1986). Hepatic microsporidiosis due to *Microgemma hepaticus* n.gen, n.sp. in juvenile grey mullet *Chelon labrosus*. *J. Fish. Dis.* (in press).

Rašín, K. (1936). *Cocconema sulci* n.sp. (Microsporidia), cizopasník vajíček jesetera malého - *Acipenser ruthenus* L. *Věst. II Sjezdu Veterin. CSR.* **1**, 20.

Rašín, K. (1949). *Cocconema sulci* n.sp. (Microsporidia), cizopasník vajíček jesetera malého (*Acipenser ruthenus* L.). *Věst. Čsl. Zool. Spol.* **13**, 295–298.

Ray, H. N. and Raghavachari, K. (1941). A note on *Encephalitozoon cuniculi* in a rabbit. *Ind. J. Vet. Sci. Anim. Husb.* **11**, 38–41.

Reddy, A. M. K. (1963). A case of *Encephalitozoon cuniculi* infection in a rabbit in Andhra Pradesh. *Ind. Vet. J.* **40**, 400–401.

Reichenbach-Klinke, H. (1952). Neue Beobachtungen über der Erreger der Neon-Fischkrankheit *Plistophora hyphessobryconis* Schäperclaus (Sporozoa, Microsporidia). *Aquar. Terrar. Z. DATZ.* **12**, 320–322.

Reichenow, E. (1932). Cnidosporidia. In Grimpe, Tierwelt der Nord-und Ostsee (Teil II-g-2/21), 49.

Reimer, L. W. and Jessen, O. (1974). Ein Beitrag zur Parasitenfauna von *Merluccius hubbsi* Marini. *Wiss. Zeitsch. Pädagog. Hochsch. Güstrow*, 53–64.

Richert, R. (1958). Über des Auftreten des Erregers der neonfischkrankheit (*Plistophora hyphessobryconis*) bei Süsswasser-fischen der gemässigten Zone. *Mikrokosmos* **47**, 327–330.

Roberts, R. J. (1975). The effects of temperature on diseases and their histopathological manifestations in fish. *In* "The Pathology of Fishes" eds. W. E. Ribelin, G. Migaki, *Univ. Wisconsin Press*, Madison, 477–496.

Robinson, J. J. (1954). Common infectious disease of laboratory rabbits questionably attributed to *Encephalitozoon cuniculi*. *Arch. Pathol.* **58**, 71–84.

Roman, E. (1955). Cercetări asupra parazitofaunei pestilor din Dunăre. *Editura Acad. Rep. Pop. Romine.*, pp.119.

Ruge, H. (1950). *Encephalitozoon* beim Meerschweinchen. *Zentr. Bakt. Parasitenk.* **156**, 543–544.

Ruiz, A. (1964). Über die Spontaninfektion mit *Encephalitozoon cuniculi* bei weissen Mausen. *Riv. Biol. Trop.* **12**, 225–227.

Sandholzer, L. A., Nostrand, T. and Young, L. (1945). Studies of an Ichthyosporidian-like parasite of ocean pout (*Zoarces anguillaris*). *Spec. Sci. Rep. U.S. Fish Wildl. Service* **31**, 1–12.

Schäperclaus, W. (1941). Eine neue Mikrosporidien-krankheit beim Neonfisch und seinen Verwandten. *Wochenschr. Aquarien Terrarienkd.* **39/40**, 381–384.

Schmidt, E. C. and Shadduck, J. A. (1983). Murine encephalitozoonosis model for studying the host-parasite relationship of a chronic infection. *Inf. Immun.* **40**, 936–942.

Schmidt, E. C. and Shadduck, J. A. (1985). Mechanisms of resistance to the intracellular protozoan *Encephalitozoon cuniculi* in mice. *Inf. Immun.* (in press).

Schrader, F. (1921). A microsporidian occurring in the smelt. *J. Parasitol.* **7**, 151–153.

Schuberg, A. (1910). Über Mikrosporidien aus dem Hoden der Barbe und durch sie verursachte Hypertrophie der Kerne. *Arb. Kaiserl. Gesundheitsamte*, Berlin, **33**, 401–434.

Schubert, G. (1969a). Ultracytologische Untersuchungen an der Spore der Mikrosporidienart, *Heterosporis finki* gen.n., sp.n.. *Z. Parasitenk.* **32**, 59–79.

Schubert, G. (1969b). Electronenmikroscopische Untersuchungen zur Sporonten und Sporenentwicklung der Mikrosporidienart *Heterosporis finki*. *Z. Parasitenk.* **32**, 80–92.

Schuetz, A. W., Selman, K. and Samson, D. (1978). Alterations in growth, function and composition of *Rana pipiens* oocytes and follicles associated with microsporidian parasites. *J. Exp. Zool.* **204**, 81–94.

Schuster, J. (1925). Über eine spontan beim Kaninchen auftretende encephalitische Erkrankung. *Klin. Wochenschr.* **4**, 550–551.

Schwartz, F. J. (1963). A new *Ichthyosporidium* parasite of the spot *Leiostomus xanthurus*: a possible answer to recent oyster mortalities. *Progr. Fish Culture* **25**, 181–184.

Šebek, Z. (1969). The finding of the genus *Nosema* Nägeli, 1857 (Microsporidia, Nosematidae) in the kidney of a guinea pig. *Folia Parasitol.* Praha, **16**, 165–169.

Seibold, H. R. and Fussell, E. N. (1973). Intestinal microsporidiosis in *Callicebus moloch*. *Lab. Anim. Sci.* **23**, 115–118.

Shadduck, J. A. (1969). *Nosema cuniculi: in vitro* isolation. *Science*, **166**, 516–517.

Shadduck, J. A., Bendele, R. and Robinson, G. T. (1978). Isolation of the causative organism of canine encephalitozoonosis. *Vet. Pathol.* **15**, 449–460.

Shadduck, J. A. and Geroulo, M. J. (1979). A simple method for the detection of antibodies to *Encephalitozon cuniculi* in rabbits. *Lab. Anim. Sci.* **29**, 330–334.

Shadduck, J. A., Kelsoe, G. and Helmke, J. (1979). A microsporidian contaminant of a non-human primate cell culture: ultrastructural comparison with *Nosema connori*. *J. Parasitol.* **65,** 185–188.

Shadduck, J. A. and Pakes, S. P. (1971). Encephalitozoonosis (Nosematosis) and Toxoplasmosis. *Am. J. Pathol.* **64,** 657–671.

Shadduck, J. A. and Polley, M. B. (1978). Some factors influencing the *in vitro* infectivity and replication of *Encephalitozoon cuniculi*. *J. Protozool.* **25,** 491–496.

Shadduck, J. A., Watson, W. T., Pakes, S. P. and Cali, A. (1979). Animal infectivity of *Encephalitozoon cuniculi*. *J. Parasitol.* **65,** 123–129.

Sheehy, D. J., Sissenwine, M. P. and Saila, S. B. (1974). Ocean pout parasites. *Marine Fish. Rev.* **36,** 29–33.

Sherburne, S. W. and Bean, L. L. (1979). Incidence and distribution of piscine erythrocytic necrosis and the microsporidian, *Glugea hertwigi* in rainbow smelt, *Osmerus mordax,* from Massachussets to the Canadian maritimes. *Fish. Bull.* **77,** 503–509.

Shulman, S., S. (1957). Parasites of fishes of the eastern part of the Baltic Sea. *Fish Diseases Conference,* Leningrad, 113–114, (abstract) (in Russian).

Shulman, S. S. (1962). Protozoa. In "Key to the parasites of freshwater fish of the U.S.S.R.", ed. B. E. Bykhovsky. Academy of Sciences Publishing House, U.S.S.R., 7–197, (in Russian).

Shulman, S. S. and Shulman-Albova, R. E. (1953). Parasites of White Sea fishes. Academy of Sciences Publishing House, U.S.S.R., pp. 199, (in Russian).

Shumilo, R. P. (1959). On parasite fauna of fish of the lower part of the river Dniestr. *Izv. Moldav. Fil. Acad. Nauk. S.S.S.R.* **8,** 31–41, (in Russian).

Sindermann, C. J. (1963). Disease in marine populations. *Trans. N. Am. Wildl. Nat. Res. Conf.* **28,** 336–356.

Sindermann, C. J. (1970). Principal diseases of marine fish and shellfish. Academic Press, New York and London, pp 369.

Singh, M., Kane, G. J., Mackinlay, L., Quaki, I. Yap, E. H., Ho, B. C., Ho, L. C. and Lim, K. C. (1982). Detection of antibodies to *Nosema cuniculi* (Protozoa: Microsporidia) in human and animal sera by the indirect fluorescent antibody technique. *Southeast Asian. J. Trop. Med. Publ. Hlth.* **13,** 110–113.

Skerry, J. B. (1952). A preliminary survey of the occurrence of *Glugea hertwigi* Weissenberg in the smelt, *Osmerus mordax* (Mitchill) of the Great Bay area, New Hampshire. M.S. Thesis, University of New Hampshire (quoted by Delisle, 1969).

Smith, T. and Florence, L. (1925). *Encephalitozoon cuniculi* as a kidney parasite in the rabbit. *J. Exp. Med.* **41,** 25–35.

Somvanshi, R., Iyer, P. K. R., Gupta, S. C. and Matanay, C. F. (1977). Spontaneous cerebral nosemiasis in a laboratory mouse. *Curr. Sci.* **46,** 272–273.

Sprague, V. (1965). *Ichthyosporidium* Caullery and Mesnil, 1905, the name of a genus of fungi or sporozoans? *Syst. Zool.* **14,** 110–114.

Sprague, V. (1966). *Ichthyosporidium* sp. Schwartz 1963, parasite of the fish *Leiostomus xanthurus,* is a microsporidian. *J. Protozool.* **13,** 356–358.

Sprague, V. (1969). Microsporidia and tumors, with particular reference to the lesion associated with *Ichthyosporidium* sp. Schwartz, 1963. *Nat. Cancer Inst. Monogr.* **31,** 231–249.

Sprague, V. (1974). *Nosema connori* n.sp., microsporidian parasite of man. *Trans. Amer. Microsc. Soc.* **93,** 400–403.

Sprague, V. (1977). Annotated list of species of microsporidia. *In* Bulla, L. A. and Cheng, T. C. eds. *Comparative Pathology* 2 Systematics of the Microsporidia. Plenum Press, New York and London, pp.333.

Sprague, V. 1982. "Microspora" *In* Parker, S. B. ed. *Synopsis and Classification of Living Organisms*. 1. McGraw-Hill, London, New York and Toronto. pp 589–594.

Sprague, V. and Hussey, K. L. (1980). Observations on *Ichthyosporidium giganteum* (Microsporidia) with particular reference to the host parasite relations during merogony. *J. Protozool.* **27**, 169–175.

Sprague, V. and Vernick, S. H. (1968a). Light and electron microscope study of a new species of *Glugea* (Microsporida, Nosematidae) in the 4-spined stickleback, *Apeltes quadracus. J. Protozool.* **15**, 547–571.

Sprague, V. and Vernick, S. H. (1968b). Observations on the spores of *Pleistophora gigantea* (Thélohan, 1895) Swellengrebel, 1911, a microsporidian parasite of the fish *Crenilabrus melops. J. Protozool.* **15**, 662–665.

Sprague, V. and Vernick, S. H. (1971). The ultrastructure of *Encephalitozoon cuniculi* (Microsporida, Nosematidae) and its taxonomic significance. *J. Protozool.* **18**, 560–569.

Sprague, V. and Vernick, S. H. (1974). Fine structure of the cyst and some sporulation stages of *Ichthyosporidium* (Microsporida). *J. Protozool.* **21**, 667–677.

Steffens, W. (1956). Weitere Beiträge zur Kenntnis der *Plistophora* - Krankheit. *Aquar. Terrar. Z.* (DATZ) **6**, 153–155.

Steffens, W. (1962). Der heutige Stand der Verbreitung von *Plistophora hyphessobryconis* Schäperclaus, 1941 (Sporozoa, Microsporidia). *Z. Parasitenk.* **21**, 535–541.

Stempell, W. (1904). Über *Nosema anomalum* Moniez. *Arch. Protistenk.* **4**, 1–42.

Stempell, W. (1919). Untersuchungen über *Leptotheca coris* n.sp. und das in dieses schmarotzende *Nosema marionis* Théloh. *Arch. Protistenk.* **40**, 113–157.

Sterba, G. (1956). Aquarienkunde. II. Urania-Verlag. Leipzig und Jena, pp. 232.

Stewart, C. G., Botha, W. S. and van Dellen, A. F. (1979). The prevalence of *Encephalitozoon* antibodies in dogs and an evaluation of the indirect fluorescent antibody test. *J. S. Af. Vet. Ass.* **50**, 169–172.

Stewart, C. G., van Dellen, A. F. and Botha, W. S. (1979). Canine encephalitozoonosis in kennels and the isolation of *Encephalitozoon* in tissue culture. *J. S. Af. Vet. Ass.* **50**, 165–168.

Stewart, C. G., van Dellen, A. F. and Botha, W. S. (1981). Antibodies to a canine isolate of *Encephalitozoon* in various species. *S. Af. J. Sci.* **17**, 572 (abstract).

Stoll, N. R. (1961). Chairman "International Code of Zoological Nomenclature adopted by the XV International Congress of Zoology" 176 pp. *Int. Trust. Zool. Nomen.*, London.

Streett, D. A. and Briggs, J. D. (1982). An evaluation of sodium dodecyl sulfate-polyacrylamide gel electrophoresis for the identification of microsporidia. *J. Invert. Pathol.* **40**, 159–165.

Stunkard, H. W. and Lux, F. E. (1965). A microsporidian infection of the digestive tract of the winter flounder *Pseudopleuronectes americanus. Biol. Bull.* **129**, 371–387.

Summerfelt, R. C. (1964). A new microsporidian parasite from the golden shiner, *Notemigonus crysoleucas. Trans. Am. Fish. Soc.* **93**, 6–10.

Summerfelt, R. C. (1972). Studies on the transmission of *Pleistophora ovariae*, the ovary parasite of the golden shiner (*Notemigonus crysoleucas*). *Final Report, Project No.4-66-R. Nat. Marine Fish. Service*, pp. 19.

Summerfelt, R. C. and Ebert, V. W. (1969). *Plistophora tahoensis* sp.n. (Microsporida, Nosematidae) in the body wall of the piute sculpin (*Cottus beldingii*) from the Lake Tahoe, California, Nevada. *Bull. Wildl. Dis. Assoc.* **5**, 330–341.

Summerfelt, R. C. and Warner, M. C. (1970a). Incidence and intensity of infection of *Plistophora ovariae*, a microsporidian parasite of the golden shiner, *Notemigonus crysoleucas. In* "A Symposium on diseases of fishes and shellfishes" ed. S. F. Snieszko, *Am. Fish. Soc., Washington, Special Publ.* **5**, 142–160.

Summerfelt, R. C. and Warner, M. C. (1970b). Geographical distribution and host parasite relationships of *Plistophora ovariae* (Microsporida, Nosematidae) in *Notemigonus crysoleucas. J. Wildl. Dis.* **6**, 457–465.

Sureau, P. (1962). Infection spontanée des souris d'élevage a Tananarive par *Encephalitozoon* et *Klossiella muris. Arch. Inst. Pasteur, Madagascar,* **31**, 125–126.

Swarczewsky, B. (1914). Über den Lebenscyclus einiger Haplosporidien. *Arch. Protistenk.* **33**, 49–108.

Sweeney, A. W., Hazard, E. I. and Graham, M. F. (1985). Intermediate host for an *Amblyospora* sp. infecting the mosquito, *Culex annulirostris. J. Invertebr. Pathol.* **46**, 98–102.

Swellengrebel, N. H. (1911). *Plistophora gigantea* Thélohan, een parasiet van *Crenilabrus melops. Verhandel. Koninklijk. Nederland. Akad. Wetensch. Afd. Natuurk.* **20**, 238–243.

Swellengrebel, N. H. (1912). The life history of *Plistophora gigantea* Thélohan (*Glugea gigantea* Thél.) *Parasitology,* **4**, 345–363.

Takahashi, S. (1978). Microsporidiosis in fish with emphasis on glugeosis of ayu. *Fish. Pathol.* **13**, 9–16, (in Japanese).

Takahashi, S. and Egusa, S. (1976). Studies on *Glugea* infection of the ayu, *Plecoglossus altivelis.* II. On the prevention and treatment - 1. Fumagillin efficacy as a treatment. *Jap. J. Fish.* **11**(2)9, 83–88, (in Japanese).

Takahashi, S. and Egusa, S. (1977a). Studies on *Glugea* infection of the ayu, *Plecoglossus altivelis.* I. Description of the *Glugea* and a proposal of a new species *Glugea plecoglossi. Jap. J. Fish.* **11**(4)3, 175–182, (in Japanese).

Takahashi, S. and Egusa, S. (1977b). Studies on *Glugea* infection in the ayu, *Plecoglossus altivelis.* III. Effect of water temperature on the development of xenoma of *Glugea plecoglossi. Jap. J. Fish.* **11**(4)3, 195–200 (in Japanese).

Takeda, S. (1933). On a new disease of rainbow trout. *Keison-iho,* **5**, 1–9, (in Japanese).

Takvorian, P. M. and Cali, A. (1981). The occurrence of *Glugea stephani* (Hagenmüller, 1899) in American winter flounder, *Pseudopleuronectes* americanus (Wallbaum) from the New York - New Jersey lower bay complex. *J. Fish. Biol.* **18**, 491–501.

Tazaki, H. (1956). Surveys on the *Toxoplasma* infection in wild rats (*Apidemus speciosus speciosus*) and sparrows (*Passer montanus staturatus*). *Keio Igaku,* **33**, 275–281. (Quoted by Matsubayashi *et al.,* 1959).

Templeman, W. (1948). The life history of the capelin (*Mallotus villosus* O. F. Müller) in Newfoundland Waters. *Res. Bull. Newfoundland Govt. Lab.* **17**, pp. 151.

Testoni, J. F. (1974). Enzootic renal nosematosis in laboratory rabbits. *Austral. Vet. J.* **50**, 159–163.

Thélohan, P. (1891). Sur deux sporozoaires nouveaux parasites des muscles des poissons. *C. R. Acad. Sci.,* Paris, **112**, 168–171.

Thélohan, P. (1892). Observations sur les myxosporidies et essai de classification de ces organismes. *Bull. Soc. Philom.,* Paris, **4**, 173–174.

Thélohan, P. (1895). Recherches sur les Myxosporidies. *Bull. Sci. Fr. Belg.* **26**, 100–394.

Thieme, H. (1952). Die Neonkrankheit. *In* "Das Leben in useren Aquarien," Berlin, 37–44.

Thieme, H. (1954). *Plistophora hyphessobryconis* Schäperclaus ihre Entwicklung und die durch den Parasiten hervorgerufenen histologischen Veranderungen des Wirtsgewebes. *Diplomarbeit Univ. Halle,* pp 61.

Thieme, H. (1956). Erkennung und Wesen der Neonkrankheit. *Aquar. Terrar.* **3**, 172–175.

Torres, C. M. (1927a). Sur une nouvelle maladie de l'homme, caractérisée par la présence d'un parasite intracellulaire, très proche de *Toxoplasma* et de l'*Encephalitozoon,* dans le tissu musculaire cardique, les muscles du squelette, le tissu cellulaire sous-cutane et le tissu nerveux. *C. R. Soc. Biol.* Paris, **97**, 1778–1781.

Torres, C. M. (1927b). Morphologie d'un nouveau parasite de l'homme, *Encephalitozoon chagasi*, n.sp. observé dans un cas de méningoencéphalomyélite congenitale avec myosite et myocardite. *C. R. Soc. Biol.* Paris, **97**, 1787–1790.

Torres, C. M. (1927c). Affinités de l'*Encephalitozoon chagasi*, agent étiologique d'une méningo-encéphalite-myélite congenitale avec myocardite et myosite chez l'homme. *C. R. Soc. Bio.* Paris, **97**, 1998–1999.

Tsuneo, C. (1960). Studies on infection and growth of *Encephalitozoon* in He La cell, L-cell and Ehrlich cancer cell. *J. Osaka Med. Centre* **9**, 2465–2477.

Twort, C. C. and Archer, H. E. (1922). Spontaneous encephalo-myelitis of rabbits, and its relation to spontaneous nephritis etc. *Vet. J.* **29**, 367–372.

Twort, J. M. and Twort, C. C. (1932). Disease in relation to carcinogenic agents among 60,000 experimental mice. *J. Pathol. Bacteriol.* **35**, 219–242.

Undeen, A. H. (1978). Spore hatching processes in some *Nosema* species with particular reference to *N. algerae* Vávra and Undeen. *Misc. Publ. Ent. Soc. Amer.* **11**, 29–40.

Undeen, A. H. and Alger, N. E. (1971). A density gradient method for fractionating microsporidian spores. *J. Invert. Pathol.* **18**, 421–424.

van Dellen, A. F., Botha, W. S., Boomker, J. and Warnes, W. E. J. (1978). Light and electron microscopical studies on canine encephalitozoonosis : cerebral vasculitis. *Onderstepoort J. Vet. Res.* **45**, 165–186.

van den Berghe, L. (1939). *Glugea microspora*, a new microsporidian species from the liver of the sand-eel, *Amnodytes lanceolatus*. *J. Parasitol.* **25** (suppl.), 238.

van den Berghe, L. (1940). *Glugea caulleryi* nom.n. for *Glugea microspora*, *J. Parasitol.* **26**, 238.

Van Duijn, C. (1956). Diseases of fishes. Water life, Poultry World Ltd., Dorset House, London, 174 pp.

Vaney, C. and Conte, A. (1901). Sur une nouvelle microsporidie, *Pleistophora mirandellae*, parasite de l'ovarie d'*Alburnus mirandella* Blanch. *C. R. Acad. Sci.* Paris, **133**, 644–646.

van Rensburg, I. B. J. and du Plessis, J. L. (1971). Nosematosis in a cat : a case report. *J. Sth. Afr. Vet. Med. Ass.* **42**, 327–331.

Vávra, J., Bedrník, P. and Činatl, J. (1972). Isolation and *in vitro* cultivation of the mammalian microsporidian *Encephalitozoon cuniculi*. *Folia Parasitol.*, Praha, **19**, 349–354.

Vávra, J., Blazěk, K., Lávička, N., Koczkova, I., Kalafa, Š. and Stehlík, M. (1971). Nosematosis in carnivores. *J. Parasitol.* **57**, 923–924.

Vavrá, J., Chalupský, J., Oktábec, J. and Bedrník, P. (1980). *Encephalitozoon cuniculi* (EC) infection in a rabbit farm: transmission and influence on body weight. *J. Protozool.* **27**(3) Suppl. 74A–75A.

Vávra, J. and Maddox, J. V. (1976). Methods in microsporidiology. *In* Bulla, L. A., and Cheng, T. C. eds. *Comparative Pathology* 1. Biology of the Microsporidia. Plenum Press, New York and London, 281–319.

Vávra, J. and Undeen, A. H. (1979). Rediscovery of the mirosporidian parasite of capelin. *J. Protozool.* (suppl.) **26**, 46 (abstract).

Vinnichenko, L. N., Zaïka, V. E., Timofeev, V. A., Shteïn, G. A. and Shulman, S. S. (1971). Parasitic protozoa of fishes from the Amur river. *Parazitol. Sbornik.* **25**, 10–40, (in Russian).

Viting, A. I. (1965). Contribution to the study of the etiology of multiple sclerosis in the light of the findings of a morphological study of the central nervous system. *J. Neuropath. Psychiat.* SS Korakov, **65**, 1641–1645, (In Russian).

Viting, A. I. (1969). The parasitic nature of disseminated or multiple sclerosis. *Parazitologiya*, Moscow, **3**, 569–577.

Vlacovich, G. P. (1866). Corpuscoli oscillanti del bombice del Gelso. *Atti. Istit. Veneto* (series 3) **11**, 1056–1058.

Voronin, V. N. (1974). Some microsporidians (Microsporidia, Nosematidae) from sticklebacks *Pungitius pungitius* and *Gasterosteus aculeatus* of the Finnish Bay. *Acta Protozool.* **13**, 211–220, (in Russian).

Voronin, V. N. (1976). Characteristics of the genus *Glugea* (Protozoa, Microsporidia) on the example of the type species *Glugea anomala* (Moniez, 1887) Gurley 1893 and its varieties. *Parazitologiya* **10**, 263–267, (in Russian).

Voronin, V. N. (1978). *Pleistophora ladogensis* sp. n. (Protozoa, Microsporidia) from muscles of *Lota lota* and *Osmerus eperlanus ladogensis*. *Parazitologiya* **12**, 453–455, (in Russian).

Voronin, V. N. (1980). On the host of *Nosema fennica* (Microsporidia). *Parastiologiya* **14**, 82–83, (in Russian).

Voronin, V. N. (1981). *Pleistophora ladogensis* infection in *Lota lota* and *Osmerus eperlanus eperlanus* natio *ladogensis*. *Parazitologiya* **15**, 259–264, (in Russian).

Wales, J. and Wolf, H. (1955). Three protozoan diseases of trout in California. *Calif. Fish. Game* **41**, 183–187.

Waller, T. (1975). Growth of *Nosema cuniculi* in established cell lines. *Lab. Anim.* **9**, 61–68.

Waller, T. (1977). The india-ink immunoreaction: a method for the rapid diagnosis of encephalitozoonosis. *Lab. Anim.* **11**, 93–97.

Waller, T. (1979). Sensitivity of *Encephalitozoon cuniculi* to various temperatures, disinfectants and drugs. *Lab. Anim.* **13**, 227–230.

Waller, T. and Bergquist, N. R. (1982). Rapid simultaneous diagnosis of toxoplasmosis and encephalitozoonosis in rabbits by carbon immunoassay. *Lab. Anim. Sci.* **32**, 515–517.

Waller, T., Lyngset, A., Elvander, M. and Morein, B. (1980). Immunological diagnosis of encephalitozoonosis from post mortem specimens. *Vet. Immun. Immunopath.* **1**, 353–360.

Waller, T., Uggla, A. and Bergquist, N. R. (1983). Encephalitozoonosis and toxoplasmosis diagnosed simultaneously by a novel rapid test : the carbon immunoassay. *Proc. 3rd Int. Symp. Vet. Lab. Diag.* (Ames, Iowa, U.S.A.) **1**, 171–178.

Walliker, D. (1966). Some protozoan parasites of the roach, *Rutilus rutilus*. *Trans. Roy. Soc. Trop. Med. Hyg.* **60**, 13 (abstract).

Weekly Epidemiological Record (1983). Parasitic Disease Surveillance. Antibody to *Encephalitozoon cuniculi* in man. No. 3–4 February, pp.30–32.

Weidner, E. (1972). Ultrastructural study of microsporidian invasion into cells. *Z. Parasitenk.* **40**, 227–242.

Weidner, E. (1975). Interactions between *Encephalitozoon cuniculi* and macrophages. *Z. Parasitenk.* **47**, 1–9.

Weidner, E. (1976a). The microsporidian invasion tube. The ultrastructure isolation and characterisation of the protein comprising the tube. *J. Cell. Biol.* **71**, 23–34.

Weidner, E. (1976b). Ultrastructure of the peripheral zone of a *Glugea*-induced xenoma. *J. Protozool.* **23**, 234–238.

Weidner, E. (1982). The microsporidian spore invasion tube. III. Tube extrusion and assembly. *J. Cell. Biol.* **93**, 976–979.

Weiser, J. (1949). Studies on some parasites of fishes. *Parasitology* **39**, 164–166.

Weiser, J. (1960). Zur Kenntnis der Krankheiten der Lurche. *Věst. Česko. Zool. Společ.* **24**, 232–233.

Weiser, J. (1964). On the taxonomic position of the genus *Encephalitozoon* Levaditi, Nicolau & Schoen, 1923 (Protozoa, Microsporidia). *Parasitology* **54**, 749–751.

Weiser, J. (1965). *Nosema muris* n.sp., a new microsporidian parasite of the white mouse (*Mus musculus* L.) *J. Protozool.* **12**, 78–83.

Weiser, J. (1964). On the taxonomic position of the genus *Encephalitozoon* Levaditi, Nicolau & Schoen, 1923 (Protozoa, Microsporidia). *Parasitology* **54**, 749–751.

Weiser, J. (1965). *Nosema muris* n.sp., a new microsporidian parasite of the white mouse (*Mus musculus* L.) *J. Protozool.* **12**, 78–83.

Weiser, J. (1976). To the identity of the microsporidia affecting man. *Věst. Českosl. Zool. Společ.* **40**, 157–159.

Weiser, J., Kalavati, C. and Sandeep, B. V. (1981). *Glugea nemipteri* sp.n. and *Nosema bengalis* sp.n., two new microsporidia of *Nemipterus japonicus* in India. *Acta. Protozool.* **20**, 201–208.

Weissenberg, R. (1909). Beiträge zur Kenntnis von *Glugea lophii* Doflein I. Über den Sitz und die Verbreitung der Mikrosporidien-cysten am Nervensystem von *Lophius piscatorius* und *budegassa*. *Sitzungsber. Ges. Natur, Freunde*, Berlin, **9**, 557–565.

Weissenberg, R. (1911a). Über Mikrosporidien aus dem Nervensystem von Fischen (*Glugea lophii* Doflein) und die Hypertrophie der befallenen Ganglienzellen. *Arch. Mikrosk. Anat.* **78**, 383–421.

Weissenberg, R. (1911b). Beiträge zur Kenntnis von *Glugea lophii* Doflein. II. Über den Bau der Cysten und die Beziehungen zwischen Parasit und Wirtsewebe. *Sitzungsber. Ges. Natur. Freunde*, Berlin, **3**, 149–157.

Weissenberg, R. (1911c). Über einigen Mikrosporidien aus Fischen (*Nosema lophii* Doflein, *Glugea anomala* Moniez, *Glugea hertwigi* nov. spec.) *Sitzungsber. Ges. Natur. Freunde*, Berlin, **8**, 344–357.

Weissenberg, R. (1913). Beiträge zur Kenntnis des Zeugungskreises der Mikrosporidien *Glugea anomala* Moniez and *hertwigi* Weissenberg. *Arch. Mikrosk. Anat.* **82**, 81–163.

Weissenberg, R. (1921). Zür Wirtsgewebsableitung des Plasmakorpers der *Glugea anomala* Cysten. *Arch. Protistenk.* **42**, 400–421.

Weissenberg. R. (1922). Mikrosporidien und Chlamydozoen als Zellparasiten von Fischen. *Verh. Deutsch. Zool. Ges.* **27**, 41–43.

Weissenberg, R. (1967). Contribution to the study of the intracellular development of the microsporidium *Glugea anomala* in the fish *Gasterosteus aculeatus*. *J. Protozool.* (suppl.) **14**, 28–29.

Weissenberg, R. (1968). Intracellular development of the microsporidian *Glugea anomala* Moniez in hypertrophying migratory cells of the fish *Gasterosteus aculeatus* L., an example of the formation of "xenoma tumors." *J. Protozool.* **15**, 44–57.

Weissenberg, R. (1976). Microsporidian interactions with host cells. *In* Bulla L. A. and Cheng T. C. eds., *Comparative Pathobiology*. 1. Biology of the Microsporidia. Plenum Press, New York and London, 203–237.

Wellings, S. R., Ashley, L. E. and McArn, G. E. (1969). Microsporidial infection of English sole, *Parophrys vetulus*. *J. Fish. Res. Board Canada* **26**, 2215–2218.

Wenyon, C. M. (1926). Protozoology. Baillière, Tindall and Cox, London, pp.1563.

Werner, H. and Pierzynski, A. (1962). Über eine neues Protozoon aus der weissen Labormaus (*Mus musculus*). *Z. Parasitenk.* **21**, 301–308.

Wilhelm, W. E. (1966). Early developmental stages of *Plistophora ovariae*, a microsporidian parasite of the golden shiner, *Notemigonus crysoleucas*. *J. Protozool.* (suppl.) **13**, 29 (abstract).

Wilson, J. M. (1979a). The biology of *Encephalitozoon cuniculi*. *Med Biol.* **57**, 84–101.

Wilson, J. M. (1979b). *Encephalitozoon cuniculi* in wild European rabbits and a fox. *Res. Vet. Sci.* **26**, 114.

Wobeser, G. and Schuh, J. C. L. (1979). Microsporidial encephalitis in muskrats. *J. Wildlife Dis.* **15**, 413–417.

Wolf, A. and Cowen, D. (1937). Granulomatous encephalomyelitis due to an *Encephalitozoon* (Encephalitozoic Encephalomyelitis) a new protosoan disease of man. *Bull. Neurol. Inst. New York* **6**, 306–371.

Wolf, A., Cowen, D. and Page, B. H. (1939). Toxoplasmic encephalomyelitis Part III. A new case of granulomatous encephalomyelitis due to a protozoan. *Amer. J. Pathol.* **15**, 657–694.

Woodcock, H. M. (1904). On Myxosporidia in flat fish. *Trans. Biol. Soc., Liverpool.* **18**, 46–62.

Wosu, N. J. Olsen, R., Shadduck, J. A., Koestner, A. and Pakes, S. P. (1977). Diagnosis of experimental encephalitozoonosis in rabbits by complement fixation. *J. Infect. Dis.* **135**, 944–948.

Wosu, N. J., Shadduck, J. A., Pakes, S. P., Frenkel, J. K., Todd, K. S. and Conroy, J. D. (1977). Diagnosis of encephalitozoonosis in experimentally infected rabbits by intradermal and immunofluorescence tests. *Lab. Anim. Sci.* **27**, 210–216.

Wright, J. H. and Craighead, E. M. (1922). Infectious motor paralysis in young rabbits. *J. Exp. Med.* **36**, 135–140.

Yost, D. H. (1958). *Encephalitozoon* infection in laboratory animals. *J. Nat. Cancer Inst.* **29**, 957–960.

Young, P. C. (1969). Parasitic cutaneous lesions in a cod (*Gadus morhua* L.). *Vet. Rec.* **84**, 99–100.

Youssef, N. N. and Hammond, D. M. (1972). Fine structure of the xenoma caused by *Glugea* sp. in the English sole. *J. Protozool.* (suppl.) **19**, 16 (abstract).

Zaika, V. E. (1965). Parasite fauna of fish from the Lake Baïkal. Nauka Publishing House, Moscow, pp. 107.

Zhukov, E. V. (1964). Parasitofauna of fishes of Chukotka III. Protozoa of marine and freshwater fishes. General conclusions. *Parasitol. Sbornik.* **22**, 224–253.

Index

Myctophum punctatum, 22, 49, 54–55
Mylopharyngodon piceus, 22, 88
Myxobolus ranae, 187
Myoxocephalus quadricornis labradoricus, 19, 22
Myoxocephalus scorpius, 22, 91

N

Natrix natrix, 174, 175–177
Negri bodies, 197
Nemipterus japonicus, 22, 63, 153
Neofelis concolor, 206
Neogobius caspius, 22, 79, 123
Neogobius cephalarges, 22, 89
Neogobius fluviatilis, 22, 81, 89
Neogobius fluviatilis pallasi, 22, 79
Neogobius kessleri gorlap, 22, 123
Neogobius melanostomus, 22, 47, 81, 89
Neogobius melanostomus affinis, 22, 79, 123
Neogobius syrman, 22, 47
Neresheimeria paradoxa, 26
Nerophis aequoreus, 49
Newt, common, 184
Nitrofurasone, 120, 165
Noemacheilus barbatulus, 22, 112
Noemacheilus malapterus longicauda, 22, 126
Nosema, 2, 5, 11, 15, 146, 153, 157, 166, 168, 170, 185, 200, 232, 237, 238
  diagnosis, 15
Nosema algerae, 240
Nosema anomala, 21, 40
Nosema anomalum, 40
Nosema bengalis, 153
Nosema bombycis, 15, 170
Nosema branchialis, 127
Nosema connori, 231, 232
Nosema cotti, 27, 154
Nosema cuniculi, 208
Nosema depressum, 87
Nosema destruens, 87
Nosema encyclometrae, 177
Nosema fennica, 59
Nosema ghigii, 177
Nosema giganteum, 147
Nosema girardini, 155
Nosema helminthorum, 234, 235

Nosema lophii, 141
Nosema marionis, 19, 153, 170
Nosema muris, 208
Nosema notabilis, 23, 73, 171
Nosema ovoideum, 156
Nosema pimephales, 64
Nosema punctiferum, 88
Nosema sauridae, 160
Nosema sp., 60, 167, 234
Nosema stephani, 68
Nosema takedai, 29, 163
Nosema tisae, 81
Nosema tritoni, 184
Nosema valamugili, 165
Nosemoides, 7, 146
Notemigonus chrysoleucas, 22, 51, 118–120

O

Octosporea, 79, 187
Octosporea machari, 79
Odontogadus merlangus, 22, 151
Oncorhynchus gorbuscha, 22, 163
Oncorhynchus keta, 22, 71, 163
Oncorhynchus kisutch, 22, 159
Oncorhynchus masou, 22, 163
Oncorhynchus nerka, 23, 132
Oncorhynchus nerka var. adonis, 23, 163
Oncorhynchus tschawytscha, 23, 133, 159, 163
Ondatra zibethica, 203
Opisthogramma seitzigi, 101
Opisthotonus, 220
Opsanus beta, 23, 171
Opsanus tau, 23, 73, 171
Ortholinea polymorpha, 23, 73, 171
Osmerus eperlanus eperlanus, 23, 60, 107
Osmerus eperlanus eperlanus var. lado-gensis, 23, 51
Osmerus eperlanus eperlanus var. spirin-chus, 23, 60
Osmerus eperlanus mordax, 49, 60, 73
Oxytetracycline, 229

P

Pansporoblast membrane, 5
Pansporoblastina, 5